Solved Problems in Analysis

AS APPLIED TO GAMMA, BETA, LEGENDRE AND BESSEL FUNCTIONS

Orin J. Farrell
Union College

Bertram Ross
New Haven College

WITHDRAWN

DOVER PUBLICATIONS, INC.
NEW YORK

This Dover edition, first published in 1971, is an unabridged and corrected republication of the work originally published by the MacMillan Company in 1963 under the title: *Solved Problems: Gamma and Beta Functions, Legendre Polynomials, Bessel Functions.*

International Standard Book Number: 0-486-62713-6
Library of Congress Catalog Card Number: 70-139975

Manufactured in the United States of America
Dover Publications, Inc.
180 Varick Street
New York, N. Y. 10014

PREFACE TO THE FIRST EDITION

This book consists of a selection of problems, each with a solution worked out in detail, dealing with the properties and applications of the Gamma function and the Beta function, the Legendre polynomials, and the Bessel functions. For those problems which involved more than mere choice of a suitable formula and appropriate use thereof, we have often endeavored to present solutions with emphasis on the considerations raised by the following questions: How does one make a start in attacking the problem? What theorems and techniques from algebra, trigonometry, analytic geometry, calculus, and the theory of functions appear applicable so as to be likely to effect a solution? How and why does one proceed from one step to the next? What clues present themselves either in the statement of the problem or in the facts which develop as the attempt at solution proceeds? What aspects of the problem must be carefully considered so that the solution will meet the demands of mathematical rigor?

Such an approach usually leads to solutions that are neither brief nor elegant. We earnestly hope, however, that the lack of brevity and elegance is compensated by what may be called a naturalness of procedure combined with a heuristic presentation that make the solutions relatively easy to follow. We hope also that the solutions presented will be found stimulating, and that they will help to develop skill in attacking and solving problems in pure and applied mathematics.

Cursory examination of this book might give the impression of an occasional haphazard choice of problem. But no problem was originated or chosen at random. Selection of problems was made so as to fulfill such purposes as exposition of suitable techniques of procedure and reasonable coverage of relevant topics. Often a problem that seems out of place in one of the chapters on the properties of the functions (Chapters I, III and V), and not closely concerned with the development of the outstanding properties of a function, will be found to serve as a useful lemma in one or more later chapters on the applications of the functions. Indeed, a goodly number of the problems and exercises in Chapters I, III and V are put to use in the chapters on applications.

References to individual texts or treatises have been used sparingly in the statements of the problems and in the solutions. However, a modest bibliography of works typical of those one would find it profitable to consult is included at the end of the book.

We gratefully make the following acknowledgments: Table III–2 is reproduced from W. E. Byerly's *Fourier's Series and Spherical Harmonics* with the permission of Ginn and Company; Tables V–2 through V–27 are printed, with slight modifications and deletions, from N. W. McLachlan's *Bessel Functions for Engineers* with the permission of Professor N. W. McLachlan and the Oxford Press; material was used from G. M. Watson's *Theory of Bessel Functions* with the permission of The Cambridge University Press; Tables V–14 through V–21 were reprinted with the permission of The Royal Society and the American Institute of Electrical Engineers.

We appreciate especially the excellent constructive criticisms and suggestions made by Dr. Melvin Hausner of New York University.

PREFACE TO THE DOVER EDITION

We have been pleased at the response to this text from students who are studying applied classical analysis for the first time, and by professors who are not only looking for ways to motivate but also for ways to bring difficult subject matter down to an understandable level. In this Dover edition, we have endeavored to correct errors in the first edition, some of which were discovered by our students. We also appreciate the very careful reading given by Professor Yoshio Matsuoka, Kagoshimashi.

CONTENTS

I   THE GAMMA

FUNCTION AND THE BETA

FUNCTION

INTRODUCTION

The Gamma function was first defined in 1729 by the great Swiss mathematician Euler. He defined the Gamma function by an infinite product:

$$\Gamma(z) = \frac{1}{z} \prod_{n=1}^{\infty} \frac{\left(1 + \dfrac{1}{n}\right)^z}{\left(1 + \dfrac{z}{n}\right)}.$$

If z be taken as the complex variable $x + iy$, Euler's product for $\Gamma(z)$ converges at every finite z except $z = 0, -1, -2, -3, \cdots$. The function defined by the product is analytic at every finite z except for the singular points just mentioned. At each of the singular points, $\Gamma(z)$ has a simple pole.

The notation $\Gamma(z)$ and the name "Gamma function" were first used by Legendre in 1814.

From Euler's infinite product for $\Gamma(z)$ can be derived the formula

$$\Gamma(z) = \int_0^\infty t^{z-1}e^{-t}dt.$$

This integral formula is convergent only when the real part of z is positive. Nevertheless this integral formula for $\Gamma(z)$ often is taken as the starting point for introductory treatments of the Gamma function. Moreover, the variable z is often confined to real values x. So shall it be in this book: unless the contrary is explicitly stated, we shall be concerned in our exercises and problems with the Gamma function of a real variable only. For positive values of x we shall take the following as our basic definition of the *Gamma function*:

$$\Gamma(x) = \int_0^\infty t^{x-1}e^{-t}dt, \qquad x > 0.$$

As is usually done, we shall extend the domain of the definition of the Gamma function into the realm of negative numbers (exclusive of negative integers) by extrapolation via the characteristic equation

$$\Gamma(x + 1) = x\Gamma(x).$$

It may be remarked that this function, namely $x\Gamma(x)$, provides an analytic function whose value at each positive integer n is $n!$.

The Gamma function itself, as set up by Euler, is such that $\Gamma(n) = (n - 1)!$ rather than $n!$ when n is a positive integer.

Although the Gamma function was devised by Euler to solve a problem in pure mathematics, here, as elsewhere in mathematics, an invention in pure mathematics has been found useful in applications of mathematics to problems in engineering and the sciences. The Gamma function is particularly useful in certain problems of probability, especially problems that involve factorials of large integers or the *incomplete Gamma function*

$$\gamma(x, \tau) = \int_0^\tau t^{x-1}e^{-t}dt, \qquad x > 0, \tau > 0.$$

Tables of values of $\Gamma(x)$ are usually given for the range $1 \leqq x < 2$. There is no need to tabulate outside a range whose spread is unity because of the fundamental property $\Gamma(x + 1) = x\Gamma(x)$. The range $1 \leqq x < 2$ is chosen because it is the interval between two successive integers whereon $\Gamma(x)$ has its lowest values for such an interval, making for economy of tabulation and interpolation.

The Beta function is a function of two arguments. As basic definition for the *Beta function* $B(x, y)$ we shall take, as is usually done, the definition

$$B(x, y) = \int_0^1 t^{x-1} (1 - t)^{y-1}dt, \qquad x > 0, \quad y > 0.$$

The Beta function is related to the Gamma function:

$$B(x, y) = \frac{\Gamma(x)\Gamma(y)}{\Gamma(x + y)}.$$

TABLE I-1

$\Gamma(x)$, $\qquad 1 \leqq x \leqq 1.99$

x	0	1	2	3	4	5	6	7	8	9
1.0	1.0000	.9943	.9888	.9835	.9784	.9735	.9687	.9642	.9597	.9555
.1	.9514	.9474	.9436	.9399	.9364	.9330	.9298	.9267	.9237	.9209
.2	.9182	.9156	.9131	.9108	.9085	.9064	.9044	.9025	.9007	.8990
.3	.8975	.8960	.8946	.8934	.8922	.8912	.8902	.8893	.8885	.8879
.4	.8873	.8868	.8864	.8860	.8858	.8857	.8856	.8856	.8857	.8859
.5	.8862	.8866	.8870	.8876	.8882	.8889	.8896	.8905	.8914	.8924
.6	.8935	.8947	.8959	.8972	.8986	.9001	.9017	.9033	.9050	.9068
.7	.9086	.9106	.9126	.9147	.9168	.9191	.9214	.9238	.9262	.9288
.8	.9314	.9341	.9368	.9397	.9426	.9456	.9487	.9518	.9551	.9584
.9	.9618	.9652	.9688	.9724	.9761	.9799	.9837	.9877	.9917	.9958

The problems worked out in this chapter are mostly exercises dealing with properties and values of the Gamma and Beta functions which can be derived directly from their definitions or which ensue from the identities

$$\Gamma(x + 1) = x\Gamma(x),$$

$$B(x, y) = \frac{\Gamma(x)\Gamma(y)}{\Gamma(x + y)}.$$

At the end of the chapter is a list of the most frequently used formulas.

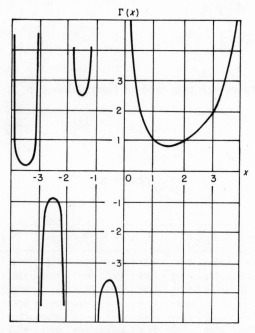

Figure I-1
The Gamma Function

Problems: Integral Expressions of $\Gamma(x)$

I-1. Show that the integral $\int_0^\infty t^{x-1}e^{-t}dt$ which defines $\Gamma(x)$ is convergent for every positive x but not convergent for any other real x. ★

We write first the integral as the sum of two integrals:

$$\int_0^\infty t^{x-1}e^{-t}dt = \int_0^m \frac{e^{-t}}{t^{1-x}}\,dt + \int_m^\infty t^{x-1}e^{-t}dt$$

where m is any positive number. Let us call the two integrals on the right A and B respectively. We see that A is proper when $x \geqq 1$. On the other hand when $x < 1$, the second factor of the integrand becomes infinite at $t = 0$, thus making the integral improper. The first factor e^{-t} does not, of course, cause us any concern in the interval $t = 0$ to $t = m$. In fact, since that factor is continuous throughout and becomes unity at $t = 0$, where the other factor becomes infinite for $x < 1$, we can conclude by the theory of improper integrals that A is convergent or not according as $\int_0^m dt/t^{1-x}$ is convergent or not. But this last-written integral we know to be convergent when and only when the exponent on t is less than unity. Thus A is convergent when and only when $1 - x < 1$, that is, when x is positive.

Integral B is improper for all x simply because the interval is infinite. The problem, then, is to determine the values of x for which it is convergent. In order to do this we first apply to B the formula for integration by parts, namely, $\int u\,dv = uv - \int v\,du$, taking $u = t^{x-1}$ and $dv = e^{-t}dt$:

$$B = \left[t^{x-1}(-e^{-t})\right]_m^\infty - \int_m^\infty (-e^{-t})(x-1)t^{x-2}dt$$

$$= -\lim_{t\to\infty}\frac{t^{x-1}}{e^t} + \frac{m^{x-1}}{e^m} + (x-1)\int_m^\infty t^{x-2}e^{-t}dt.$$

Now we know by the theory of indeterminate forms (by successive applications of L'Hospital's Rule) that in the race to infinity e^t will always win out over any constant power of t to such an extent that $\lim_{t\to\infty} [t^{x-1}/e^t] = 0$ for every x. Thus convergence of B now hinges on the convergence of our last-written integral, in which we observe that the exponent on t is less by unity than what it was in B. We keep applying integration by parts to the remaining integral until the exponent on t is nonpositive. (Incidentally, we would not have to do any integrating by parts when $x \leq 1$.) In any event we finally get for B a finite sum of numbers added to a polynomial in x times an integral of the form

$$\int_m^\infty \frac{1}{t^p} \frac{1}{e^t}\, dt, \qquad p \text{ positive.}$$

If m be taken sufficiently large, the first factor $1/t^p$ in this last integral is less than unity for all $t \geq m$, which makes the curve $y = 1/t^p e^t$ lie *under* the curve $y = 1/e^t$ for $t \geq m$. But we can see that $\int_m^\infty dt/e^t$ is convergent by actual integration. Therefore, our final integral is convergent for every x, which in turn makes B convergent for every x.

The Gamma integral, then, is convergent for those values of x, and *only* those, for which *both* A and B are convergent, namely for all *positive* x:

$$\Gamma(x) = \int_0^\infty t^{x-1} e^{-t} dt, \qquad x > 0. \tag{I-1}$$

I-2. Show that

$$\Gamma(x) = \int_0^1 \left[\log_e \frac{1}{u} \right]^{x-1} du, \qquad x > 0. \qquad \star$$

We start with $\Gamma(x) = \displaystyle\int_0^\infty t^{x-1} e^{-t} dt, \qquad x > 0$ from Eq. (I-1).

Let

$$u = e^{-t}; \text{ then } 1/u = e^t, \ \log_e(1/u) = t, \ -(1/u) du = dt, \text{ and}$$
$$[\log_e(1/u)]^{x-1} = t^{x-1}.$$

For our limits of integration: when $t = 0$, $u = 1$; and when $t = \infty$, $u = 0$. Then,

$$\int_0^\infty t^{x-1} e^{-t} dt = -\int_1^0 \left[\log_e\left(\frac{1}{u}\right) \right]^{x-1} u \cdot \frac{1}{u}\, du = \int_0^1 \left[\log_e\left(\frac{1}{u}\right) \right]^{x-1} du.$$

I-3. Show that $\Gamma(x) = 2 \int_0^\infty m^{2x-1} e^{-m^2} dm, \qquad x > 0.$ ★

It is worth noting that here, as in many other cases, our starting point is the definition $\Gamma(x) = \int_0^\infty t^{x-1} e^{-t} dt, \qquad x > 0$ in Eq. (I-1). Let $t = m^2$, then $dt = 2m\, dm$. Our limits of integration remain the same. So we have

$$\Gamma(x) = \int_0^\infty t^{x-1} e^{-t} dt = \int_0^\infty m^{2(x-1)} e^{-m^2} 2m\, dm$$

$$= 2 \int_0^\infty m^{2x-1} e^{-m^2} dm, \qquad x > 0.$$

Problem: Properties of Γ(x)

I-4. Establish the fundamental identity $\Gamma(x + 1) = x\Gamma(x)$.   ★

Before proving the identity directly from the Gamma integral for all positive x we note that this identity is used to *define* the Gamma function first for $-1 < x < 0$ by writing it in the form $\Gamma(x) = \Gamma(x + 1)/x$, thence for $-2 < x < -1$ by the same formula, and so on for all nonintegral negative values of x. It remains, then, to show that $\Gamma(x + 1) = x\Gamma(x)$ for every positive x.

Letting x be any positive number, we write the Gamma integral for the argument $x + 1$:

$$\Gamma(x + 1) = \int_0^\infty t^{(x+1)-1} e^{-t} dt$$

$$= \int_0^\infty t^x e^{-t} dt.$$

Next we apply integration by parts, namely $\int u\, dv = uv - \int v\, du$, to this latter integral, taking $dv = e^{-t} dt$ and $u = t^x$:

$$\Gamma(x + 1) = t^x(-e^{-t}) \Big|_0^\infty - \int_0^\infty (-e^{-t}) x t^{x-1} dt$$

$$= -\lim_{t \to \infty} \frac{t^x}{e^t} + \frac{0^x}{e^0} + x \int_0^\infty t^{x-1} e^{-t} dt.$$

The limit indicated in the first term on the right we know to be zero by the treatment of the indeterminate form ∞/∞ as learned in introductory calculus, using L'Hospital's Rule (once or twice or several or many times in succession according to the size of x). The second term on the right vanishes, while the third term is none other than $x\Gamma(x)$. Thus we have

$$\Gamma(x + 1) = x\Gamma(x). \tag{I-4.1}$$

REMARK.   It is often found convenient to apply the fundamental identity in one or other of the following forms:

$$\Gamma(x) = \frac{\Gamma(x + 1)}{x}, \tag{I-4.2}$$

$$\Gamma(x) = (x - 1)\Gamma(x - 1), \tag{I-4.3}$$

$$\Gamma(-x) = \frac{\Gamma(1 - x)}{-x}, \; x \neq 0, 1, 2, \cdots. \tag{I-4.4}$$

Problems: Specific Evaluations of $\Gamma(x)$

I-5. Evaluate $\Gamma(.37)$.   ★

We have merely to increase the argument from .37 to 1.37 via the identity $\Gamma(x) = \Gamma(x + 1)/x$ so that we can use Table I-1 where $\Gamma(x)$ is tabulated for $1 \leq x < 2$.

$$\Gamma(0.37) = \frac{\Gamma(.37 + 1)}{.37} = \frac{\Gamma(1.37)}{.37} \cong \frac{.8893}{.37} \cong 2.404.$$

I-6. Evaluate $\Gamma(9/4)$.   ★

Here we have to decrease the argument from $\frac{9}{4}$ to $\frac{5}{4}$ using the fundamental identity in the form $\Gamma(x) = (x - 1)\Gamma(x - 1)$:

$$\Gamma\left(\frac{9}{4}\right) = \left(\frac{5}{4}\right)\Gamma\left(\frac{5}{4}\right) = \left(\frac{5}{4}\right)\Gamma(1.25) \cong \frac{5}{4}(.9064) \cong 1.133.$$

I-7. Evaluate $\Gamma(4.6)$.   ★

This requires three applications in succession of the identity $\Gamma(x) = (x - 1)\Gamma(x - 1)$:

$$\Gamma(4.6) = (3.6)\Gamma(3.6) = (3.6)(2.6)\Gamma(2.6) = (3.6)(2.6)(1.6)\Gamma(1.6)$$

$$\cong (14.976)(.8935) \cong 13.38.$$

I-8. Evaluate $\Gamma(-1.3)$.   ★

This requires three successive increases of argument by unity via the fundamental identity in the form $\Gamma(x) = \Gamma(x + 1)/x$:

$$\Gamma(-1.3) = \frac{\Gamma(-.3)}{-1.3} = \frac{\Gamma(.7)}{(-1.3)(-.3)} = \frac{\Gamma(1.7)}{(-1.3)(-.3)(.7)}$$

$$\cong \frac{.9086}{.2730} \cong 3.328.$$

I-9. Show that $\Gamma(1) = 1$.   ★

Putting 1 for x in $\int_0^\infty t^{x-1}e^{-t}dt$, we have

$$\Gamma(1) = \int_0^\infty e^{-t}dt$$

$$= \lim_{N \to \infty} \left\{ \left[-e^{-t} \right]_0^N \right\}$$

$$= \lim_{N \to \infty} \left\{ -\frac{1}{e^N} + 1 \right\}$$

$$= 1.$$

Problem: Properties of $\Gamma(x)$

I-10. If n be any positive integer ≥ 2, show that

$$\Gamma(n) = (n - 1)!.$$ ★

We start with the result of Prob. I-9: $\Gamma(1) = 1$. Then by Prob. I-4 we have $\Gamma(2) = \Gamma(1 + 1) = (1)\Gamma(1) = 1$. Similarly, by continuing to apply the fundamental identity $\Gamma(x + 1) = x\Gamma(x)$, we get

$$\Gamma(3) = 2\Gamma(2) = 2 \cdot 1$$
$$\Gamma(4) = 3\Gamma(3) = 3 \cdot 2 \cdot 1.$$

At this point we perceive the truth of the formula we have to establish. To *prove* the formula true we have yet to apply the second stage of the method of proof by mathematical induction. Assume the formula true for an arbitrary integer $n \geq 2$. Then for the next integer $m = n + 1$ we have

$$\Gamma(m) = \Gamma(n + 1) = n\Gamma(n) = n(n - 1)! = n! = (m - 1)!.$$

Thus, the formula also holds for $n + 1$. But we already know from our work above that it holds for $n = 2$, $n = 3$, and $n = 4$. Consequently, it must hold for the next integer $n = 5$ and for the next after 5, namely 6, and so on *ad infinitum*.

REMARKS 1.   It is by virtue of the formula just established that the convention of defining and accepting a value for the factorial of zero, namely $0! = 1$, came to be adopted. For, if we apply the formula *formally* with $n = 1$, we have $\Gamma(1) = (1 - 1)! = 0!$. (The exclamation point here may be considered, if you will, as having double significance.) But we know that $\Gamma(1) = 1$. So, we agree that zero shall be considered as having a factorial which shall be taken as unity. With this convention for $0!$ we have

$$\Gamma(n) = (n - 1)!, \qquad n = 1, 2, 3, \cdots. \qquad \text{(I-10.1)}$$

2. Since the Gamma function provides a smooth interpolation function relative to the factorials of the positive integers, it is sometimes used as a means of defining $x!$ when x is nonintegral, i.e., $x! = \Gamma(x + 1)$. For example $(3.6)! = \Gamma(4.6) \cong 13.38$ by Prob. I-7.

Problem: Specific Evaluation of $\Gamma(x)$

I-11. Evaluate $\Gamma(\tfrac{1}{2})$.   ★

We put $\tfrac{1}{2}$ for x in Eq. (I-1) and get $\Gamma(\tfrac{1}{2}) = \int_0^\infty t^{-1/2}e^{-t}dt$. But we do not see any way to evaluate this integral. There is no use trying to integrate by parts because the exponent on t is not a positive integer.

Let us try again. If we change the variable of integration via $t = u^2$, we get $\Gamma(\tfrac{1}{2}) = \int_0^\infty 2e^{-u^2}du$. This looks a little better. The integrand is not as complicated as before. But we are still baffled when we try to integrate. What to do? We begin at this point to suspect that we may have to resort to some *indirect* method for evaluating $\Gamma(\tfrac{1}{2})$. But what? The following scheme appears to be without motivation. Indeed, its discoverer was surely a person of great mathematical ingenuity.

We write our second trial above for $\Gamma(\tfrac{1}{2})$ *twice*: once with x as the variable of integration, then with y as the variable of integration. Then we multiply the results:

$$\Gamma^2(\tfrac{1}{2}) = \left[\int_0^\infty 2e^{-x^2}dx\right]\left[\int_0^\infty 2e^{-y^2}dy\right].$$

Now, although the right side is the *product* of two integrals, its appearance suggests an *iterated* integral. Indeed, in this instance we may actually write

$$\Gamma^2(\tfrac{1}{2}) = 4\int_0^\infty \left[\int_0^\infty e^{-y^2}dy\right]e^{-x^2}dx,$$

because the integral in y yields a mere *constant* to carry over into the integral in x. May we now equate the iterated integral to a *double* integral? Yes, we may:

$$\Gamma^2(\tfrac{1}{2}) = 4\iint_Q e^{-(x^2+y^2)}dA,$$

where Q denotes the entire first quadrant of the cartesian xy-plane.

This improper double integral over the entire first quadrant is, of course,

$$\lim_{\substack{M \to \infty \\ N \to \infty}} \iint_R e^{-(x^2+y^2)} dA.$$

where R denotes the rectangular region $0 \leq x \leq M$, $0 \leq y \leq N$, and is, therefore equivalent to

$$\lim_{M \to \infty} \int_0^M \left[\lim_{N \to \infty} \int_0^N e^{-y^2} dy \right] e^{-x^2} dx.$$

Now, our double integral may likewise be equated to an iterated integral in polar coordinates:

$$\Gamma^2(\tfrac{1}{2}) = 4 \int_0^{\pi/2} \left[\int_0^\infty e^{-r^2} r \, dr \right] d\theta,$$

which in turn (by the same argument as used before with xy-coordinates) may be expressed as a *product* of two integrals (since the integral in r yields only a constant value independent of θ). We now have

$$\Gamma^2(\tfrac{1}{2}) = 4 \left[\int_0^{\pi/2} d\theta \right] \left[\int_0^\infty e^{-r^2} r \, dr \right]$$

$$= 4 \left(\frac{\pi}{2} \right) \left(\frac{1}{2} \right) = \pi,$$

making

$$\Gamma(\tfrac{1}{2}) = \sqrt{\pi}. \tag{I-11.1}$$

Motivation for the ingenious scheme of evaluation is now apparent. It was the presence of the factor r in the iterated polar-coordinates integral that made the integration possible, and that is what suggested the original multiplication whereby we got from e^{-x^2} to $e^{-(x^2+y^2)} = e^{-r^2}$.

Problems: Properties of $\Gamma(x)$

I-12. Show that if n be a positive integer, then

$$\Gamma(n + \tfrac{1}{2}) = \frac{(2n - 1)(2n - 3)(2n - 5) \cdots (3)(1)\sqrt{\pi}}{2^n}. \qquad \star \quad \text{I-12.1}$$

The argument $n + \tfrac{1}{2}$ can be written $(2n + 1)/2$. If we recall the property of Eq. (I-4.3), namely $\Gamma(x) = (x - 1)\Gamma(x - 1)$ and if we take $x = (2n + 1)/2$, then we have

$$\Gamma\left(\frac{2n + 1}{2}\right) = \left(\frac{2n + 1}{2} - 1\right) \Gamma\left(\frac{2n + 1}{2} - 1\right),$$

that is,

$$\Gamma(n + \tfrac{1}{2}) = \left(\frac{2n - 1}{2}\right) \Gamma\left(\frac{2n - 1}{2}\right).$$

The process of decreasing the argument by unity is repeated for $\Gamma[(2n - 1)/2]$:

$$\Gamma\left(\frac{2n - 1}{2}\right) = \left(\frac{2n - 3}{2}\right) \Gamma\left(\frac{2n - 3}{2}\right).$$

So far we have

$$\Gamma(n + \tfrac{1}{2}) = \left(\frac{2n - 1}{2}\right)\left(\frac{2n - 3}{2}\right) \Gamma\left(\frac{2n - 3}{2}\right).$$

We recall that $\Gamma(\tfrac{1}{2}) = \sqrt{\pi}$; consequently, we want to repeat the process of changing the argument until we reach $\Gamma(\tfrac{1}{2})$. If we take $\Gamma(\tfrac{7}{2})$, for instance, we can write

$$\Gamma\left(\frac{7}{2}\right) = \frac{5}{2} \cdot \frac{3}{2} \cdot \frac{1}{2} \Gamma\left(\frac{1}{2}\right).$$

We see that we obtained $\Gamma(\tfrac{1}{2})$ multiplied by three factors; moreover, this number three is the same integer as occurs in the argument of $\Gamma(\tfrac{7}{2})$ when written $\Gamma(3 + \tfrac{1}{2})$. This 3 corresponds to the n of $\Gamma(n + \tfrac{1}{2})$.

Consequently, we can say that if we start with $\Gamma(n + \frac{1}{2})$ we have to continue decreasing the argument by unity n times in order to reach $\frac{1}{2}$. We also observe that each time the process is repeated there is another 2 in the denominator. Since we have determined that the process is to be repeated n times, we can collect the 2's and write 2^n. We then have

$$\Gamma(n + \tfrac{1}{2}) = \frac{(2n - 1)(2n - 3)(2n - 5) \cdots (3)(1)\sqrt{\pi}}{2^n}. \quad \text{(I-12.2)}$$

I-13. Show that, if n be a positive integer, then

$$\Gamma(n + \tfrac{3}{2}) = \frac{(2n + 1)(2n - 1)(2n - 3) \cdots (3)(1)\sqrt{\pi}}{2^{n+1}}. \quad \star \quad \text{(I-13.1)}$$

Applying Eq. (I-4.1), we have

$$\Gamma(n + \tfrac{3}{2}) = (n + \tfrac{1}{2})\Gamma(n + \tfrac{1}{2})$$
$$= \frac{2n + 1}{2}\,\Gamma(n + \tfrac{1}{2}).$$

Thus, by Prob. I-12, we have

$$\Gamma(n + \tfrac{3}{2}) = \frac{2n + 1}{2}\,\frac{(2n - 1)(2n - 3) \cdots (3)(1)\sqrt{\pi}}{2^n}$$
$$= \frac{(2n + 1)(2n - 1)(2n - 3) \cdots (3)(1)\sqrt{\pi}}{2^{n+1}}.$$

I-14. Show that, if n be a positive integer,

$$\Gamma(n - \tfrac{1}{2}) = \frac{(2n - 3)(2n - 5) \cdots (3)(1)\sqrt{\pi}}{2^{n-1}}. \quad \star$$

Applying Eq. (I-4.1) to $\Gamma(n + \frac{1}{2})$, we have

$$\Gamma(n + \tfrac{1}{2}) = (n - \tfrac{1}{2})\Gamma(n - \tfrac{1}{2})$$
$$= \frac{2n - 1}{2}\,\Gamma(n - \tfrac{1}{2}).$$

Thus,

$$\Gamma(n - \tfrac{1}{2}) = \frac{2}{2n - 1}\,\Gamma(n + \tfrac{1}{2}).$$

Then, by Prob. I-12, we can write

$$\Gamma(n - \tfrac{1}{2}) = \frac{2}{2n - 1}\,\frac{(2n - 1)(2n - 3)\cdots(3)(1)\sqrt{\pi}}{2^n}$$

$$= \frac{(2n - 3)(2n - 5)\cdots(3)(1)\sqrt{\pi}}{2^{n-1}}.$$

I-15. If n is a positive integer and $x - n \neq 0, -1, -2, \cdots,$ evaluate

$$\frac{\Gamma(x + n)}{\Gamma(x - n)}. \qquad \star$$

We apply Eq. (I-4.1) to the numerator of the given fraction:

$$\Gamma(x + n) = (x + n - 1)\Gamma(x + n - 1).$$

By repeated application of Eq. (I-4.1) to the numerator, we have

$$\frac{\Gamma(x + n)}{\Gamma(x - n)} = \frac{(x + n - 1)(x + n - 2)(x + n - 3)\cdots(x + n - 2n)\Gamma(x + n - 2n)}{\Gamma(x - n)}$$

$$= (x + n - 1)(x + n - 2)\cdots(x - n).$$

I-16. Show that $2 \cdot 4 \cdot 6 \cdots 2n = 2^n\Gamma(n + 1).$ $\qquad \star$

$$2 \cdot 4 \cdot 6 \cdots 2n = (2 \cdot 1)(2 \cdot 2)(2 \cdot 3)\cdots(2 \cdot n)$$

$$= 2^n n!$$

$$= 2^n\Gamma(n + 1).$$

REMARK. If $n - 1$ were substituted in place of n, we would have

$$2 \cdot 4 \cdot 6 \cdots (2n - 2) = 2^{n-1}\Gamma(n).$$

I-17. Show that

$$1 \cdot 3 \cdot 5 \cdots (2n - 1) = \frac{2^{1-n}\Gamma(2n)}{\Gamma(n)}. \qquad \star$$

We insert unity in the left side of the given equation in the form

$$\frac{2 \cdot 4 \cdot 6 \cdots (2n - 2)}{2 \cdot 4 \cdot 6 \cdots (2n - 2)}.$$

We have then

$$1 \cdot 3 \cdot 5 \cdots (2n - 1) = \frac{1 \cdot 2 \cdot 3 \cdot 4 \cdot 5 \cdots (2n - 2)(2n - 1)}{2 \cdot 4 \cdots (2n - 2)}. \quad \text{(I-17.1)}$$

The numerator of Eq. (1) is $(2n - 1)!$, which is $\Gamma(2n)$ by Eq. (I-4.1). The denominator of Eq. (1) is $2^{n-1}\Gamma(n)$ by Prob. I-16. Thus we have

$$1 \cdot 3 \cdot 5 \cdots (2n - 1) = \frac{2^{1-n}\Gamma(2n)}{\Gamma(n)}.$$

By property (I-4.2) this last equation may be written alternatively as

$$1 \cdot 3 \cdot 5 \cdots (2n - 1) = \frac{2^{1-n}n\Gamma(2n)}{\Gamma(n + 1)}.$$

I-18. Show that $\sqrt{\pi}\Gamma(2n + 1) = 2^{2n}\Gamma(n + \tfrac{1}{2})\Gamma(n + 1)$ for every positive integer n. $\star$

In the right side of the given equation we substitute the value of $\Gamma(n + \tfrac{1}{2})$ found in Prob. I-12. Then we have for the right side

$$2^{2n}\Gamma(n + \tfrac{1}{2})\Gamma(n + 1)$$

$$= \frac{2^{2n}(2n - 1)(2n - 3) \cdots 3 \cdot 1\sqrt{\pi}\Gamma(n + 1)}{2^n}. \qquad \text{(I-18.1)}$$

Multiplying numerator and denominator of the right side of Eq.
(I-18.1) by $2n(2n-2)(2n-4)\cdots(4)(2)$, we have

$$2^{2n}\Gamma(n+\tfrac{1}{2})\Gamma(n+1)$$
$$=\frac{2^{2n}(2n)(2n-1)(2n-2)\cdots4\cdot3\cdot2\cdot1\cdot\sqrt{\pi}\Gamma(n+1)}{2^{n}(2n)(2n-2)(2n-4)\cdots(4)(2)}.\qquad\text{(I-18.2)}$$

The numerator in the right side of Eq. (I-18.2) is

$$2^{2n}(2n)!\sqrt{\pi}\Gamma(n+1).\qquad\text{(I-18.3)}$$

The denominator is $2^{n}\cdot2^{n}n!$. If now we write $(2n)!=\Gamma(2n+1)$
and $n!=\Gamma(n+1)$, then we have for the right side of Eq. (I-18.2)

$$\frac{2^{2n}\sqrt{\pi}\Gamma(2n+1)\Gamma(n+1)}{2^{2n}\Gamma(n+1)}.$$

When the factors common to the numerator and denominator are
canceled out we have remaining the left side of the given equation,
thus demonstrating the identity.

REMARK. Using $2n\Gamma(2n)$ for $\Gamma(2n+1)$ and $n\Gamma(n)$ for $\Gamma(n+1)$
we can write the given equation as

$$\sqrt{\pi}\Gamma(2n)=2^{2n-1}\Gamma(n)\Gamma\!\left(n+\frac{1}{2}\right).\qquad\text{(I-18.4)}$$

This is known as *Legendre's duplication formula*. (See also Prob.
II-15b.)

It is at times convenient to use this formula in the ratio form

$$\frac{\Gamma(2n)}{\Gamma(n)}=\frac{\Gamma(n+\tfrac{1}{2})}{\sqrt{\pi}2^{1-2n}}.\qquad\text{(I-18.5)}$$

I-19. Show that

$$\frac{1\cdot3\cdot5\cdots(2n-1)}{2\cdot4\cdot6\cdots2n}=\frac{\Gamma(n+\tfrac{1}{2})}{\sqrt{\pi}\Gamma(n+1)}.\qquad\star$$

Replacing the numerator and denominator on the left by the expressions found for them in Prob. I-16 and I-17 and then replacing the ratio $\Gamma(2n)/\Gamma(n)$ by the equivalent ratio from Eq. (I-18.5), we have at once the right side of the equation to be demonstrated.

I-20. Show that $\Gamma(x)$ is differentiable at every real x except $x = 0, -1, -2, -3, \cdots$.   *

There is, of course, no possibility of differentiability at $x = 0$, $-1, -2, -3, \cdots$, since the absolute value of $\Gamma(x)$ becomes infinite at each of these values of x.

We consider first only positive values of x for which we have $\Gamma(x)$ defined by the integral in Prob. I-1. May we apply Leibniz's rule for differentiation of an integral (with respect to a parameter in the integrand when the integral is *improper*, as in Prob. I-1)? For the problem at hand this question means that we are concerned with determining, if possible, an interval of values of x for which the following equation is valid:

$$\frac{d\Gamma}{dx} = \int_0^\infty \frac{\partial}{\partial x}\left[t^{x-1}e^{-t} \right] dt$$

$$= \int_0^\infty t^{x-1}(\log_e t)e^{-t}dt. \qquad (I\text{-}20.1)$$

Let us, first of all, consider the question of convergence of the integral in (I-20.1). An examination of the integrand indicates that, for $x > 1$, we can demonstrate convergence of this integral by the following considerations. Choosing any positive number b, we write

$$\int_0^\infty t^{x-1}(\log_e t)e^{-t}dt$$

$$= \int_0^b t^{x-1}(\log_e t)e^{-t}dt + \int_b^\infty t^{x-1}(\log_e t)e^{-t}dt. \qquad (I\text{-}20.2)$$

We have, for $x > 1$,

$$
\begin{aligned}
\lim_{t \to 0} \left[t^{x-1}(\log_e t)e^{-t} \right] &= \lim_{t \to 0} \left[t^{x-1} \log_e t \right] \\
&= \lim_{t \to 0} \left[\frac{\log_e t}{t^{1-x}} \right] \\
&= \lim_{t \to 0} \left[\frac{1/t}{(1-x)t^{-x}} \right] \quad \text{by L'Hospital's Rule} \\
&= \frac{1}{1-x} \lim_{t \to 0} \left[t^{x-1} \right] \\
&= 0, \quad x > 1.
\end{aligned}
$$

(I-20.3)

Thus, as $t \to 0$, the integrand not only remains finite for $x > 1$; it approaches the limit zero.

On the other hand, as $t \to \infty$, we have

$$
\begin{aligned}
\lim_{t \to \infty} \left[\frac{t^{x-1}(\log_e t)e^{-t}}{t^x e^{-t}} \right] &= \lim_{t \to \infty} \left[\frac{\log_e t}{t} \right] \\
&= \lim_{t \to \infty} \left[\frac{1/t}{1} \right] \quad \text{by L'Hospital's Rule} \\
&= 0.
\end{aligned}
$$

(I-20.4)

But it is well known that the integral

$$
\int_b^\infty t^x e^{-t} dt, \quad b > 0
$$

(I-20.5)

is convergent for every x. And since, as we have just shown, the integrand in (I-20.2) bears ratio to the integrand in (I-20.5) which approaches a finite limit, namely zero, as $t \to \infty$, it follows that the second integral on the right in (I-20.2) is convergent. The first integral on the right in (I-20.2) was found to be a proper integral for $x > 1$ by virtue of (I-20.3). It follows that the integral in (I-20.1) is convergent at least for all x such that $x > 1$.

When $0 < x \leqq 1$, the first integral on the right in (I-20.2) is no longer proper, since the integrand becomes negatively infinite as $t \to 0$. But direct application of the definition for convergence of such an integral shows, in the usual manner, that this integral is convergent for $0 < x \leqq 1$. Accordingly, we find that the integral in (I-20.1) is convergent for every positive x. We will, however, content ourselves with the convergence of this integral only for $x \geqq 1$. This restriction, as we shall see, will make it relatively easy to establish validity of application of Leibniz's rule for such values of x.

Choosing any $x > 1$ and taking Δx so that $|\Delta x| < x - 1$ and tentatively assuming existence of the indicated limits, we have

$$\frac{d\Gamma}{dx} = \lim_{\Delta x \to 0} \frac{\Delta \Gamma}{\Delta x}$$

$$= \lim_{\Delta x \to 0} \left\{ \frac{1}{\Delta x} \left[\int_0^\infty t^{x+\Delta x-1} e^{-t} dt - \int_0^\infty t^{x-1} e^{-t} dt \right] \right\}$$

$$= \lim_{\Delta x \to 0} \int_0^\infty \frac{t^{x+\Delta x-1} e^{-t} - t^{x-1} e^{-t}}{\Delta x} dt$$

$$= \lim_{\Delta x \to 0} \int_0^\infty \frac{\Delta_x (t^{x-1} e^{-t})}{\Delta x} dt$$

$$= \lim_{\Delta x \to 0} \int_0^\infty \frac{\left\{ \left[\frac{\partial}{\partial x} (t^{x-1} e^{-t}) \right]_{at(x^*,t)} \right\} \Delta x}{\Delta x} dt,$$

with x^* somewhere between x and $x + \Delta x$. It will be recognized that the replacement of $\Delta_x (t^{x-1} e^{-t})$ by the derivative (with respect to x) of $t^{x-1} e^{-t}$ evaluated at (x^*, t) was made by application of the Theorem of the Mean for first derivative.

Evaluating the partial derivative of $t^{x-1} e^{-t}$ at (x^*, t), we have

$$\frac{d\Gamma}{dx} = \lim_{\Delta x \to 0} \int_0^\infty t^{x^*-1} (\log_e t) e^{-t} dt$$

$$= \lim_{\Delta x \to 0} \left[\int_0^b t^{x^*-1} (\log_e t) e^{-t} dt + \int_b^\infty t^{x^*-1} (\log_e t) e^{-t} dt \right]. \quad \text{(I-20.6)}$$

We observe that $x^* \to x$ as $\Delta x \to 0$, since x^* is always *between* x and $x + \Delta x$.

If now a positive ϵ be arbitrarily assigned, we can take b so large that the second integral on the right in (I-20.6) will differ by less than $\epsilon/2$, simultaneously for all x^*, from the integral

$$\int_b^\infty t^{x-1}(\log_e t)e^{-t}dt.$$

With b so chosen, we find then that the first integral on the right in (I-20.6) will differ, for all Δx sufficiently near zero, from the integral

$$\int_0^b t^{x-1}(\log_e t)e^{-t}dt$$

by less than $\epsilon/2$, since the function

$$f(x, t) = \begin{cases} 0, & t = 0 \\ t^{x-1}(\log_e t)e^{-t}, & t \neq 0, \end{cases}$$

is continuous and hence *uniformly* continuous for

$$1 < x_1 \leqq x \leqq x_2, \qquad 0 \leqq t \leqq b.$$

It follows that the limit indicated on the right in (I-20.6) *does* exist for every $x > 1$, which is to say that $\Gamma(x)$ is differentiable for $x > 1$ and that, for $x > 1$, the derivative $d\Gamma/dx$ is given by (I-20.1).

And now we can readily show that $\Gamma(x)$ is differentiable also for $x \leqq 1$ except, of course, at $x = 0, -1, -2, -3, \cdots$, by making use of the characteristic identity (I-4.2), namely

$$\Gamma(x) = \frac{1}{x}\Gamma(x + 1).$$

Since we have shown that $\Gamma(x)$ is differentiable for $x > 1$, it follows that $\Gamma(x + 1)$ is differentiable for $x > 0$. And $1/x$ is certainly differentiable for all $x > 0$. Consequently, the product $(1/x)\Gamma(x + 1)$ is differentiable for all $x > 0$, which is to say that $\Gamma(x)$ is differentiable for $x > 0$.

For x negative we recall that $\Gamma(x)$ is *defined* by successive extrapolations by way of the formula $\Gamma(x) = \Gamma(x + 1)/x$, first to the range $-1 < x < 0$, thence to the range $-2 < x < -1$, and so on. Accordingly, we find by repetition of the argument used in the preceding paragraph that $\Gamma(x)$ is also differentiable at every non-integral negative x.

REMARKS. 1. The Gamma function is continuous at every x except 0 and the negative integers. This is an immediate consequence of the differentiability of $\Gamma(x)$.

2. The second derivative $\Gamma''(x)$ is found in similar manner to exist for all x except zero and the negative integers. For x positive it is found that

$$\Gamma''(x) = \int_0^\infty t^{x-1}(\log_e t)^2 \, e^{-t}dt.$$

We observe that $\Gamma''(x)$ is positive for every $x > 0$, since the integrand is positive for all $t > 0$. Consequently, the curve $y = \Gamma(x)$ is concave upward for all positive values of x.

I-21. Prove that $\Gamma(x) \to +\infty$ as $x \to 0$ through positive values and as $x \to +\infty$; also that $\Gamma(x)$ becomes alternatively negatively infinite and positively infinite (as shown in Fig. I-1) at the negative integers. ⋆

We know that $\Gamma(1) = 1$ from Prob. I-9; also that $\Gamma(x)$ is a continuous function for all positive x by Prob. I-20. From Prob. I-4 we have $\Gamma(x) = \Gamma(x + 1)/x$. Now, letting $x \to 0$ through positive values, we merely have to observe that the numerator of $\Gamma(x + 1)/x$ approaches unity, namely $\Gamma(1)$, as $x \to 0$ while the denominator of the fraction is approaching zero as limit. Thus, $\Gamma(x) \to +\infty$ as $x \to 0^+$.

To prove that $\Gamma(x) \to +\infty$ as $x \to +\infty$ we note first that the *particular* sequence $\Gamma(1)$, $\Gamma(2)$, $\Gamma(3)$, $\Gamma(4)$, $\cdots$, $\Gamma(n)$, $\cdots$, *does* have the property we are trying to establish because $\Gamma(n) = (n - 1)!$ by

Prob. I-10. Our task, however, is to prove that $\Gamma(x)$ becomes positively infinite for *all* positive sequences x_1, x_2, x_3, $\cdots$, in which $x_n \to +\infty$ as $n \to \infty$. We shall do this by establishing a property of the Gamma function indicated by Fig. I-1, namely, that the derivative of $\Gamma(x)$ is positive for all values of x beyond a certain x. The graph indicates such a place between $x = 1$ and $x = 2$. We find it convenient (as well as sufficient) to demonstrate the positiveness of the derivative for all $x > e + 1$, where as usual e denotes the natural base 2.71828 . . . for logarithms.

The property to be proved will follow at once from the fact that for $x > e + 1$ the Γ-function always increases and takes on successively the factorial values 2!, 3!, 4!, 5!, $\cdots n!$, $\cdots$, as x equals respectively, the integers 3, 4, 5, $\cdots$. From Prob. I-20 we take the derivative of $\Gamma(x)$ and write it as the sum of two integrals:

$$\Gamma'(x) = \int_0^1 t^{x-1}(\log_e t)e^{-t}dt + \int_1^\infty t^{x-1}(\log_e t)e^{-t}dt.$$

Let us call these two integrals A and B respectively. We choose any $x > e + 1$. In A we let p denote the positive exponent $x - 1$. As $t \to 0$ we find by L'Hospital's Rule that the indeterminate form $t^p \log_e t$ approaches zero. So, at $t = 0$ we assign the value zero to our integrand in A. Since $\log_e t$ is negative for $0 < t < 1$ while the other two factors are positive, we see that A has a negative value. We shall appraise this negative value and show that it is outweighed by the positive value of B. For $0 < t < 1$ we have

$$\left| t^p \log_e t \, e^{-t} \right| < \left| t^p \log_e t \right|$$

because $e^{-t} < 1$ in that range of values of t. Consequently,

$$\left| \int_0^1 t^p(\log_e t)e^{-t}dt \right| < \left| \int_0^1 t^p \log_e t \, dt \right|.$$

But when this last integral is evaluated, using integration by parts and L'Hospital's Rule, we find its value to be $- 1/(p + 1)^2$.

We turn now to examination of B in which the integrand is positive over the whole range $1 \leqq t < \infty$. Thus, B is positive. Moreover, we have $B > C = \int_e^\infty t^p(\log_e t)e^{-t}dt$, where, as before, we put p for $x - 1$, remembering that we fixed upon any $x > e + 1$. The integral C is easily appraised as follows. First of all, $C > \int_e^\infty t^p e^{-t}dt$ because $\log_e t > 1$ for all $t > e$. Applying integration by parts to this last integral together with L'Hospital's Rule, we have

$$C > e^p e^{-e} + \int_e^\infty e^{-t}pt^{p-1}dt.$$

We do not try to evaluate this last integral. We merely note that it has a positive value. Thus we find that $C > e^{p-e}$. If now $p > e$, that is if $x > e + 1$, we have C greater than 1 whereas the absolute of integral A is less than unity. So, the derivative $\Gamma'(x)$ is positive for all $x > e + 1$.

Do you wonder how we knew in advance to take $x > e + 1$? Well, let us confess it: we did not know in advance. But we felt from a look at Fig. I-1 that we ought to be able to prove $\Gamma'(x)$ to be positive for all x greater than *some one x*. So we started on our proof and as we got along with our analysis of integral B we kept our eyes open and finally we saw that the proof would complete itself easily by taking that x to be $> e + 1$.

Finally, to show that $\Gamma(x)$ becomes alternately negatively infinite and positively infinite at the negative integers, we merely have to take the result of the first part of the present problem, namely that $\Gamma(x) \to +\infty$ as $x \to 0^+$, together with the Fundamental Identity $\Gamma(x) = \Gamma(x + 1)/x$ which is used to define $\Gamma(x)$ for x negative.

REMARK. Since $\Gamma(x) \to +\infty$ both as $x \to 0^+$ and as $x \to +\infty$ and since the curve $y = \Gamma(x)$ is concave upward (as was observed in Prob. I-20) for all positive x, it follows that $\Gamma(x)$ attains a single minimum as x ranges over all positive numbers. This minimum occurs between $x = 1$ and $x = 2$, as indicated in Fig. I-1.

Problem: Specific Evaluations of Γ(*x*)

I-22. Find the value of Γ(1.15) given only the following data:

$$\Gamma(1.1) = .9514, \quad \Gamma(1.2) = .9182, \quad \Gamma(1.3) = .8975. \quad \star$$

Ordinary linear interpolation will give the value of Γ(1.15) equal to .9348. In Table I-1 it is seen that the value is given as .9330. Even though the curve is shallow on the range from $x = 1$ to 2, linear interpolation is not as accurate as the following method of interpolation which is treated in detail in the branch of mathematics called the Calculus of Finite Differences.

Let $y(x)$ be the function to be evaluated.

Let $y_1 = \Gamma(1.1) = .9514,$ Let $x_1 = 1.1,$

$y_2 = \Gamma(1.2) = .9182,$ $x = 1.15.$

$y_3 = \Gamma(1.3) = .8975.$

Let h denote the equally spaced difference between 1.1, 1.2, 1.3. The following formula from the Calculus of Finite Differences is applicable.

$$y(x) = y_1 + \frac{(x - x_1)(y_2 - y_1)}{h}$$

$$+ \frac{(x - x_1)(x - x_1 - h)(y_3 - 2y_2 + y_1)}{2h^2} + \cdots.$$

Thus,

$$\Gamma(1.15) \cong .9514 + \frac{(.05)(-.0332)}{.1} + \frac{(.05)(-.05)(.0125)}{2(.01)}$$

$$\cong .9514 - .0166 - .00156$$

$$\cong .9332.$$

This result differs from the entry to be found in tables by .0002.

I-23. Compute $\Gamma(1.2)$ approximately by means of the asymptotic expansion

$$\log_{10} \Gamma(x) \sim .3991 + (x - \tfrac{1}{2}) \log_{10} (x)$$
$$- .4343 \left[x - \frac{1}{12x} + \frac{1}{360x^3} - \frac{1}{1260x^5} + \cdots \right],$$

which, although not convergent for any x, has the property that the error committed in taking a partial sum is numerically less than the last term retained.   ★

We have

$$\log_{10} \Gamma(1.2) \cong .3991 + (.7) \log_{10} (1.2)$$
$$- .4343 \left[1.2 - \frac{1}{(12)(1.2)} + \frac{1}{(360)(1.728)} \right]$$
$$\cong 9.9629 - 10,$$

whence

$$\Gamma(1.2) \cong .9181.$$

REMARKS. 1. The four-place entry for $\Gamma(1.2)$ in Table I-1 is .9182.

2. All the entries in Table I-1 can thus be computed by use of the asymptotic expansion given above. Retention of the term $-1/1260x^5$ will provide four-place accuracy.

Problem: Infinite Product Expression of $\Gamma(x)$

I-24. A fundamental formula, due to Euler, for $\Gamma(z)$ states that

$$\Gamma(z) = \lim_{n \to \infty} \frac{1 \cdot 1 \cdot 2 \cdot 3 \cdots (n - 1)}{z(z + 1)(z + 2) \cdots (z + n - 1)} \, n^z. \quad \text{(I-24.1)}$$

The limit on the right exists for all real and complex values of z other than $z = 0, -1, -2, -3, \cdots$, and equals the value given by

Eq. (I-1) when z is real and positive. From Euler's formula (I-24.1) derive Weierstrass's formula

$$\frac{1}{\Gamma(z)} = ze^{Cz}\prod_{n=1}^{\infty}\left(1 + \frac{z}{n}\right)e^{-z/n},\qquad\text{(I-24.2)}$$

where C denotes Euler's constant, namely,

$$\lim_{n\to\infty}\left(1 + \frac{1}{2} + \frac{1}{3} + \cdots + \frac{1}{n} - \log_e n\right).\qquad\star$$

First we write the fraction in (I-24.1) as

$$\frac{1}{z}\prod_{k=1}^{n-1}\frac{k}{z+k} = \frac{1}{z}\prod_{k=1}^{n-1}\frac{1}{1 + \dfrac{z}{k}}.\qquad\text{(I-24.3)}$$

If, now, we put $n - 1 = m$, we can write (I-24.1) as

$$\Gamma(z) = \frac{1}{z}\lim_{m\to\infty}\left[(m+1)^z\prod_{k=1}^{m}\frac{1}{1 + \dfrac{z}{k}}\right],\quad z \neq 0, -1, -2, \cdots.$$

This last equation is equivalent to

$$\frac{1}{\Gamma(z)} = z\lim_{m\to\infty}\left[e^{-z\,\log_e(m+1)}\prod_{k=1}^{m}\left(1 + \frac{z}{k}\right)\right].\qquad\text{(I-24.4)}$$

In order to get from (I-24.4) to (I-24.2) we use a rather unusual way of expressing unity, namely,

$$1 = \lim_{m\to\infty}\left[e^{\left(1 + \frac{1}{2} + \frac{1}{3} + \cdots + \frac{1}{m}\right)z}\prod_{k=1}^{m-1}e^{-\frac{z}{k}}\right]\qquad\text{(I-24.5)}$$

which is readily seen to be true because the right side telescopes into nothing more than $\lim_{m\to\infty} e^{z/m}$.

Multiplying the left side and the right side of (I-24.4) each by the corresponding side of (I-24.5), we get

$$\frac{1}{\Gamma(z)} = z \lim_{m \to \infty} \left[e^{\left(1 + \frac{1}{2} + \frac{1}{3} + \cdots + \frac{1}{m} - \log_e(m+1)\right)z} \prod_{k=1}^{m}\left(1 + \frac{z}{k}\right) \prod_{k=1}^{m-1} e^{-\frac{z}{k}} \right]$$

$$= z \lim_{m \to \infty} \left[e^{\left(1 + \frac{1}{2} + \frac{1}{3} + \cdots + \frac{1}{m} + \frac{1}{m+1} - \log_e(m+1)\right)z} \right] \cdot \lim_{m \to \infty} e^{-\frac{z}{m+1}}$$

$$\cdot \lim_{m \to \infty} \prod_{k=1}^{m}\left(1 + \frac{z}{k}\right) e^{-\frac{z}{k}} \cdot \lim_{m \to \infty} e^{\frac{z}{m}}.$$

If you wonder where the extra two limits here, namely, $\lim e^{-z/(m+1)}$ and $\lim e^{z/m}$, came from, look very sharply and you will see that an extra factor $e^{z/(m+1)}$ was put in the first limit and also that, in combining the two products we *had* to put in an additional factor $e^{-(z/m)}$ in the second product. These two had to be counterbalanced by multiplying by $e^{-z/(m+1)}$ and $e^{z/m}$.

The first of our four limits in our last equation we see to be e^{Cz}, where C is *Euler's constant*. The second and fourth limits are both unity at every finite z. The third limit does exist for every finite z. So, we have Weierstrass's formula

$$\frac{1}{\Gamma(z)} = z e^{Cz} \prod_{n=1}^{\infty}\left(1 + \frac{z}{n}\right) e^{-z/n}.$$

OBSERVATION. We were careful to *combine* the two products in our last step before using the theorem

$$\lim_{m \to \infty}(A_1 A_2 \cdots A_n) = \lim_{m \to \infty} A_1 \cdot \lim_{m \to \infty} A_2 \cdots \lim_{m \to \infty} A_n,$$

because the product

$$\prod_{k=1}^{m}\left(1+\frac{z}{k}\right)$$

does *not* converge in general as $m \to \infty$. For instance, when z is taken at unity, this product diverges to infinity as $m \to \infty$.

Problem: Integral Expressions of B(x, y)

I-25. Show that the Beta integral

$$\int_0^1 t^{x-1}(1-t)^{y-1}dt$$

defines a function of x and y as real arguments when and only when x and y are both positive.   ★

When x and y are both greater than unity, the integral is a proper Riemann integral because the integrand is then a continuous function of t for $0 \leq t \leq 1$.

If $x < 1$ or if $y < 1$ or if both x and y are less than unity, then the integral is improper because at least one of the factors t^{x-1}, $(1-t)^{y-1}$ becomes infinite at an endpoint of the interval $0 \leq t \leq 1$. Let us examine the case when x and y are both less than unity. The other two cases are handled in a similar manner and are even a bit simpler to analyze. We write

$$\int_0^1 t^{x-1}(1-t)^{y-1}dt = \int_0^c \frac{1}{t^{1-x}}\frac{1}{(1-t)^{1-y}}\,dt$$

$$+ \int_c^1 \frac{1}{t^{1-x}}\frac{1}{(1-t)^{1-y}}dt, \qquad 0 < c < 1.$$

The second fraction in the integrand of the first integral on the right is continuous for $0 \leq t \leq c < 1$. Consequently, the first integral on the right is convergent when and only when the integral

$$\int_0^c \frac{1}{t^{1-x}}\,dt$$

is convergent, namely, when and only when $1 - x < 1$, that is, when $x > 0$ and only when $x > 0$. In similar manner one finds that the second integral on the right is convergent when and only when $y > 0$. We conclude, therefore, that the integral on the left is convergent when and only when x and y are both positive:

$$B(x, y) = \int_0^1 t^{x-1}(1 - t)^{y-1}dt, \qquad x > 0, \qquad y > 0. \quad \text{(I-25.1)}$$

Problem: Properties of $B(x, y)$

I-26. Show that $B(y, x) = B(x, y)$. ★

By the definition (I-25.1) we have

$$B(y, x) = \int_0^1 t^{y-1}(1 - t)^{x-1}dt, \qquad y > 0, \qquad x > 0.$$

Upon changing the variable of integration by taking $1 - t = s$, we get

$$B(y, x) = -\int_1^0 (1 - s)^{y-1}s^{x-1}ds = \int_0^1 s^{x-1}(1 - s)^{y-1}ds,$$

which is the *same* integral as in (I-25.1) even though a different letter now represents the variable of integration. Thus,

$$B(y, x) = B(x, y).$$

Problem: Integral Expressions of $B(x, y)$

I-27. Show that $B(x, y) = \int_0^{\pi/2} 2 \sin^{2x-1} \theta \cos^{2y-1} \theta \, d\theta.$ ★

We start again with (I-25.1). Let $t = \sin^2 \theta$. This makes $dt = 2 \sin \theta \cos \theta \, d\theta$. As t ranges over the interval of integration from

$t = 0$ to $t = 1$ we may have θ ranging over any interval on which $\sin \theta$ increases continuously from 0 to 1, say $\theta = 0$ to $\theta = \pi/2$. Then (I-25.1) is transformed into

$$B(x, y) = \int_0^{\pi/2} (\sin^2 \theta)^{x-1} (\cos^2 \theta)^{y-1} \, 2 \sin \theta \cos \theta \, d\theta$$

$$= \int_0^{\pi/2} 2 \sin^{2x-1} \theta \cos^{2y-1} \theta \, d\theta.$$

I-28. Show that $B(x, y)$ can be expressed as an integral whose interval of integration is from zero to infinity.   ★

The problem first of all is to seek a change of the variable of integration, $t = f(u)$, so that as t, in the integral $B(x, y) = \int_0^1 t^{x-1}(1 - t)^{y-1} dt$ courses through the interval zero to unity, the new variable u will vary continuously and monotonically from zero to infinity. We also require the relation $t = f(u)$ to have continuous derivative for all $u \geq 0$. Since we have to stretch the interval $0 \leq t < 1$ out to the interval $0 \leq u < \infty$, we see after a little reflection that a *fractional* formula will do the trick:

Let $t = u/(u + 1)$. Then

$$\frac{dt}{du} = \frac{1}{(u + 1)^2}, \qquad 1 - t = \frac{1}{u + 1}.$$

Then (I-25.1) becomes

$$B(x, y) = \int_0^\infty \frac{u^{x-1} du}{(u + 1)^{x+y}}, \qquad x > 0, \quad y > 0. \qquad \text{(I-28.1)}$$

Properties of B(x, y)

I-29. Show that

$$B(x, y) = \frac{\Gamma(x)\Gamma(y)}{\Gamma(x + y)}. \qquad ★$$

When confronted with this problem it is but natural for one to ask, "How would one suspect a relationship between the Beta function and the Gamma function?" Well, that is why Prob. I-28 preceded this problem. Having learned in Prob. I-28 that the Beta function can be defined by an integral over the interval 0 to $+\infty$, we at least *begin* to suspect the possibility of a relationship to the Gamma function. The very next thought that comes to our minds is concerned with the *kind* of relationship that might exist and with a search for a means of establishing a suspected relation. But it is not easy to get started. In fact, it seems to us that the relation would be more likely to result as a *by-product* of experiment with the Gamma integral somewhat along the following line of thought.

In (I-11) we hit upon the idea of multiplying $\Gamma(\tfrac{1}{2})$ by $\Gamma(\tfrac{1}{2})$ and changing the product first to an iterated integral, thence to the equivalent double integral in rectangular coordinates, thence to polar coordinates, and finally to a product of integrals. Let us ask ourselves, "If that scheme was successful in evaluating the product of $\Gamma(\tfrac{1}{2})\Gamma(\tfrac{1}{2})$, might we not expect it to yield something useful not only for a particular such product but also for the general product $\Gamma(x)\Gamma(y)$?" At any rate it is worth trying to see what happens.

We start with the product $\Gamma(p)\Gamma(q)$ instead of $\Gamma(x)\Gamma(y)$, because we shall find it convenient and familiar in doing this problem to use the letters x and y as variables of integration both in single integrals and in a double integral taken over a quadrant of the cartesian xy-plane. Then by Prob. (I-3) we have

$$\Gamma(p) = 2\int_0^\infty x^{2p-1}e^{-x^2}dx, \qquad \Gamma(q) = 2\int_0^\infty y^{2q-1}e^{-y^2}dy,$$

whence by multiplication

$$\Gamma(p)\Gamma(q) = 4\left[\int_0^\infty x^{2p-1}e^{-x^2}dx\right]\left[\int_0^\infty y^{2q-1}e^{-y^2}dy\right].$$

Now the right side here is a *product* of two integrals, except for the factor 4. But its *appearance* suggests the following line of thought.

The *double* integral

$$\iint_Q x^{2p-1}y^{2q-1}e^{-(x^2+y^2)}dA$$

in which Q denotes the entire first quadrant of the cartesian xy-plane, though improper, is seen to be convergent by considering it as the limit of the double integral over a closed rectangular region R for which $0 \leq x \leq M, \quad 0 \leq y \leq N$ as M and N both $\rightarrow \infty$. The double integral over R can be evaluated by the *iterated* integral

$$\int_0^M x^{2p-1}e^{-x^2}dx \int_0^N y^{2q-1}e^{-y^2}dy.$$

But the integral in y here has an integrand free of x and the limits of integration are free of x. So the integral in y will yield merely a constant to carry forward into the integral in x. Thus the iterated integral is equivalent to the *product*

$$\left[\int_0^M x^{2p-1}e^{-x^2}dx\right]\left[\int_0^N y^{2q-1}e^{-y^2}dy\right]$$

whose limit as M and N both $\rightarrow \infty$ is the product of integrals on the right side of our equation above which we set up for $\Gamma(p)\Gamma(q)$. We conclude, therefore, that

$$\Gamma(p)\Gamma(q) = 4\iint_Q x^{2p-1}y^{2q-1}e^{-(x^2+y^2)}dA.$$

The exponent on e here suggests the use of polar coordinates: $x^2 + y^2 = r^2, \quad x = r \cos \theta, \quad y = r \sin \theta$. Then we have

$$\Gamma(p)\Gamma(q) = 4\iint_Q \sin^{2p-1}\theta \cos^{2q-1}\theta \, e^{-r^2}r^{2p+2q-2}dA$$

$$= \int_0^{\pi/2} 2\sin^{2p-1}\theta \cos^{2q-1}\theta \, d\theta \int_0^\infty 2r^{2p+2q-1}e^{-r^2}dr.$$

Now by the same argument that we used with xy-coordinates we can replace the iterated integral here by a product of integrals:

$$\Gamma(p)\Gamma(q) = \left[\int_0^{\pi/2} 2 \sin^{2p-1} \theta \cos^{2q-1} \theta \, d\theta\right]\left[\int_0^{\infty} 2 \, r^{2(p+q)-1} e^{-r^2} \, dr\right].$$

By Prob. (I-27) and Prob. (I-3) these last two integrals are recognized respectively to be $B(p, q)$ and $\Gamma(p + q)$. Finally, then

$$\Gamma(p)\Gamma(q) = B(p, q)\Gamma(p + q),$$

whence, upon replacing p and q respectively by x and y and dividing both sides by $\Gamma(p + q)$, we have

$$B(x, y) = \frac{\Gamma(x)\Gamma(y)}{\Gamma(x + y)}. \qquad (\text{I-29.1})$$

REMARK.   It can be shown via (I-29.1) that

$$B(x, y) = B(x + 1, y) + B(x, y + 1).$$

For we have

$$B(x + 1, y) = \frac{\Gamma(x + 1)\Gamma(y)}{\Gamma(x + 1 + y)} = \frac{x\Gamma(x)\Gamma(y)}{(x + y)\Gamma(x + y)} = \frac{x}{x + y} B(x, y)$$

by Eq. (I-29.1) and Prob. I-4. In similar manner we find that

$$B(x, y + 1) = \frac{y}{x + y} B(x, y).$$

Therefore,

$$B(x + 1, y) + B(x, y + 1) = B(x, y)\left[\frac{x}{x + y} + \frac{y}{x + y}\right].$$

The sum in brackets in the last equation equals unity.

I-30. Evaluate $B(p, 1 - p)$ when p is any positive number less than unity.   ★

Taking $x = p$ with $0 < p < 1$ and $y = 1 - p$ in $\displaystyle\int_0^\infty \frac{u^{x-1}du}{(u+1)^{x+y}}$ of Eq. (I-28.1), we get

$$B(p, 1 - p) = \int_0^\infty \frac{u^{p-1}du}{u + 1}, \qquad 0 < p < 1. \qquad \text{(I-30.1)}$$

What do we do now? If the interval of integration were from $-\infty$ to $+\infty$ we might be able to evaluate the resulting integral by the usual method of contour integration on the plane of the complex variable $z = x + iy$, where $i = \sqrt{-1}$. Well let us *make* the interval of integration stretch from $-\infty$ to $+\infty$ by the transformation $u = e^x$, which puts x at $-\infty$ when $u = 0$ and puts x at $+\infty$ when u is at $+\infty$. Not'ng that $du/dx = e^x$, we have

$$B(p, 1 - p) = \int_{-\infty}^{+\infty} \frac{e^{px}}{1 + e^x} dx, \qquad 0 < p < 1. \qquad \text{(I-30.2)}$$

To evaluate this last integral we consider the integral

$$\oint_C \frac{e^{pz}dz}{1 + e^z}, \qquad z = x + iy,$$

taken in the positive sense around the rectangle C lying on the complex plane and having its vertices at $z = -R$, $z = R$, $z = R + 2\pi i$, $z = -R + 2\pi i$. The integrand is analytic everywhere on the complex plane except at $z = \pm\pi i$, $\pm\pi 3i$, $\cdots$. At each of these points the integrand has a simple pole. Our contour encloses just one such point, namely $z = \pi i$. To verify, for instance, that the integrand has a simple pole at $z = \pi i$ we have but to write out the Taylor series expansion for the denominator in powers of $z - \pi i$:

$$1 + e^z = 0 + (-1)(z - \pi i) + \frac{-1}{2!}(z - \pi i)^2 + \cdots$$
$$= (z - \pi i)G(z),$$

where $G(z)$ is analytic at $z = \pi i$ with $G(\pi i) = -1 \neq 0$. Thus, the denominator $1 + e^z$ has a simple zero at $z = \pi i$. And, since the numerator e^{pz} does not vanish at $z = \pi i$, the integrand has a simple pole there. Accordingly, the residue of the integrand at $z = \pi i$ is equal to the value $e^{p\pi i}$ of the numerator there divided by the value there of the derivative of the denominator, namely $e^{\pi i} = -1$. Thus, the residue at our lone pole within the contour is $-e^{p\pi i}$. Then, by Cauchy's residue theorem we have

$$\oint_C \frac{e^{pz} dz}{1 + e^z} = 2\pi i(- e^{p\pi i}).$$

We now write this contour integral as the sum of four line integrals, one along each side of the rectangle. Then we compute the sum of the limits approached by the four integrals as we make $R \to \infty$. Since we are holding the width of the rectangle constant at 2π all the while that we are stretching the length beyond all bounds, there will always be just one pole of the integrand within the contour. Consequently, the sum of the four limits, provided we show that each of the four limits exists individually, must be $-2\pi i e^{p\pi i}$.

The integral along the base of the rectangle, where $z = x$, has as its limit the integral on the right in Eq. (I-30.2).

The integral going from east to west along the top of the rectangle, where $z = x + 2\pi i$, becomes

$$\int_R^{-R} \frac{e^{px} e^{2p\pi i} \, dx}{1 + e^x e^{2\pi i}} = - e^{2p\pi i} \int_{-R}^{R} \frac{e^{px}}{1 + e^x} \, dx.$$

Hence, the limit of the integral along the top is $-e^{2p\pi i}$ times the integral in Eq. (I-30.2)

The integral taken upward along the right side of the rectangle where $z = R + iy$ from $y = 0$ to $y = 2\pi$ is appraised as to its absolute value by the well known appraisal theorem which says that the absolute value of an integral along a curve C of a continuous function of z does not exceed the product of the length of C by the maximum of the absolute value of the integrand on C. For the integral along the right side of the rectangle we thus have

$$\left| \int_0^{2\pi} \frac{e^p R e^{ipy}}{1 + e^{R+iy}} dy \right| \leq (2\pi) \frac{e^p R}{e^R - 1}.$$

Since the power of e in the denominator of the fraction on the right side of this appraisal is greater than the power of e in the numerator inasmuch as $0 < p < 1$, we conclude that the limit of the appraisal as $R \rightarrow + \infty$ is zero. Likewise one finds that the limit of the integral along the left side of the rectangle is also zero. Adding together the four limits, two of which vanish, we have

$$(1 - e^{2p\pi i}) \int_{-\infty}^{+\infty} \frac{e^{px}}{1 + e^x} dx = - 2\pi i e^{p\pi i},$$

whence by Eq. (I-30.2)

$$B(p, q) = \frac{- 2\pi i e^{p\pi i}}{1 - e^{2p\pi i}}, \qquad 0 < p < 1, \quad q = 1 - p.$$

Now we multiply numerator and denominator of this result by $-e^{-p\pi i}/2i$ and we find that the new denominator, which is now under π, is none other than $\sin p\pi$ in the Euler exponential form. At last we have our evaluation, namely

$$B(p, 1 - p) = \frac{\pi}{\sin p\pi}, \qquad 0 < p < 1. \tag{I-30.3}$$

Problem: Properties of $\Gamma(x)$

I-31. Apply the fundamental identity $\Gamma(x + 1) = x\Gamma(x)$ and its variant $\Gamma(-x) = \Gamma(1 - x)/(-x)$ and the identity $B(x, y) = \Gamma(x)\Gamma(y)/\Gamma(x + y)$ together with $B(p, 1 - p) = \pi/\sin p\pi, 0 < p < 1$ to show that

$$\Gamma(p)\Gamma(1 - p) = \frac{\pi}{\sin p\pi}, \qquad p \text{ nonintegral.} \qquad \star$$

The reason for the restriction that p be nonintegral is, of course, evident. An integral value for p could not be used in the denominator of the formula to be established, for the denominator would then be zero.

We start with $0 < p < 1$ and apply Eq. (I-29.1) together with Prob. I-9:

$$\frac{\Gamma(p)\Gamma(1-p)}{\Gamma(p+1-p)} = B(p, 1-p),$$

$$\Gamma(p)\Gamma(1-p) = \frac{\pi}{\sin p\pi}, \qquad 0 < p < 1.$$

Next, taking $h = p + 1$ and using Eq. (I-4.1) and Eq. (I-4.4) in conjunction with the result just obtained, we have

$$
\begin{aligned}
\Gamma(h)\Gamma(1-h) &= \Gamma(p+1)\Gamma(-p) \\
&= p\Gamma(p)\frac{\Gamma(1-p)}{-p} \\
&= \frac{-\pi}{\sin p\pi}, \qquad 0 < p < 1 \\
&= \frac{-\pi}{\sin(h-1)\pi} \\
&= \frac{-\pi}{-\sin h\pi} \\
&= \frac{\pi}{\sin h\pi}, \qquad 1 < h < 2.
\end{aligned}
$$

Similarly one shows that the formula holds for $2 < h < 3$ and then by mathematical induction for every positive nonintegral h.

Starting again with $0 < p < 1$ and taking now $h = p - 1$, we get in the same manner as before

$$\Gamma(h)\Gamma(1-h) = \frac{\pi}{\sin h\pi}, \qquad -1 < h < 0.$$

Again, as before, one finds that the formula holds for $-2 < h < -1$ and then by mathematical induction for every negative nonintegral h.

Finally, we conclude that

$$\Gamma(p)\Gamma(1 - p) = \frac{\pi}{\sin p\pi}$$

is true for every nonintegral p, positive or negative.

Problem: $\Gamma'(1) = $ *Negative of Euler's Constant*

I-32. Show that

$$\log_e x = \int_0^\infty \frac{e^{-t} - e^{-xt}}{t}dt, \qquad x > 0; \qquad \text{(I-32.1)}$$

and then from Eq. (I-32.1) obtain

$$\Gamma'(1) = -\gamma, \qquad \text{(I-32.2)}$$

where γ denotes *Euler's constant*, namely

$$\gamma = \lim_{n \to \infty}\left[1 + \frac{1}{2} + \frac{1}{3} + \cdots + \frac{1}{n} - \log_e n\right]. \qquad \star \qquad \text{(I-32.3)}$$

The denominator of the integral in Eq. (I-32.1) vanishes at $t = 0$. But by L'Hospital's Rule we find that the integrand approaches the limit $x - 1$ as $t \to 0$. We therefore assign this limit value $x - 1$ to the integrand at $t = 0$, thereby making the integrand continuous for $t \geqq 0$.

Examination of the integral in Eq. (I-32.1) for convergence in the manner employed in Prob. I-1 shows that the integral is convergent for every positive value of x, thus defining a function of x for $x > 0$. But how do we see that it defines the function $\log_e x$? One way of answering this question is to show that the derivative of the integral is $1/x$, which is done by differentiating the integral with respect to x

by Leibniz's Rule for differentiating under the integral sign as in Prob. I-20:

$$\frac{d}{dx}\int_0^\infty \frac{e^{-t} - e^{-xt}}{t}\,dt = \int_0^\infty \frac{\partial}{\partial x}\left[\frac{e^{-t} - e^{-xt}}{t}\right]dt \qquad \text{(I-32.4)}$$

$$= \int_0^\infty \frac{te^{-xt}}{t}\,dt \qquad \text{(I-32.5)}$$

$$= \lim_{N\to\infty}\left[-\frac{1}{xe^{xt}}\right]_{t=0}^{t=N} \qquad \text{(I-32.6)}$$

$$= \frac{1}{x}. \qquad \text{(I-32.7)}$$

So, the integral in Eq. (I-32.1) equals $\log_e x + C$, where C is a constant. To find C we evaluate the integral with $x = 1$, getting zero because the integrand is then identically zero. Thus $C = 0$, and our integral in Eq. (I-32.1) does equal $\log_e x$.

Another way of establishing Eq. (I-32.1) is to start from the equation

$$\frac{1}{u} = \int_0^\infty e^{-ut}\,dt, \quad u > 0, \qquad \text{(I-32.8)}$$

which is apparent from the right side of Eqs. (I-32.5) and (I-32.7). Integrating both sides of Eq. (I-32.8) from $u = 1$ to $u =$ an arbitrary positive x, we get

$$\int_1^x \frac{1}{u}\,du = \int_1^x\left[\int_0^\infty e^{-ut}\,dt\right]du, \qquad \text{(I-32.9)}$$

$$\log_e x = \int\!\!\int_R e^{-ut}\,dA, \qquad \text{(I-32.10)}$$

where R denotes the rectangular plane region comprised of all the points (u, t) for which $t > 0$ and for which $1 \leq u \leq x$ or $u \leq x \leq 1$. We may evaluate the double integral in Eq. (I-32.10) by an iterated integral in which the order of integration is reversed from the order in the iterated integral in Eq. (I-32.9):

$$\log_e x = \int_0^\infty \left[\int_1^x e^{-ut} du \right] dt, \qquad \text{(I-32.11)}$$

$$= \int_0^\infty \frac{e^{-t} - e^{-xt}}{t} dt, \qquad \text{(I-32.12)}$$

assigning the value $x - 1$ to the integrand when $t = 0$, as in the first paragraph.

Now to the second part of our problem: to show that the derivative of $\Gamma(x)$ at $x = 1$ is the negative of Euler's Constant. Since we wish to evaluate the derivative of $\Gamma(x)$ for a particular x, it seems that a natural point of departure is the integral formula for the derivative $\Gamma'(x)$ established in Prob. I-20:

$$\Gamma'(x) = \int_0^\infty (\log_e t) t^{x-1} e^{-t} dt, \qquad \text{(I-32.13)}$$

$$\Gamma'(1) = \int_0^\infty (\log_e t) e^{-t} dt. \qquad \text{(I-32.14)}$$

The next step is perhaps not so apparent nor so easily motivated. But its exploitation will indicate how to proceed to our goal. It is to replace $\log_e t$ in Eq. (I-32.14) by the integral expression for $\log_e t$ as established in Eq. (I-32.1), namely,

$$\log_e t = \int_0^\infty \frac{e^{-u} - e^{-tu}}{u} du. \qquad \text{(I-32.15)}$$

Thus, Eq. (I-32.14) becomes

$$\Gamma'(1) = \int_0^\infty \left[\int_0^\infty \frac{e^{-u} - e^{-tu}}{u} du \right] e^{-t} dt. \qquad \text{(I-32.16)}$$

And now, as we did in the first part of the present problem, we may replace the iterated integral in Eq. (I-32.16) by a double integral over the first quadrant in the tu-plane and then replace the double integral by an iterated integral in which the order of integration is reversed from what it is in Eq. (I-32.16):

$$\Gamma'(1) = \int_0^\infty \left[\int_0^\infty (e^{-u} - e^{-tu})e^{-t}dt \right] \frac{du}{u} \qquad \text{(I-32.17)}$$

$$= \int_0^\infty \left[e^{-u} \int_0^\infty t^0 e^{-t}dt - \int_0^\infty t^0 e^{-(u+1)t}dt \right] \frac{du}{u}. \qquad \text{(I-32.18)}$$

The integrals within the brackets are respectively $\Gamma(1)$ and $\Gamma(1)/(u+1)$ by Prob. II-3. But $\Gamma(1) = 1$ by Prob. I-9. Accordingly, Eq. (I-32.18) becomes

$$\Gamma'(1) = \int_0^\infty \frac{-1}{u}\left(\frac{1}{u+1} - e^{-u} \right)du, \qquad \text{(I-32.19)}$$

wherein we assign the value zero to the integrand at $u = 0$, for this is the limit value which the integrand (by L'Hospital's Rule) approaches as $u \to 0$.

And now we have

$$\Gamma'(1) = -\lim_{\substack{n \to \infty \\ h \to 0}} \left[\int_h^n \frac{du}{u} - \int_h^n \frac{e^{-u}}{u}du - \int_h^n \frac{du}{u+1} \right]. \qquad \text{(I-32.20)}$$

Since the limit on the right in Eq. (I-32.20) exists, namely $\Gamma'(1)$, and is approached as n and h vary simultaneously and independently with $n \to \infty$ and $h \to 0$, we may and we will let $n \to \infty$ through the sequence of positive integers:

$$n = 1, 2, 3, \cdots.$$

We will also take advantage of the appraisal

$$0 \leq e^{-u} - \left(1 - \frac{u}{n} \right)^n \leq e^{-u}u^2/n \qquad \text{(I-32.21)}$$

established in Whittaker and Watson's *Modern Analysis* (4th ed.), page 242. This allows us to write

$$e^{-u} = \left(1 - \frac{u}{n}\right)^n + R, \qquad 0 \leq R \leq e^{-u}u^2/n.$$

Making this replacement for e^{-u} in the second integral in Eq. (I-32.20) and evaluating the third integral, we get

$$\Gamma'(1) = -\lim_{\substack{n\to\infty \\ h\to 0}} \left[\int_h^n \frac{1 - \left(1 - \frac{u}{n}\right)^n}{u}\, du - \log_e n + \left\{\log_e(h+1)\right.\right.$$

$$\left.\left. - \log_e\left(1 + \frac{1}{n}\right) - \int_h^n \frac{R\,du}{u}\right\}\right].$$

$$(I\text{-}32.22)$$

Each of the three terms in the braces in Eq. (I-32.22) approaches the limit zero as $h \to 0$ and $n \to \infty$. Accordingly, Eq. (I-32.22) reduces to

$$\Gamma'(1) = -\lim_{\substack{n\to\infty \\ h\to 0}} \left[\int_h^n \frac{1 - \left(1 - \frac{u}{n}\right)^n}{u}\, du - \log_e n\right]. \qquad (I\text{-}32.23)$$

To simplify the evaluation of the integral in Eq. (I-32.23) we let $u/n = t$. Then we have

$$\Gamma'(1) = -\lim_{\substack{n\to\infty \\ h\to 0}} \left[\int_{h/n}^1 \frac{1 - (1-t)^n}{t}\, dt - \log_e n\right]$$

$$= -\lim_{\substack{n\to\infty \\ h\to 0}} \left[\int_0^1 \frac{1 - (1-t)^n}{t}\, dt - \int_0^{h/n} \frac{1 - (1-t)^n}{t}\, dt - \log_e n\right].$$

$$(I\text{-}32.24)$$

The first integral on the right in Eq. (I-32.24) is readily evaluated by letting $1 - t = w$. Thus,

$$\int_0^1 \frac{1 - (1 - t)^n}{t}\, dt = \int_0^1 \frac{1 - w^n}{1 - w}\, dw$$

$$= \int_0^1 (1 + w + w^2 + \cdots + w^{n-1})dw$$

$$= \left[1 + \frac{1}{2}w + \frac{1}{3}w^3 + \cdots + \frac{1}{n}w^n \right]_0^1$$

$$= 1 + \frac{1}{2} + \frac{1}{3} + \cdots + \frac{1}{n}.$$

The second integral on the right in Eq. (I-32.24) looks as if its value might approach the limit zero as $h \to 0$ while $n \to \infty$. Accordingly, we shall try to *appraise* this integral rather than actually evaluate it. Applying the binomial theorem to $(1 - t)^n$, subtracting the unit terms and then replacing all negative terms by their absolute values, we get

$$0 < \int_0^{h/n} \frac{1 - (1 - t)^n}{t}dt < \int_0^{h/n} \frac{1}{t}\left[nt + \frac{n(n-1)}{2!}t^2 + \frac{n(n-1)(n-2)}{3!}t^3 \right.$$

$$\left. + \cdots + nt^{n-1} + t^n \right]dt.$$

The integral on the right in this inequality equals

$$h + \frac{(1)\left(1 - \dfrac{1}{n}\right)}{2!}\frac{h^2}{2} + \frac{(1)\left(1 - \dfrac{1}{n}\right)\left(1 - \dfrac{2}{n}\right)}{3!}\frac{h^3}{3} + \cdots + \frac{h^{n-1}}{(n-1)n^{n-2}}$$

$$+ \frac{h^n}{n^{n+1}},$$

which, for $n > 1$, is seen to be less than

$$h + h^2 + h^3 + \cdots + h^{n-1} + h^n = \frac{(h)(1 - h^n)}{1 - h}.$$

But $\lim_{h \to 0} \left[h(1 - h^n)/(1 - h) \right] = 0$. It follows that

$$\lim_{\substack{n \to \infty \\ h \to 0}} \int_0^{h/n} \frac{1 - (1 - t)^n}{t} \, dt = 0.$$

Thus, Eq. (I-32.24) now becomes

$$\Gamma'(1) = - \lim_{n \to \infty} \left[1 + \frac{1}{2} + \frac{1}{3} + \cdots + \frac{1}{n} - \log_e n \right] = - \gamma. \quad \text{(I-32.25)}$$

REMARKS. 1. An approximate value for Euler's constant γ is .5772157.

2. We can generalize the result of this problem by taking $x = n = $ a positive integer ≥ 2 in Eq. (I-32.13). Then paralleling the steps taken in Eqs. (I-32.14) through (I-32.25), we find that

$$\frac{\Gamma'(n)}{\Gamma(n)} = - \gamma + 1 + \frac{1}{2} + \frac{1}{3} + \cdots + \frac{1}{n - 1}, \quad n = 2, 3, \cdots.$$

This ratio $\Gamma'(n)/\Gamma(n)$ is, of course, the derivative of $\log_e \Gamma(x)$ at $x = n$. Evaluation of $\Gamma'(n)/\Gamma(n)$ for $n = 2, 3, \ldots$ is carried out in Prob. I-33.

Problem: Logarithmic Derivative of $\Gamma(x)$ *for Positive Integers*

I-33. The logarithmic derivative of $\Gamma(x)$ is often denoted by $\psi(x)$:

$$\psi(x) = \frac{d}{dx} \left[\log_e \Gamma(x) \right] = \frac{\Gamma'(x)}{\Gamma(x)}. \quad \text{(I-33.1)}$$

Evaluate $\psi(n)$ where n denotes a positive integer ≥ 2. $\star$

Starting with the formula for $\Gamma'(x)$ in Eq. (I-20.1) and carrying through as in Prob. I-32, we get

$$\Gamma'(n) = \int_0^\infty (\log_e x) x^{n-1} e^{-x} dx \tag{I-33.2}$$

$$= \int_0^\infty \left[\int_0^\infty \frac{e^{-t} - e^{-xt}}{t} dt \right] x^{n-1} e^{-x} dx \tag{I-33.3}$$

$$= \int_0^\infty \left[\int_0^\infty (e^{-t} - e^{-tx}) x^{n-1} e^{-x} dx \right] \frac{dt}{t} \tag{I-33.4}$$

$$= \int_0^\infty \left[e^{-t} \int_0^\infty x^{n-1} e^{-x} dx - \int_0^\infty x^{n-1} e^{-(t+1)x} dx \right] \frac{dt}{t}. \tag{I-33.5}$$

As in Prob. I-32 we observe that the first integral within the brackets in Eq. (I-33.5) is $\Gamma(n)$ by Eq. (I-1) and that the second integral within the brackets is $\Gamma(n)/(t+1)^n$ by Prob. II-3. Taking out the constant factor $\Gamma(n)$, we find that Eq. (I-33.5) becomes

$$\psi(n) = \frac{\Gamma'(n)}{\Gamma(n)} = \int_0^\infty \frac{1}{t} \left[e^{-t} - \frac{1}{(t+1)^n} \right] dt, \tag{I-33.6}$$

assigning the value $n-1$ to the integrand at $t=0$, as in Prob. I-32, inasmuch as $n-1$ is the limit which we find (by L'Hospital's Rule) to be approached by the integrand as $t \to 0$.

Our integrand, except for the exponent n on the binomial $t+1$, is the same as the integrand in Eq. (I-32.19). This suggests that we transform our integrand so as to follow through with the steps taken in Eqs. (I-32.20) through (I-32.25). So we change Eq. (I-33.6) of the present problem to

$$\psi(n) = \int_0^\infty \left[\frac{1}{t} \left(\frac{1}{t+1} - \frac{1}{(t+1)^n} \right) - \left(\frac{1}{t} \frac{1}{t+1} - \frac{e^{-t}}{t} \right) \right] dt \tag{I-33.7}$$

$$= \int_0^\infty \frac{1}{t} \left(\frac{1}{t+1} - \frac{1}{(t+1)^n} \right) dt + \int_0^\infty \frac{-1}{t} \left(\frac{1}{t+1} - e^{-t} \right) dt. \tag{I-33.8}$$

It is permissible to write the integral in Eq. (I-33.7) as the sum of the two integrals in Eq. (I-33.8) because in Prob. I-32 we established convergence of the second integral in Eq. (I-33.8) and found its value to be $\Gamma'(1) = -\gamma$. Since the integral for $\psi(n)$ in Eq. (I-33.6) is convergent and since the second integral in Eq. (I-33.8) is also convergent, it follows that the first integral in Eq. (I-33.8) must be convergent. Eq. (I-33.8) becomes

$$\psi(n) = -\gamma + \int_0^\infty \frac{1}{t}\left(\frac{1}{t+1} - \frac{1}{(t+1)^n}\right) dt. \qquad \text{(I-33.9)}$$

The integrand in Eq. (I-33.9) is undefined at $t = 0$. But we find by L'Hospital's Rule that the integrand approaches the limit $n - 1$ as $t \to 0$. So we assign this value to the integrand at $t = 0$. To simplify the evaluation of the integral in Eq. (I-33.9) we let $t + 1 = b$. Then

$$\psi(n) = -\gamma + \int_1^\infty \frac{1}{b-1}\left(\frac{1}{b} - \frac{1}{b^n}\right) db \qquad \text{(I-33.10)}$$

$$= -\gamma + \int_1^\infty \frac{1}{b-1}\left(\frac{b^{n-1} - 1}{b^n}\right) db \qquad \text{(I-33.11)}$$

$$= -\gamma + \int_1^\infty \frac{1}{b^n}(b^{n-2} + b^{n-3} + \cdots + b + 1) db \qquad \text{(I-33.12)}$$

$$= -\gamma + \int_1^\infty (b^{-2} + b^{-3} + \cdots + b^{-n}) db. \qquad \text{(I-33.13)}$$

We have thus transformed the integral in Eq. (I-33.9) into a form where we can carry out the integration. The result is seen to be

$$\psi(n) = -\gamma + 1 + \frac{1}{2} + \frac{1}{3} + \cdots + \frac{1}{n-1}, \qquad \text{(I-33.14)}$$

where n denotes any integer ≥ 2, $\psi(n)$ denotes the logarithmic derivative $\Gamma'(n)/\Gamma(n)$, and γ denotes Euler's constant:

$$\gamma = \lim_{n\to\infty}\left(1 + \frac{1}{2} + \frac{1}{3} + \cdots + \frac{1}{n} - \log_e n\right).$$

TABLE I-2   FORMULAS

$$\Gamma(x) = \int_0^\infty t^{x-1}e^{-t}dt, \quad x > 0.$$

$$\Gamma(x) = \int_0^\infty 2u^{2x-1}e^{-u^2}du, \quad x > 0.$$

$$\Gamma(x) = \int_0^1 \log\left(\frac{1}{v}\right)^{x-1}dv, \quad x > 0.$$

$$\Gamma(x) = \frac{\Gamma(x+1)}{x}, \quad x \neq 0, -1, -2, \cdots .$$

$$\Gamma(x) = (x-1)\Gamma(x-1), \quad x \neq 0, -1, -2, \cdots .$$

$$\Gamma(-x) = \frac{\Gamma(1-x)}{-x}, \quad x \neq 0, 1, 2, \cdots .$$

$$\Gamma(n) = (n-1)!, \quad n = 1, 2, 3, \cdots, \text{ with the convention } 0! = 1.$$

$$\Gamma\left(\frac{1}{2}\right) = \sqrt{\pi}$$

$$\Gamma\left(n+\frac{1}{2}\right) = \frac{1 \cdot 3 \cdot 5 \cdots (2n-1)\sqrt{\pi}}{2^n}, \quad n = 1, 2, 3, \cdots .$$

$$\Gamma(x)\Gamma(1-x) = \frac{\pi}{\sin x\pi}, \quad x \neq 0, \pm 1, \pm 2, \pm 3, \cdots .$$

$$n! = \left(\frac{n}{e}\right)^n\sqrt{2\pi n} + h, \quad n = 1, 2, 3, \cdots, 0 < \frac{h}{n!} < \frac{1}{12n}.$$

$$\int_0^\infty t^a e^{-bt^c}dt = \frac{\Gamma\left(\dfrac{a+1}{c}\right)}{cb^{(a+1)/c}}, \quad a > -1, \quad b > 0, \quad c > 0.$$

$$B(x, y) = \int_0^1 t^{x-1}(1-t)^{y-1}dt, \quad x > 0, \quad y > 0.$$

$$B(x, y) = \int_0^{\pi/2} 2 \sin^{2x-1} \theta \cos^{2y-1} \theta \, d\theta, \qquad x > 0. \quad y > 0.$$

$$B(x, y) = \int_0^\infty \frac{u^{x-1}}{(u + 1)^{x+y}} du, \qquad x > 0, \quad y > 0.$$

$$B(x, y) = \frac{\Gamma(x)\Gamma(y)}{\Gamma(x + y)}.$$

$$B(x, y) = B(y, x).$$

$$B(x, 1 - x) = \frac{\pi}{\sin x\pi}, \qquad 0 < x < 1.$$

2 APPLICATIONS

OF THE GAMMA FUNCTION &

THE BETA FUNCTION

INTRODUCTION

In this chapter we present a modest selection of worked out problems involving applications of the Gamma and Beta Functions to (1) the evaluation of certain definite integrals, (2) the development of Wallis's infinite product for $\pi/2$, (3) the derivation and application of formulas for certain geometrical and physical magnitudes, (4) the derivation of Stirling's approximation formula for $n!$, and (5) the application thereof to a few problems in probability.

Problems: Evaluation of Integrals

II-1. Show

$$\int_0^\infty t^a e^{-bt^c} dt = \frac{\Gamma\left(\dfrac{a+1}{c}\right)}{cb^{(a+1)/c}},$$

where b and c are positive constants and a is a constant such that $a > -1$. ★

Let $bt^c = x$, from which we can write various powers of t:

$$t = \frac{x^{1/c}}{b^{1/c}}, \quad t^a = \frac{x^{a/c}}{b^{a/c}}, \quad t^{c-1} = \frac{x^{(c-1)/c}}{b^{(c-1)/c}} \text{ or } \frac{1}{t^{c-1}} = \frac{bx^{1/c-1}}{b^{1/c}}.$$

Also $cbt^{c-1}dt = dx$. Then

$$dt = \left(\frac{1}{t^{c-1}}\right)\frac{dx}{cb}$$

$$= \frac{x^{1/c-1}}{cb^{1/c}}dx.$$

We note that the change in variable did not change the interval of integration. Hence,

$$\int_0^\infty t^a e^{-bt^c}dt = \int_0^\infty \frac{x^{a/c}(e^{-x})}{b^{a/c}} \frac{x^{1/c-1}}{cb^{1/c}}dx.$$

By rearranging the integrand and taking the constants out in front of the integral sign we get

$$\int_0^\infty t^a e^{-bt^c}dt = \frac{1}{cb^{\frac{(a+1)}{c}}} \int_0^\infty x^{\frac{a+1}{c}-1} e^{-x}dx.$$

The integral on the right hand side we recognize to be $\Gamma[(a + 1)/c]$ by Eq. (I-1).

II-2. If n is a nonnegative integer and p is a positive constant, show that

$$\int_0^\infty t^n p^{-t}dt = \frac{n!}{(\log_e p)^{n+1}}. \qquad \star$$

We recall that $p^{-t} = (e^{\log_e p})^{-t} = e^{-(\log_e p)t}$. Thus

$$\int_0^\infty t^n p^{-t}dt = \int_0^\infty t^n e^{-(\log_e p)t}dt.$$

We can now apply Prob. II-1, taking $a = n$, $b = \log_e p$, $c = 1$:

$$\int_0^\infty t^n p^{-t} dt = \frac{\Gamma(n+1)}{(\log_e p)^{n+1}} = \frac{n!}{(\log_e p)^{n+1}}$$

by Eq. (I-10.1).

II-3. Show that

$$\int_0^\infty t^{n-1} e^{-(a+1)t} dt = \frac{\Gamma(n)}{(a+1)^n}, \qquad n > 0, \quad a > -1. \qquad \star$$

Let $t = (a+1)^{-1} y$. Then $dt = (a+1)^{-1} dy$. Thus we have

$$\int_0^\infty t^{n-1} e^{-(a+1)t} dt = \int_0^\infty [(a+1)^{-1} y]^{n-1} e^{-y} (a+1)^{-1} dy$$

$$= (a+1)^{-n} \int_0^\infty y^{n-1} e^{-y} dy$$

$$= \frac{\Gamma(n)}{(a+1)^n} \qquad \text{by Eq. (I-1).}$$

II-4. Evaluate $\int_0^\infty e^{-x^2} dx.$ $\star$

This is the probability integral. Writing the integral as

$$\int_0^\infty x^0 e^{-x^2} dx$$

and using the result of Prob. II-1 with $a = 0$, $b = 1$, $c = 2$, we find that

$$\int_0^\infty e^{-x^2} dx = \frac{\Gamma\left(\dfrac{a+1}{c}\right)}{cb^{(a+1)/c}} = \frac{\Gamma\left(\dfrac{0+1}{2}\right)}{(2)1^{\frac{1}{4}}}.$$

But $\Gamma(\frac{1}{2}) = \sqrt{\pi}$ by Prob. I-11. Consequently, the value of the integral is $\sqrt{\pi}/2$.

It is worth noting that if the interval of integration were $-\infty$ to $+\infty$, then the result would be $\sqrt{\pi}$.

II-5. Evaluate $\displaystyle\int_0^\infty x^{\frac{1}{2}}e^{-x^{\frac{1}{3}}}dx$. ★

Using the result of Prob. II-1 with $a = \frac{1}{2}$, $b = 1$, $c = \frac{1}{3}$, we find that the value of the integral to be evaluated is

$$\frac{\Gamma\left(\dfrac{a+1}{c}\right)}{cb^{(a+1)/c}} = \frac{\Gamma\left(\dfrac{\frac{1}{2}+1}{\frac{1}{3}}\right)}{(\frac{1}{3})(1)} = 3\Gamma\left(\frac{9}{2}\right).$$

Then by Prob. I-6 we can change the argument to a number between 1 and 2. Thus the value of our integral is

$$3\Gamma\left(\frac{9}{2}\right) = (3)\left(\frac{7}{2}\right)\left(\frac{5}{2}\right)\left(\frac{3}{2}\right)\Gamma\left(\frac{3}{2}\right).$$

From Table I-1, $\Gamma(1.5) \cong .8862$, so that our result reduces to the numerical result 34.9 to the nearest tenth. If we take $\Gamma(\frac{3}{2}) = (\frac{1}{2})\Gamma(\frac{1}{2})$ by Prob. I-6 and replace $\Gamma(\frac{1}{2})$ by $\sqrt{\pi}$ by Prob. I-11, we find the exact value of the integral to be $315\sqrt{\pi}/16$.

II-6. Evaluate

$$\int_0^1 x^2 \left(\log_e \frac{1}{x}\right)^3 dx. \quad ★$$

Let $\log_e (1/x) = t$. Then $1/x = e^t$, $x = e^{-t}$, $dx = -e^{-t}dt$. The interval of integration changes: when $x = 0$, $t = \infty$ and when $x = 1$, $t = 0$. Then we have

$$\int_0^1 x^2 \left(\log_e \frac{1}{x}\right)^3 dx = \int_\infty^0 t^3(-e^{-2t})e^{-t}dt$$

$$= \int_0^\infty t^3 e^{-3t}dt.$$

Applying Prob. II-1 with $a = 3$, $b = 3$, $c = 1$, we get

$$\frac{\Gamma\left(\frac{3+1}{1}\right)}{(1)(3^4)} = \frac{3!}{3^4} = \frac{2}{27}$$

as the value of the given integral.

II-7. Show that

$$\int_0^1 x^m \left(\log_e \frac{1}{x}\right)^n dx = \frac{\Gamma(n+1)}{(m+1)^{n+1}}, \qquad m > -1, \quad n > -1. \qquad \star$$

The method of solving this problem is the same as in Prob. II-6.
Let $\log_e (1/x) = t$. Then $1/x = e^t$, $x = e^{-t}$, $x^m = e^{-mt}$, $dx = -e^{-t}dt$. As in Prob. II-6, the interval of integration changes. We note that the negative sign in the integrand is removed by reversing the limits of integration. Then we have

$$\int_0^1 x^m \left(\log_e \frac{1}{x}\right)^n dx = \int_0^\infty t^n e^{-(m+1)t}dt.$$

Evaluating the integral on the right by Prob. II-1 with $a = n$, $b = m + 1$, $c = 1$, we find the value of the given integral to be $\Gamma(n + 1)/(m + 1)^{n+1}$.

II-8. Evaluate

$$\int_0^\infty \frac{t^5 dt}{(1 + t)^9}. \qquad \star$$

We have only to apply Eq. (I-28.1) with $u = t$, $x - 1 = 5$, and $x + y = 9$. This makes $x = 6$ and $y = 3$. Accordingly the value of the integral before us is $B(6, 3)$ which by Eq. (I-29.1) equals $\frac{\Gamma(6)\Gamma(3)}{\Gamma(9)}$; and this by Eq. (I-10.1) equals $\frac{5!2!}{8!} = \frac{1}{168}$.

II-9. Evaluate $\displaystyle\int_0^\infty \frac{x^{p-1}dx}{1+x}$, where $0 < p < 1$, in terms of the Gamma function.   ★

We can evaluate by Eq. (I-30.1). Accordingly, we have

$$\int_0^\infty \frac{x^{p-1}dx}{1+x} = B(p,\, 1-p) = \frac{\Gamma(p)\Gamma(1-p)}{\Gamma(1)} = \Gamma(p)\Gamma(1-p)$$

by Probs. I-29 and I-9.

II-10. Show that

$$\int_0^\infty \frac{x^a\,dx}{(m+x^b)^c} = \frac{m^{(a+1)/b-c}}{b}\left[\frac{\Gamma\!\left(\dfrac{a+1}{b}\right)\Gamma\!\left(c-\dfrac{a+1}{b}\right)}{\Gamma(c)}\right],$$

$$a > -1,\quad b > 0,\quad m > 0,\quad c > (a+1)/b. \qquad ★$$

Let $x^b = m\tan^2\theta$. Then $m + x^b = m\sec^2\theta$, $\quad (m+x^b)^c = m^c\sec^{2c}\theta$.
$x = m^{1/b}\tan^{2/b}\theta$, $\quad x^a = m^{a/b}\tan^{2a/b}\theta$, $\quad dx = \dfrac{2}{b}m^{1/b}\tan^{(2/b)-1}\theta\sec^2\theta\,d\theta$.
Then

$$\int_0^\infty \frac{x^a}{(m+x^b)^c}dx = \int_0^{\pi/2} \frac{m^{\frac{a}{b}}\tan^{\frac{2a}{b}}\theta}{m^c\sec^{2c}\theta}\frac{2}{b}\,m^{\frac{1}{b}}\tan^{\frac{2}{b}-1}\theta\,\sec^2\theta\,d\theta.$$

$$= \frac{m^{\frac{a+1}{b}-c}}{b}\int_0^{\pi/2} 2\sin^{\frac{2a+2}{b}-1}\theta\cos^{2c-\frac{2a+2}{b}-1}\theta\,d\theta.$$

Prob. I-27 applies:

$$2x - 1 = \frac{2a+2}{b} - 1, \qquad 2y - 1 = 2c - \frac{2a+2}{b} - 1,$$

$$x = \frac{a+1}{b}, \qquad\qquad y = c - \frac{a+1}{b}.$$

Thus

$$
\int_0^\infty \frac{x^a}{(m + x^b)^c} \, dx = \frac{m^{\frac{a+1}{b} - c}}{b} B\left(\frac{a + 1}{b}, c - \frac{a + 1}{b}\right)
$$

$$
= \frac{m^{\frac{a+1}{b} - c}}{b} \left[\frac{\Gamma\left(\dfrac{a + 1}{b}\right)\Gamma\left(c - \dfrac{a + 1}{b}\right)}{\Gamma(c)} \right],
$$

$$
a > -1, \quad b > 0, \quad m > 0, \quad c > \frac{a + 1}{b}
$$

by Prob. I-29.

II-11. Show that

$$
\int_0^\infty \frac{w^m}{w^n + a} \, dw = \frac{1}{na^{(n-m-1)/n}} \Gamma\left(\frac{m + 1}{n}\right)\Gamma\left(1 - \frac{(m + 1)}{n}\right)
$$

where the constants m, n, a are such that $a > 0$ and $n > m + 1 > 0$.   ★

Let $w^n = a \tan^2\theta$. Then $w = a^{1/n} \tan^{2/n}\theta$, from which $w^m = a^{m/n} \tan^{2m/n}\theta$ and $dw = a^{1/n}(2/n) \tan^{(2/n)-1}\theta \sec^2\theta \, d\theta$. When $w = 0$, $\theta = 0$ and when $w = \infty$, $\theta = \pi/2$. So we have

$$
\int_0^\infty \frac{w^m}{w^n + a} \, dw = \int_0^{\pi/2} \frac{\left[a^{m/n} \tan^{2m/n}\theta \right] a^{1/n} \left(\dfrac{2}{n}\right) \tan^{2/n-1}\theta \sec^2\theta \, d\theta}{a \tan^2\theta + a}.
$$

$$
\text{(II-11.1)}
$$

The denominator becomes $a \sec^2\theta$. The constants are all collected except for the number 2 and are taken out in front of the integral sign. The other terms are combined or are canceled out. Thus, we have for the right side of Eq. (II-11.1)

$$
\frac{1}{na^{(n-m-1)/n}} \int_0^{\pi/2} 2 \tan^{\frac{2(m+1)}{n} - 1}\theta \, d\theta,
$$

which may be written as

$$\frac{1}{na^{(n-m-1)/n}} \int_0^{\pi/2} 2 \sin^{\frac{2(m+1)}{n}-1} \theta \cos^{-\frac{2(m+1)}{n}+1} \theta\, d\theta.$$

We recognize this last integral as the Beta integral in trigonometric form, as was established in Prob. I-27. Letting $2x - 1$ and $2y - 1$ equal the exponents of $\sin\theta$ and $\cos\theta$ respectively, we get $x = (m + 1)/n$ and $y = 1 - (m + 1)/n$. Thus, the given integral equals

$$\frac{1}{na^{(n-m-1)/n}} B(x, y) = \frac{1}{na^{(n-m-1)/n}} \frac{\Gamma([m + 1]/n)\Gamma(1 - [m + 1]/n)}{\Gamma(1)}$$

$$= \frac{1}{na^{(n-m-1)/n}} \Gamma([m + 1]/n)\Gamma(1 - [m + 1]/n)$$

by Probs. I-29 and I-9.

Problems: $\Gamma(\tfrac{1}{2}) = \sqrt{\pi}$

II-12. Use the definition $B(p, q) = \int_0^{\pi/2} 2 \sin^{2p-1}\theta \cos^{2q-1}\theta\, d\theta$ to show $\Gamma(\tfrac{1}{2}) = \sqrt{\pi}$.   ★

We let $p = q = \tfrac{1}{2}$. Then we have

$$B\left(\frac{1}{2}, \frac{1}{2}\right) = \int_0^{\pi/2} 2 \sin^0\theta \cos^0\theta\, d\theta$$

$$= \int_0^{\pi/2} 2\, d\theta$$

$$= \left[2\theta\right]_0^{\pi/2} = \pi.$$

But

$$B\left(\frac{1}{2}, \frac{1}{2}\right) = \frac{\Gamma\left(\dfrac{1}{2}\right)\Gamma\left(\dfrac{1}{2}\right)}{\Gamma\left(\dfrac{1}{2} + \dfrac{1}{2}\right)} = \left[\Gamma\left(\frac{1}{2}\right)\right]^2$$

by Probs. I-29 and I-9.

Then we have

$$\left[\Gamma\!\left(\frac{1}{2}\right)\right]^2 = \pi,$$

$$\Gamma\!\left(\frac{1}{2}\right) = \sqrt{\pi}.$$

II-13. Apply the formula $\Gamma(x)\Gamma(1-x) = \pi/\sin x\pi$ to evaluate $\Gamma(\frac{1}{2})$.   ★

Let $x = \frac{1}{2}$. Then we have

$$\Gamma\!\left(\frac{1}{2}\right)\Gamma\!\left(\frac{1}{2}\right) = \frac{\pi}{\sin\dfrac{\pi}{2}},$$

$$\left[\Gamma\!\left(\frac{1}{2}\right)\right]^2 = \pi,$$

$$\Gamma\!\left(\frac{1}{2}\right) = \sqrt{\pi}.$$

REMARK. This way of evaluating $\Gamma(\frac{1}{2})$ is indeed much shorter and simpler than either of the two previous ways used in Prob. I-11 and II-12. But in all fairness we must admit, on the other hand, that the derivation in Probs. I-30 and I-31 of the formula applied here in II-13 was neither short nor simple.

Problems: Evaluation of Integrals

II-14. Evaluate $\int_0^{\pi/2} \sqrt{\cos\theta}\, d\theta$.   ★

Inserting unity in the integrand in the form of $2/2$ and $\sin^0\theta$ we have

$$\int_0^{\pi/2} \sqrt{\cos\theta}\, d\theta = \frac{1}{2} \int_0^{\pi/2} 2 \sin^0\theta \cos^{\frac{1}{2}}\theta\, d\theta.$$

We can now apply Prob. I-27 with $2x - 1 = 0$ and $2y - 1 = 1/2$.

Thus,

$$\int_0^{\pi/2} \sqrt{\cos\theta}\, d\theta = \frac{1}{2} B(\tfrac{1}{2}, \tfrac{3}{4})$$

$$= \frac{\Gamma(\tfrac{1}{2})\Gamma(\tfrac{3}{4})}{2\Gamma(\tfrac{5}{4})} \qquad \text{by Prob. I-27}$$

$$= \frac{\sqrt{\pi}\,\Gamma(\tfrac{3}{4})}{2\Gamma(\tfrac{5}{4})} \qquad \text{by Prob. I-11.} \qquad \text{(II-14.1)}$$

From Prob. I-31 we have

$$\Gamma(\tfrac{1}{4})\Gamma(\tfrac{3}{4}) = \frac{\pi}{\sin\tfrac{1}{4}\pi} = \pi\sqrt{2}, \qquad \text{(II-14.2)}$$

whence

$$\Gamma(\tfrac{3}{4}) = \frac{\pi\sqrt{2}}{\Gamma(\tfrac{1}{4})}.$$

Moreover, by Eq. (I-4.1)

$$\Gamma(\tfrac{5}{4}) = \tfrac{1}{4}\Gamma(\tfrac{1}{4}). \qquad \text{(II-14.3)}$$

Substitution in (II-14.1) from (II-14.2) and (II-14.3) yields

$$\int_0^{\pi/2} \sqrt{\cos\theta}\, d\theta = \frac{(2\pi)^{3/2}}{[\Gamma(\tfrac{1}{4})]^2}.$$

REMARKS. 1. The integral to be evaluated in this problem can also be evaluated as an elliptic integral as follows.

Let $\theta = 2t$. Then $\cos\theta = 1 - 2\sin^2 t$. And we have

$$\int_0^{\pi/2} \sqrt{\cos\theta}\, d\theta = 2\int_0^{\pi/4} \sqrt{1 - 2\sin^2 t}\, dt.$$

Now let $\sqrt{2}\sin t = \sin\phi$. Then

$$\sqrt{2}\cos t\, dt = \cos\phi\, d\phi$$

$$dt = \frac{\cos\phi\, d\phi}{\sqrt{2}\cos t} = \frac{\cos\phi\, d\phi}{\sqrt{2}\sqrt{1 - \sin^2 t}}$$

$$= \frac{\cos\phi\, d\phi}{\sqrt{2}\sqrt{1 - \tfrac{1}{2}\sin^2\phi}}.$$

Accordingly,

$$\int_0^{\pi/2} \sqrt{\cos\theta}\, d\theta = \sqrt{2} \int_0^{\pi/2} \frac{\cos^2\phi\, d\phi}{\sqrt{1 - \frac{1}{2}\sin^2\phi}}$$

$$= \sqrt{2} \int_0^{\pi/2} \frac{1 - \sin^2\phi}{\sqrt{1 - \frac{1}{2}\sin^2\phi}}\, d\phi$$

$$= \sqrt{2} \int_0^{\pi/2} \frac{2(1 - \frac{1}{2}\sin^2\phi) - 1}{\sqrt{1 - \frac{1}{2}\sin^2\phi}}\, d\phi$$

$$= \sqrt{2}\left(2\int_0^{\pi/2} \sqrt{1 - \tfrac{1}{2}\sin^2\phi}\, d\phi - \int_0^{\pi/2} \frac{d\phi}{\sqrt{1 - \frac{1}{2}\sin^2\phi}}\right)$$

$$= \sqrt{2}[2E(\sqrt{\tfrac{1}{2}}) - K(\sqrt{\tfrac{1}{2}})],$$

where $E(\sqrt{\frac{1}{2}})$ denotes the value of the complete elliptic integral of the second kind with modulus $\sqrt{\frac{1}{2}}$ and $K(\sqrt{\frac{1}{2}})$ denotes the value of the complete elliptic integral of the first kind with modulus $\sqrt{\frac{1}{2}}$.

2. $\int_0^{\pi/2} \sqrt{\sin\theta}\, d\theta = \int_0^{\pi/2} \sqrt{\cos\theta}\, d\theta.$

II-15. Evaluate $\int_0^{1/2} (t - t^2)^{x-1} dt$, $x > 0$.   ★

(a)   The value of this integral evidently depends on choice of x. In other words the evaluation of this integral will yield us an integration formula available for all such integrals in which the exponent appearing in the integrand is greater than -1. The given integral may be written as

$$\int_0^{1/2} t^{x-1}(1 - t)^{x-1} dt.$$

In this form it looks like the Beta integral (I-25.1) except for two things: (1) the second exponent is not $y - 1$ but is the same, namely

$x - 1$, as the first exponent, (2) the interval of integration extends only to $t = \frac{1}{2}$ instead of to $t = 1$. When we ponder these two facts in conjunction with each other we become aware of the consequence that

$$\int_0^{1/2} t^{x-1}(1 - t)^{x-1}dt = \int_{1/2}^1 t^{x-1}(1 - t)^{x-1}dt,$$

because, as t ranges from $\frac{1}{2}$ to 1, the values taken by the product of t^{x-1} and $(1 - t)^{x-1}$ are the very same values as are taken by the product of $(1 - t)^{x-1}$ and t^{x-1} when t ranges from 0 to $\frac{1}{2}$. It follows that

$$\int_0^{1/2} (t - t^2)^{x-1}dt = \frac{1}{2}\int_0^1 t^{x-1}(1 - t)^{x-1}dt$$

$$= \frac{1}{2} B(x, x) \qquad \text{by Eq. (I-25.1)}$$

$$= \frac{\Gamma(x)\Gamma(x)}{2\Gamma(2x)} \qquad \text{by Eq. (I-29.1).}$$

In the alternative solution (**b**) another approach is employed.

(**b**) Alternative solution. Let $t^2 - t = -u/4$. By completing the square we get $t = \pm (1 - u)^{1/2}/2 + \frac{1}{2}$. We take $t = -\frac{1}{2}(1 - u)^{1/2} + \frac{1}{2}$. Then $dt = \frac{1}{4}(1 - u)^{-1/2}du$. When $t = 0$, $u = 0$; $t = \frac{1}{2}$, $u = 1$. Then

$$\int_0^{1/2} (t - t^2)^{x-1}dt = \int_0^1 \left(\frac{u}{4}\right)^{x-1} \frac{1}{4}(1 - u)^{-1/2}du$$

$$= \frac{1}{4^x}\int_0^1 u^{x-1}(1 - u)^{-1/2}du.$$

Now let $u = \sin^2\theta$. Then $du = 2 \sin \theta \cos \theta \, d\theta$ and $(1 - u)^{-1/2} = (\cos \theta)^{-1}$.

When $u = 0$, $\theta = 0$; when $u = 1$, $\theta = \pi/2$. And the given integral becomes

$$\frac{1}{2^{2x}} \int_0^{\pi/2} \sin^{2x-2}\theta(\cos\theta)^{-1}(2\sin\theta\cos\theta\,d\theta) = 2^{1-2x}\frac{1}{2}\int_0^{\pi/2} 2\sin^{2x-1}\theta\cos^0\theta\,d\theta$$

$$= \frac{2^{1-2x}}{2}B(x, \tfrac{1}{2})$$

$$= \frac{2^{1-2x}\Gamma(x)\sqrt{\pi}}{2\Gamma(x+1/2)}$$

by Probs. I-27 and I-29.

If we equate this result to the value obtained in solution (a) for the integral of the present problem, namely $\Gamma(x)\Gamma(x)/2\Gamma(2x)$, we find that

$$\frac{\Gamma(2x)}{\Gamma(x)} = \frac{\Gamma(x+1/2)}{2^{1-2x}\sqrt{\pi}},$$

which establishes Legendre's duplication formula as given in Eq. (I-18.5) not only when x is any positive integer, as was done in Prob. I-18, but when x is any real positive number.

II-16. Evaluate in terms of the Gamma function, all arguments being between 1 and 2:

$$\int_0^1 x^2(1 - x^4)^{-1/3}dx. \qquad \star$$

Let $x^2 = \sin\theta$. Then $x^4 = \sin^2\theta$, $dx = (1/2)\sin^{-1/2}\theta\cos\theta\,d\theta$. The interval of integration changes: when $x = 0$, $\theta = 0$ and when $x = 1$, $\theta = \pi/2$. Then

$$\int_0^1 x^2(1 - x^4)^{-1/3}dx = \int_0^{\pi/2}\frac{(\sin\theta)\sin^{-1/2}\theta\cos\theta\,d\theta}{2(1 - \sin^2\theta)^{1/3}}$$

$$= \int_0^{\pi/2}\frac{1}{2}\sin^{1/2}\theta\cos^{1/3}\theta\,d\theta$$

$$= \frac{1}{4} B\left(\frac{3}{4}, \frac{2}{3}\right) = \frac{1}{4} \frac{\Gamma\left(\frac{3}{4}\right)\Gamma\left(\frac{2}{3}\right)}{\Gamma\left(\frac{17}{12}\right)} \qquad \text{by Probs.} \atop \text{I-27 and I-29}$$

$$= \frac{1}{2} \frac{\Gamma\left(\frac{7}{4}\right)\Gamma\left(\frac{5}{3}\right)}{\Gamma\left(\frac{17}{12}\right)} \qquad \text{by Eq. (I-4.2).}$$

This integral can also be evaluated by use of Prob. II-17, which is a generalization of the present problem.

II-17. Evaluate $\int_0^a x^b(a^c - x^c)^d dx$ where a, b, c, d are constants with $a > 0$, $c > 0$, $b > -1$, $d > -1$, $b + cd > -1$.   ★

This integral certainly looks like the Beta integral in disguise. We penetrate the disguise by letting $x^c = a^c u$, which makes the given integral equivalent to

$$c^{-1}a^{b+cd+1}\int_0^1 u^{\frac{b+1}{c}-1}(1 - u)^d du.$$

The new integral is just what we expected: the Beta integral. Applying Eq. (I-25.1) and Eq. (I-29.1), we get

$$\int_0^a x^b(a^c - x^c)^d dx = \frac{a^{b+cd+1}\Gamma\left(\frac{b+1}{c}\right)\Gamma(d+1)}{(b + cd + 1)\Gamma\left(\frac{b+cd+1}{c}\right)} \cdot \qquad \text{(II-17.1)}$$

II-18. Evaluate $\int_0^2 x^2(2 - x)^{13} dx$.   ★

We have merely to apply (II-17.1) with $a = 2$, $b = 2$, $c = 1$, $d = 13$. We find at once that the value of the integral is

$$\frac{2^{16}\Gamma(3)\Gamma(14)}{16\Gamma(16)} = \frac{4096}{105}.$$

II-19. Evaluate

$$\int_0^1 \frac{x^5 dx}{\sqrt[3]{(1 - x^4)}}. \qquad \star$$

This integral, like the integral in the preceding problem, is quickly evaluated by application of the formula obtained in Prob. II-17. We have $a = 1$, $b = 5$, $c = 4$, $d = -\frac{1}{3}$, making the value of the integral to be

$$\frac{\Gamma\left(\frac{3}{2}\right)\Gamma\left(\frac{2}{3}\right)}{\left(\frac{14}{3}\right)\Gamma\left(\frac{7}{6}\right)}.$$

This is as far as we shall go here. To get a numerical result, one would apply (I-4.2), to replace $\Gamma(\frac{2}{3})$ by $\frac{3}{2}\Gamma(\frac{5}{3})$, then use Table I-1.

REMARK.   The following integral, namely

$$\int_0^1 \frac{dx}{\sqrt{1 - x^{1/n}}},$$

where n denotes an arbitrary positive integer can also be solved via Prob. II-17, taking $a = 1$, $b = 0$, $c = 1/n$, $d = -1/2$. It may be of interest, however, to show how this integral can be transformed into a Beta integral:

Let $x^{1/n} = \sin^2\theta$. Then $x = \sin^{2n}\theta$, $dx = 2n \sin^{2n-1}\theta \cos \theta \, d\theta$. The interval of integration changes. When $x = 0$, $\theta = 0$ and when $x = 1$, $\theta = \pi/2$. We have, then,

$$\int_0^1 \frac{dx}{\sqrt{1 - x^{1/n}}} = n \int_0^{\pi/2} 2 \sin^{2n-1}\theta \cos^0\theta \, d\theta$$

$$= nB(n, \tfrac{1}{2}) \qquad \text{by Prob. I-27}$$

$$= \frac{n\Gamma(n)\Gamma(\tfrac{1}{2})}{\Gamma(n + \tfrac{1}{2})} \qquad \text{by Prob. I-29}$$

$$= \frac{\Gamma(n + 1)\sqrt{\pi}}{\Gamma(n + \tfrac{1}{2})} \qquad \text{by Eqs. (I-4.1) and (I-11.1)}$$

$$= \frac{2 \cdot 4 \cdot 6 \cdots 2n}{1 \cdot 3 \cdot 5 \cdots (2n - 1)} \qquad \text{by Prob. I-19.}$$

II-20. Evaluate $I = \int_0^\infty e^{-a^2x^2} \cos bx\, dx$, $ab \neq 0$. ★

We differentiate with respect to b by Leibniz's Rule:

$$\frac{d}{db}\left[\int_{x_1}^{x_2} f(x, b)dx\right] = \int_{x_1}^{x_2} \frac{\partial f}{\partial b}dx.$$

In the problem at hand this formula holds even with $x_2 = \infty$. Then we have

$$\frac{dI}{db} = \int_0^\infty e^{-a^2x^2}(-x \sin bx)dx. \tag{II-20.1}$$

If the use of Leibniz's Rule for differentiation of an integral did not occur to you in trying to solve this problem do not be dismayed. Knowing when to use this technique requires considerable mathematical experience. In the solution of this problem you should see how and why this technique works. Then you may find yourself in a position to try it in similar situations.

Now integrate by parts: $\int u\,dv = uv - \int v\,du$. We take

$$u = \sin bx, \quad du = b \cos bx\, dx;$$

$$dv = -\, xe^{-a^2x^2}dx, \quad v = \frac{e^{-a^2x^2}}{2a^2}.$$

Then we have

$$\frac{dI}{db} = \underbrace{\left[\frac{(\sin bx)e^{-a^2x^2}}{2a^2}\right]_{x=0}^{x=\infty}}_{(A)} - \underbrace{\int_0^\infty \frac{e^{-a^2x^2}(b \cos bx)}{2a^2}dx}_{(B)}.$$

The expression marked (A) turns out to be zero; and when we take the constants out in front of the integral marked (B), we see we have the very same integral we started with. Thus,

$$\frac{dI}{db} = \frac{-b}{2a^2}\int_0^\infty e^{-a^2x^2} \cos bx\, dx = -\frac{b}{2a^2}I. \tag{II-20.2}$$

Eq. (II-20.2) may be written as $dI/I = -b\, db/2a^2$, whose solution is seen to be

$$I = ke^{-b^2/4a^2}. \tag{II-20.3}$$

Our problem will be solved as soon as we can determine the value of k. This we can do as follows. We observe in Eq. (II-20.3) that if we take $b = 0$, then $I = k$. But when $b = 0$, then we have

$$I = \int_0^\infty e^{-a^2x^2}dx$$

$$= \int_0^\infty x^0 e^{-a^2x^2}dx$$

$$= \frac{\Gamma(\frac{1}{2})}{2(a^2)^{1/2}} = \sqrt{\pi}/2a \qquad \text{by Probs. II-1 and I-11.}$$

Thus we have $k = \sqrt{\pi}/2a$; and our problem is solved:

$$\int_0^\infty e^{-a^2x^2} \cos bx \, dx = (\sqrt{\pi}/2a)e^{-b^2/4a^2}.$$

REMARK.   Justification for application of Leibniz's Rule in Eq. (II-20.1) may be made by arguments quite similar as those used in Prob. I-20.

II-21. Evaluate $\int_a^b (x-a)^m (b-x)^n dx$, $\quad m > -1, \quad n > -1$, $b > a$.   ★

The form of the integrand suggests the possibility of transformation to the basic Beta integral. To effect such a transformation we need a change of variable $x = f(t)$ which will change the interval of integration from $a \leq x \leq b$ to $0 \leq t \leq 1$. Accordingly, we let $x - a = (b-a)t$, which makes $b - x = (b-a)(1-t)$ and $dx = (b-a)dt$. Then we have

$$\int_a^b (x-a)^m (b-x)^n dx = \int_0^1 [(b-a)t]^m [(b-a)(1-t)]^n (b-a)dt$$

$$= (b-a)^{m+n+1} \int_0^1 t^m (1-t)^n dt.$$

The transformed integral is the Beta integral as in Prob. I-25 with $x - 1 = m$ and $y - 1 = n$. Its value, therefore, is $B(m + 1, n + 1)$. Expressing the Beta function in terms of the Gamma function by Eq. (I-29.1), we have

$$\int_a^b (x - a)^m (b - x)^n dx = (b - a)^{m+n+1} \frac{\Gamma(m + 1)\Gamma(n + 1)}{\Gamma(m + n + 2)}.$$

II-22. Evaluate the integrals

$$\int_0^\infty e^{-t \cos \phi} t^{b-1} \sin(t \sin \phi)dt, \qquad \int_0^\infty e^{-t \cos \phi} t^{b-1} \cos(t \sin \phi)dt,$$

$$b > 0, \quad -\frac{\pi}{2} < \phi < \frac{\pi}{2}. \qquad \star$$

We observe that the restriction $b > 0$ is made to insure convergence of the integrals, for if b were zero or negative the exponent of the factor t^{b-1} of the integrands would be ≤ -1 thereby causing the integrands, as $t \to 0$, to become infinite "like" $1/t^p$ with $p \geq 1$, making the integrals divergent. The observation just made does not, however, apply to one exceptional situation, namely when $\phi = 0$. Then the factor $\sin(t \sin \phi)$ in the integrand of the first integral is zero for all t, making that integral convergent and worth zero for all b.

The restriction $- \pi/2 < \phi < \pi/2$ is made also to insure convergence of the integrals. In that range of values of ϕ the cosine of ϕ is positive, making the exponent on e in the integrands negative. If the exponent on e were positive, then as $t \to + \infty$, the absolute value of the integrands would become infinite no matter what the exponent of the factor t^{b-1} would be, thus making the integrals divergent. Similar reasoning shows convergence when the exponent on e is negative.

Since the value of each of these integrals depends on choice of b and choice of ϕ, keeping b and ϕ restricted as just explained, we shall designate them with functional notation like this:

$$G(b, \phi) = \int_0^\infty e^{-t\cos\phi} t^{b-1} \sin(t\sin\phi)dt,$$

$$H(b, \phi) = \int_0^\infty e^{-t\cos\phi} t^{b-1} \cos(t\sin\phi)dt.$$

Now that we have convinced ourselves of the need for the restrictions on b and ϕ together with realization of convergence of the integrals when b and ϕ are so restricted, and have expressed the integrals as functions of their parameters, we cast about for some means of evaluating the integrals in terms of known functions. Surely $G(b, \phi)$ and $H(b, \phi)$ must be related to each other somehow, for they differ only in the third factor of the integrand. And their third factors are related, being sine and cosine of the same argument respectively. Presuming, then, that there is a relationship between G and H, let us try to squeeze it out of the relationship between sine and cosine. We think first of the identity $\sin^2 x + \cos^2 x = 1$. But that does not look very hopeful. We cannot see how to use its connection with the two integrals. On the other hand the *derivative* relation between sine and cosine looks hopeful inasmuch as we can differentiate G and H with respect to either parameter, since each integral is *uniformly* convergent for all pairs (b, ϕ) that lie on any *closed* region taken within the open two-dimensional region $-\pi/2 < \phi < \pi/2, b > 0$. We get

$$\frac{\partial H}{\partial \phi} = \int_0^\infty t^{b-1} \left\{ e^{-t\cos\phi}[-\sin(t\sin\phi)]t\cos\phi \right.$$

$$\left. + [\cos(t\sin\phi)]e^{-t\cos\phi}t\sin\phi \right\} dt.$$

We suspect that $\partial H/\partial \phi$ is related to G. But $\partial H/\partial \phi$ has *two* parts in its integrand added together. Now if we can express G with a two-part integrand we may find out something. A little reflection shows that this would come about as a result of integration by parts:

$$G = \left[e^{-t\cos\phi}\sin(t\sin\phi) \right]\frac{t^b}{b}\Big|_{t=0}^{t=\infty} - \int_0^\infty \frac{t^b}{b}\frac{d}{dt}\left\{ e^{-t\cos\phi}\sin(t\sin\phi) \right\} dt$$

$$= \text{zero} - \frac{1}{b}\int_0^\infty t^b \left\{ e^{-t\cos\phi}[\cos(t\sin\phi)]\sin\phi \right.$$

$$\left. + [\sin(t\sin\phi)]e^{-t\cos\phi}(-\cos\phi) \right\} dt.$$

And there we have it! Comparison of $\partial H/\partial\phi$ with G shows immediately that $\partial H/\partial\phi = -bG$.

In exactly the same manner one finds that $\partial G/\partial\phi = bH$. Out of these two first-order differential equations we can get a second-order differential equation in H alone, like this:

$$\frac{\partial H}{\partial\phi} = -bG,$$

$$\frac{\partial^2 H}{\partial\phi^2} = -b\frac{\partial G}{\partial\phi} = -b\cdot bH,$$

$$\frac{\partial^2 H}{\partial\phi^2} + b^2 H = 0.$$

So when b is held constant H must be such a function of ϕ as to satisfy the well-known equation for simple harmonic motion. It follows that

$$H = A\sin b\phi + B\cos b\phi + f(b)$$

where A and B are constants and $f(b)$ is a function of b alone. In order for the second-order differential equation to be satisfied identically in b and ϕ [identically, that is, for value pairs (b, ϕ) satisfying $b > 0$, $-\pi/2 < \phi < \pi/2$], we must have $f(b)$ identically zero:

$$H = A\sin b\phi + B\cos b\phi.$$

To determine the constant coefficients A and B, we first take $\phi = 0$. Then this last equation becomes

$$\int_0^\infty e^{-t}t^{b-1}dt = B,$$

which says that $B = \Gamma(b)$ by Eq. (I-1). To determine A, we use the relation $\partial H/\partial\phi = -bG$:

$$Ab \cos b\phi - Bb \sin b\phi = -bG,$$

which, when $\phi = 0$, gives us

$$Ab = -b\int_0^\infty e^{-t}t^{b-1}(\text{zero})dt = 0,$$

$$A = 0.$$

We have now $H = \Gamma(b) \cos b\phi$. To get G, we use the relation

$$\frac{\partial H}{\partial\phi} = -bG,$$

$$\Gamma(b)[-b \sin b\phi] = -bG,$$

whence $G = \Gamma(b) \sin b\phi$. Our problem is solved. Let us display the solution, expressing G and H as originally given:

$$\int_0^\infty e^{-t\cos\phi}t^{b-1}\sin(t \sin \phi)dt = \Gamma(b)\sin b\phi, \qquad b > 0, \ -\frac{\pi}{2} < 0 < \frac{\pi}{2},$$

$$\int_0^\infty e^{-t\cos\phi}t^{b-1}\cos(t \sin \phi)dt = \Gamma(b)\cos b\phi, \qquad b > 0, \ -\frac{\pi}{2} < 0 < \frac{\pi}{2}.$$

II-23. Evaluate

$$\int_0^\infty t^{b-1}\cos t \, dt, \qquad 0 < b < 1,$$

$$\int_0^\infty t^{b-1}\sin t \, dt, \qquad 0 < b < 1. \qquad \star$$

Let us first check the restriction in range imposed on the parameter b. Because of the unceasing oscillation, as $t \to \infty$, between -1 and 1 of the factor $\cos t$ or $\sin t$, we evidently require the exponent $b - 1$ to be *negative* in order to make the magnitude of oscillation of the integrand decrease toward zero as limit as $t \to \infty$. When this is done, either integral taken over an interval $t = h$ to $t = \infty$, with $h > 0$, is convergent, as may be seen by considering the integral as a sum of an infinite series of alternating positive and negative terms (areas between the t-axis and the arches of the curve $y = t^{b-1}\cos t$ or $y = t^{b-1}\sin t$) such that each term is less in magnitude than its immediate predecessor and the limit of the nth term is zero as $n \to \infty$. Then, for convergence of the given integrals we must have $b - 1 < 0$, that is, $b < 1$.

We must yet examine each integral taken over an interval $t = 0$ to $t = h > 0$ when $b < 1$. The factor $\cos t$ in the integrand of the first integral becomes unity at $t = 0$. So, that integrand is "like" its first factor t^{b-1} as $t \to 0$. Consequently, the cosine integral taken over an interval $t = 0$ to $t = h > 0$ converges when and only when $b - 1 > -1$, that is, when $b > 0$. Our conclusion, then, is that the cosine integral as given from $t = 0$ to $t = \infty$ is convergent when and only when $0 < b < 1$.

The situation presented by the sine integral taken over an interval $t = 0$ to $t = h > 0$ is a little different. We let $b - 1 = -c$ where $c > 0$. Then we write the integrand like this:

$$t^{b-1}\sin t = \frac{\sin t}{t} \frac{1}{t^{c-1}}.$$

From here on the analysis of the situation proceeds as in the preceding paragraph, since the ratio $\sin t$ to t approaches the limit unity as $t \to 0$. Our conclusion is that the sine integral is convergent when and only when we have both $b < 1$ and $c - 1 < 1$, that is, when $-1 < b < 1$. However, we shall evaluate the sine integral only for $0 < b < 1$ because we want to consider the two integrals together in the light of Prob. II-22 where we found that we had a common interval of convergence for the two integrals concerned.

In Prob. II-22 we found the integrals to be convergent for all positive values of b when $-\pi/2 < \phi < \pi/2$. In the present problem we do not have the factor $e^{-t \cos \phi}$ (with $-\pi/2 < \phi < \pi/2$) to outweigh t^{b-1} as $t \to \infty$. That is why we now require $b < 1$. But inasmuch as the integrals of Prob. II-22 become those of the present problem when $\phi = \pi/2$, we see that the results of Prob. II-22 hold not only for $-\pi/2 < \phi < \pi/2$ when b is any positive number but also for $\phi = \pi/2$ when $0 < b < 1$. Accordingly we have

$$\int_0^\infty t^{b-1} \cos t \, dt = \Gamma(b) \cos (b\pi/2), \qquad 0 < b < 1,$$

$$\int_0^\infty t^{b-1} \sin t \, dt = \Gamma(b) \sin (b\pi/2), \qquad 0 < b < 1.$$

II-24. Evaluate $\displaystyle\int_0^{\pi/2} \tan^h\theta \, d\theta, \qquad 0 < h < 1.$ ★

We have

$$\int_0^{\pi/2} \tan^h\theta \, d\theta = \frac{1}{2} \int_0^{\pi/2} 2 \sin^h\theta \cos^{-h}\theta \, d\theta$$

$$= \frac{1}{2} B\left(\frac{h+1}{2}, \frac{-h+1}{2}\right) \qquad \text{by Prob. I-27.}$$

We observe that each of the arguments $(h+1)/2$ and $(-h+1)/2$ lies between 0 and 1, also that their sum is unity. Accordingly Eq. (I-30.3) is applicable with $p = (h+1)/2$. This gives us

$$\int_0^{\pi/2} \tan^h\theta \, d\theta = \frac{1}{2} \frac{\pi}{\sin\left(\dfrac{h+1}{2}\pi\right)}$$

$$= \frac{\pi}{2 \sin\left(\dfrac{h\pi}{2} + \dfrac{\pi}{2}\right)}$$

$$= \frac{\pi}{2 \cos\left(\dfrac{h\pi}{2}\right)}, \qquad 0 < h < 1.$$

II-25. Show that

$$\int_0^{\pi/2} \sin^{2n+1}\theta \, d\theta = \frac{2 \cdot 4 \cdots (2n)}{1 \cdot 3 \cdots (2n+1)}, \qquad n = 1, 2, 3, \cdots . \qquad \star$$

We have

$$\int_0^{\pi/2} \sin^{2n+1}\theta \, d\theta = \frac{1}{2}\int_0^{\pi/2} 2\sin^{2n+1}\theta\cos^0\theta \, d\theta.$$

This is the Beta integral, as shown in Prob. I-27. Letting $2x - 1$ and $2y - 1$ equal the exponents of $\sin\theta$ and $\cos\theta$ respectively, we get $x = n + 1$ and $y = \frac{1}{2}$. Thus we have

$$\int_0^{\pi/2} \sin^{2n+1}\theta \, d\theta = \frac{1}{2}B\left(n+1, \frac{1}{2}\right)$$

$$= \frac{1}{2}\frac{\Gamma(n+1)\Gamma(\frac{1}{2})}{\Gamma(n+\frac{3}{2})} \qquad \text{by Probs. I-27 and I-29}$$

However,

$$\Gamma(n + \tfrac{3}{2}) = \frac{1 \cdot 3 \cdot 5 \cdots (2n+1)\sqrt{\pi}}{2^{n+1}} \qquad \text{by Prob. I-13;}$$

and

$$\Gamma(n+1) = \frac{2 \cdot 4 \cdots (2n)}{2^n} \qquad \text{by Prob. I-16.}$$

Thus we have

$$\int_0^{\pi/2} \sin^{2n+1}\theta \, d\theta = \frac{1}{2}\left[\frac{2 \cdot 4 \cdots (2n)}{2^n}\right]\left[\frac{2^{n+1}}{1 \cdot 3 \cdot 5 \cdots (2n+1)\sqrt{\pi}}\right](\sqrt{\pi})$$

$$= \frac{2 \cdot 4 \cdots (2n)}{1 \cdot 3 \cdots (2n+1)}, \qquad n = 1, 2, 3, \cdots .$$

II-26. Show that

$$\int_0^{\pi/2} \sin^{2n}\theta \, d\theta = \frac{1 \cdot 3 \cdots (2n-1)}{2 \cdot 4 \cdots (2n)}\left(\frac{\pi}{2}\right), \qquad n = 1, 2, 3, \cdots . \qquad \star$$

We have

$$\int_0^{\pi/2} \sin^{2n} \theta \, d\theta = \frac{1}{2} \int_0^{\pi/2} 2 \sin^{2n} \theta \cos^0 \theta \, d\theta.$$

This is the Beta integral, as shown in Prob. I-27. Letting $2x - 1$ and $2y - 1$ equal the exponents on $\sin \theta$ and $\cos \theta$ respectively, we get $x = n + \frac{1}{2}$ and $y = \frac{1}{2}$. Thus, we have

$$\int_0^{\pi/2} \sin^{2n} \theta \, d\theta = \frac{1}{2} B\left(n + \frac{1}{2}, \frac{1}{2}\right) = \left(\frac{1}{2}\right) \frac{\Gamma\left(n + \frac{1}{2}\right)\Gamma\left(\frac{1}{2}\right)}{\Gamma(n + 1)}$$

by Probs. I-27 and I-29.

From Prob. I-19 we have

$$\frac{\Gamma\left(n + \frac{1}{2}\right)}{\Gamma(n + 1)} = \frac{1 \cdot 3 \cdot 5 \cdots (2n - 1)\sqrt{\pi}}{2 \cdot 4 \cdots (2n)}.$$

Therefore,

$$\int_0^{\pi/2} \sin^{2n} \theta \, d\theta = \frac{1 \cdot 3 \cdot 5 \cdots (2n - 1)}{2 \cdot 4 \cdots (2n)}\left(\frac{\pi}{2}\right), \qquad n = 1, 2, 3, \cdots.$$

II-27. Show that

$$\int_0^{\pi/2} \sin^{2n-1} \theta \, d\theta = \frac{2 \cdot 4 \cdots (2n - 2)}{1 \cdot 3 \cdots (2n - 1)}, \qquad n = 1, 2, 3, \cdots. \qquad \star$$

We have

$$\int_0^{\pi/2} \sin^{2n-1} \theta \, d\theta = \frac{1}{2} \int_0^{\pi/2} 2 \sin^{2n-1} \theta \cos^0 \theta \, d\theta.$$

We apply Prob. I-27. Letting $2x - 1$ and $2y - 1$ equal the exponents on $\sin\theta$ and $\cos\theta$ respectively, we get $x = n$ and $y = \frac{1}{2}$. Thus we have

$$\int_0^{\pi/2} \sin^{2n-1}\theta \; d\theta = \frac{1}{2} B\left(n, \frac{1}{2}\right) = \left(\frac{1}{2}\right) \frac{\Gamma(n)\Gamma\left(\dfrac{1}{2}\right)}{\Gamma\left(n + \dfrac{1}{2}\right)}. \qquad \text{(II-27.1)}$$

From Prob. I-12 we have

$$\Gamma\left(n + \frac{1}{2}\right) = \frac{1 \cdot 3 \cdot 5 \cdots (2n-1)\sqrt{\pi}}{2^n}.$$

From Prob. I-16 we have

$$\Gamma(n) = \frac{2 \cdot 4 \cdot 6 \cdots (2n-2)}{2^{n-1}}.$$

Then, substituting these values for $\Gamma(n)$ and $\Gamma(n + \frac{1}{2})$ in Eq. (II-27.1), we obtain the equation that was to be demonstrated.

Problems: Infinite Product Expression of $\pi/2$

II-28. Show that

$$\frac{\pi}{2} = \prod_{n=1}^{\infty} \frac{4n^2}{4n^2 - 1}. \qquad \star$$

This is known as *Wallis's product for* $\pi/2$. We observe first that the notation on the right side of the equation denotes an unending product (also called infinite product) of factors in which $n = 1$ in the first factor, $n = 2$ in the second factor, and so on ad infinitum. The infinite product in question may also be written

$$\prod_{n=1}^{\infty} \frac{(2n)^2}{(2n-1)(2n+1)} = \frac{(2)(2)}{(1)(3)} \cdot \frac{(4)(4)}{(3)(5)} \cdot \frac{(6)(6)}{(5)(7)} \cdots.$$

If now we write *only a part* of the whole product, namely

$$\frac{2^2 \cdot 4^2 \cdot 6^2 \cdots (2n)^2}{3^2 \cdot 5^2 \cdots (2n-1)^2(2n+1)},$$

we observe some resemblance of this expression to the expressions involved in Probs. II-25, II-26, II-27. This resemblance gives us the clue to the proof. We note first that, for $0 < \theta < \pi/2$ and for every positive integer n, we have

$$\sin^{2n+1}\theta < \sin^{2n}\theta < \sin^{2n-1}\theta.$$

It follows that the integrals of such powers of $\sin\theta$ over the interval $0 \leq \theta \leq \pi/2$ will be unequal in the same order:

$$\int_0^{\pi/2} \sin^{2n+1}\theta \, d\theta < \int_0^{\pi/2} \sin^{2n}\theta \, d\theta < \int_0^{\pi/2} \sin^{2n-1}\theta \, d\theta.$$

Then by Probs. II-25, II-26, II-27 we have

$$\frac{2 \cdot 4 \cdots (2n)}{1 \cdot 3 \cdots (2n+1)} < \frac{1 \cdot 3 \cdots (2n-1)}{2 \cdot 4 \cdots (2n)} \frac{\pi}{2} < \frac{2 \cdot 4 \cdots (2n-2)}{1 \cdot 3 \cdots (2n-1)}.$$

Upon multiplying each term of the threefold inequality by the reciprocal of the first term we get

$$1 < \frac{1^2 \cdot 3^2 \cdots (2n-1)^2(2n+1)}{2^2 \cdot 4^2 \cdots (2n)^2} \frac{\pi}{2} < \frac{2n+1}{2n}.$$

Now as $n \to \infty$, the fraction on the right approaches unity as limit. And since the term in the middle is always between unity and the fraction on the right, it follows that the middle term is forced by the squeeze play to approach unity as a limit. This in turn means that the fraction multiplying $\pi/2$ in the middle term must approach $2/\pi$ as limit, that is, the *reciprocal* of the fraction multiplying $\pi/2$ must approach $\pi/2$ as limit as $n \to \infty$:

$$\frac{\pi}{2} = \lim_{n\to\infty} \left[\frac{2 \cdot 2 \cdot 4 \cdot 4 \cdots (2n)(2n)}{1 \cdot 1 \cdot 3 \cdot 3 \cdots (2n-1)(2n+1)} \right].$$

If, now, we cast away the first factor in the denominator and group the remaining factors in pairs, we get

$$\frac{\pi}{2} = \lim_{n\to\infty}\left[\frac{(2)^2}{1\cdot 3}\cdot\frac{(4)^2}{3\cdot 5}\cdot\frac{(6)^2}{5\cdot 7}\cdots\frac{(2n)^2}{(2n-1)(2n+1)}\right],$$

which is seen to be what we set out to prove.

II-29. Get an approximation for π via the product of the first several terms of

$$\prod_{n=1}^{\infty}\frac{4n^2}{4n^2-1}. \qquad \star$$

Let us take, say, eight terms. Then we have

$$\frac{\pi}{2}\cong\frac{4}{3}\cdot\frac{16}{15}\cdot\frac{36}{35}\cdot\frac{64}{63}\cdot\frac{100}{99}\cdot\frac{144}{143}\cdot\frac{196}{195}\cdot\frac{256}{255}.$$

Carrying out the multiplication on the slide rule, we get

$$\pi \cong 3.05.$$

Not so good! We must admit, then, that Wallis's product for $\pi/2$ is a slowly convergent one, requiring many terms even for accuracy to two decimal places, when a partial product is taken as an approximation for $\pi/2$.

Problems: Evaluation of Certain Geometrical Magnitudes

II-30. Express in terms of Gamma Functions the area enclosed by the curve $x^{b/c} + y^{b/c} = a^{b/c}$, where a is a positive constant, c is a positive odd integer, b is a positive, even integer. Apply your resulting formula to compute the area enclosed by the curve when the exponent b/c is 2/3. $\star$

Before carrying out the computation, we pause to observe that the family of curves represented by the given equation has as one of its members the familiar four-pointed star or asteroid $x^{2/3} + y^{2/3} = a^{2/3}$. Indeed, whenever $b < c$, we have such a concave star with four sharp points (cusps). On the other hand when $b > c$, the curve is convex; and when $c = 1$ and b is a large even integer the curve is almost a square. In any case the curve is symmetric to both co-ordinate axes, so that we shall compute only that part of the area which is in the first quadrant and multiply the result by 4 to get the whole area within the curve.

Thus we have

$$\text{Area} = 4 \iint_R dA,$$

where R denotes the region in the first quadrant bounded by the curve and the coordinate axes. We set the double integral equal to an equivalent iterated integral:

$$A = 4 \int_0^a dx \int_0^{\left(a^{\frac{b}{c}} - x^{\frac{b}{c}}\right)^{\frac{c}{b}}} dy.$$

To evaluate the integral in y we make the change of variable

$$y = av^{\frac{c}{b}}, \qquad dy = a\left(\frac{c}{b}\right)v^{\frac{c}{b}-1}dv.$$

While we are at it we may as well make the same kind of change in variable for x:

$$x = au^{c/b}, \qquad dx = a\left(\frac{c}{b}\right)u^{c/b-1}du.$$

These changes in variable carry with them, of course, corresponding changes in limits of integration, so that we get

$$A = \frac{4a^2c^2}{b^2} \int_0^1 u^{c/b-1}du \int_0^{1-u} v^{c/b-1}dv.$$

Carrying out the integration only with respect to v, we have

$$A = 4a^2\left(\frac{c}{b}\right)\int_0^1 u^{c/b-1}(1-u)^{c/b}du.$$

Here we observe that the resulting integral is none other than the Beta integral, so that

$$A = 4\left(\frac{c}{b}\right)a^2 B\left(\frac{c}{b}, \frac{c}{b}+1\right) \qquad \text{by Prob. I-25}$$

$$= 4\left(\frac{c}{b}\right)a^2\left[\frac{\Gamma\left(\frac{c}{b}\right)\Gamma\left(\frac{c}{b}+1\right)}{\Gamma\left(\frac{2c}{b}+1\right)}\right] \qquad \text{by Prob. I-29.}$$

We can obtain a neater form for our formula upon replacing

$\Gamma\left(\dfrac{c}{b}+1\right)$ by $\dfrac{c}{b}\Gamma\left(\dfrac{c}{b}\right)$ and $\Gamma\left(\dfrac{2c}{b}+1\right)$ by $\dfrac{2c}{b}\Gamma\left(\dfrac{2c}{b}\right)$ by Eq. (I-4.1).

Then

$$A = \frac{\Gamma\left(\frac{c}{b}\right)\Gamma\left(\frac{c}{b}\right)}{\Gamma\left(\frac{2c}{b}\right)}\left(\frac{2ca^2}{b}\right). \tag{II-30.1}$$

Applying this formula to the asteroid when $b = 2$ and $c = 3$, we have

$$A = \frac{\Gamma\left(\frac{3}{2}\right)\Gamma\left(\frac{3}{2}\right)}{\Gamma(3)}(2)\left(\frac{3}{2}\right)a^2 = \frac{3a^2\pi}{8}$$

by Eqs. (I-4.1), (I-10.1), and (I-11.1).

II-31. Apply the results of Prob. II-30 to the curve $x^{20} + y^{20} = a^{20}$. ★

This is an "almost-a-square" curve where the common exponent b/c is a fairly large even integer. By Prob. II-30 we have

$$A = \frac{\Gamma\!\left(\dfrac{c}{b}\right)\Gamma\!\left(\dfrac{c}{b}\right)}{\Gamma\!\left(\dfrac{2c}{b}\right)}(2)\!\left(\frac{c}{b}\right)a^2 = \frac{\Gamma\!\left(\dfrac{1}{20}\right)\Gamma\!\left(\dfrac{1}{20}\right)}{\Gamma\!\left(\dfrac{2}{20}\right)}(2)\!\left(\frac{1}{20}\right)a^2$$

$$= \frac{400\Gamma\!\left(\dfrac{21}{20}\right)\Gamma\!\left(\dfrac{21}{20}\right)}{10\Gamma\!\left(\dfrac{11}{10}\right)}(2)\!\left(\frac{1}{20}\right)a^2$$

$$\cong 4\left[\frac{(.9735)(.9735)}{.9514}\right]a^2.$$

The quantity in brackets to four figures is .9961. Thus, we have

$$A \cong 4a^2.$$

II-32. Show that if the exponent b/c in Prob. II-30 be a positive even integer $2m$, the enclosed area A approaches the limit $4a^2$ as $m \to \infty$.   ★

This result is what we naturally expect from consideration of the shape and position of the curves of the family $x^{2m} + y^{2m} = a^{2m}$. We have from (II-30.1)

$$\lim_{m \to \infty} A = a^2 \lim_{m \to \infty} \frac{\Gamma\!\left(\dfrac{1}{2m}\right)\Gamma\!\left(\dfrac{1}{2m}\right)}{m\Gamma\!\left(\dfrac{1}{m}\right)}.$$

Applying (I-4.1), we get

$$\lim_{m \to \infty} A = a^2 \lim_{m \to \infty} \frac{2m\Gamma\!\left(\dfrac{1}{2m} + 1\right)2m\Gamma\!\left(\dfrac{1}{2m} + 1\right)}{m \cdot m\Gamma\!\left(\dfrac{1}{m} + 1\right)}$$

$$= 4a^2\Gamma(1) = 4a^2 \qquad \text{by Prob. I-9.}$$

Problem: Evaluation of Integrals

II-33. Compute in terms of the Gamma function the value of $\iint\limits_{A} x^{m-1}y^{n-1}dA$ where A denotes the region of the cartesian xy-plane bounded by the coordinate axes and that portion of the curve $(x/a)^p + (y/b)^q = 1$ which lies in the first quadrant, each of the letters m, n, p, q, a, b denoting a positive constant.   ★

This problem, except for the restriction that A lie in the first quadrant, is a generalization of Prob. II-30. When $0 < m < 1$ or $0 < n < 1$ or both, the integral is improper in that the integrand becomes infinite along a coordinate axis (or along both). But if neither exponent is as much negative as minus one, the integral is convergent even when improper. If we were to allow one or the other or both of the exponents on x and y to be more negative than minus one, the integral would not be convergent. That is why we require m and n to be positive. If we were to allow p or q or both to be negative, region A would not be limited but would extend infinitely far off.

If a or b or both were negative, there might not be any part of the curve $(x/a)^p + (y/b)^q = 1$ lying in the first quadrant. On the other hand the constants m, n, p, q, a, b might be such that the integral could be taken over an area covering four symmetric portions, one in each quadrant as in Prob. II-30.

At any rate we shall confine the present problem to the first quadrant and shall even omit most of the details of the solution since the details will parallel closely those of Prob. II-30.

To begin with, we have

$$\iint\limits_{A} x^{m-1}y^{n-1}dA = \int_0^a dx \int_0^{b\left[1-\left(\frac{x}{a}\right)^p\right]^{\frac{1}{q}}} x^{m-1}y^{n-1}dy.$$

As in Prob. II-30 we make the following change of variables:

$$x = au^{1/p} \qquad \text{and} \qquad y = bv^{1/q},$$

$$dx = a \cdot \frac{1}{p} u^{1/p-1} du \quad \text{and} \quad dy = b \cdot \frac{1}{q} v^{1/q-1} dv,$$

which results in

$$\iint_A x^{m-1} y^{n-1} dA = \frac{a^m b^n}{pq} \frac{q}{n} \int_0^1 u^{m/p-1} (1 - u)^{n/q} du$$

$$= \frac{a^m b^n}{pq} \frac{q}{n} B\left(\frac{m}{p}, \frac{n}{q} + 1\right) \qquad \text{by Prob. I-25}$$

$$= \frac{a^m b^n}{pq} \frac{q}{n} \frac{\Gamma\left(\frac{m}{p}\right) \frac{n}{q} \Gamma\left(\frac{n}{q}\right)}{\Gamma\left(\frac{m}{p} + \frac{n}{q} + 1\right)} \qquad \begin{array}{l} \text{by Probs.} \\ \text{I-29 and I-4.} \end{array}$$

Thus the solution of our problem is

$$\iint_A x^{m-1} y^{n-1} dA = \frac{a^m b^n}{pq} \frac{\Gamma\left(\frac{m}{p}\right) \Gamma\left(\frac{n}{q}\right)}{\Gamma\left(\frac{m}{p} + \frac{n}{q} + 1\right)}. \qquad \text{(II-33.1)}$$

Problem: Evaluation of Certain Geometrical Magnitudes

II-34. Compute the area A in the first quadrant bounded by the coordinate axes and the curve $y^2 = (1 - x)^3$.   ★

Letting R denote the region whose area we are to compute, we have

$$A = \iint_R dA = \iint_R x^0 y^0 dA,$$

which means that our problem is a special case of Prob. II-33 with $m = 1$, $n = 1$. In order to ascertain the values to be taken for

a, b, p, q, we must write the equation of the curve $y^2 = (1 - x)^3$ in the form $x + y^{2/3} = 1$. Thus we have $a = 1$, $b = 1$, $p = 1$, $q = \frac{2}{3}$. Applying Eq. (II-33.1), we have

$$A = \frac{(1)(1)}{\frac{2}{3}} \frac{\Gamma(1)\Gamma\left(\frac{3}{2}\right)}{\Gamma\left(1 + \frac{3}{2} + 1\right)}$$

$$= \frac{2}{5} \qquad \text{by Probs. I-9 and I-4.}$$

Problem: Evaluation of Integrals

II-35. Compute in terms of the Gamma function the value of the integral $H = \iiint\limits_{R} x^{h-1}y^{m-1}z^{n-1}dV$ where R denotes the region of space bounded by the coordinate planes and that portion of the surface $(x/a)^p + (y/b)^q + (z/c)^k = 1$ which lies in the first octant, each of the letters h, m, n, p, q, k, a, b, c denoting a positive constant.　　★

The solution of this problem will naturally proceed along the same lines as the solution of Prob. II-33. We have

$$H = \int_0^a x^{h-1}dx \int_0^{b\left[1-\left(\frac{x}{a}\right)^p\right]^{\frac{1}{q}}} y^{m-1}dy \int_0^{c\left[1-\left(\frac{x}{a}\right)^p-\left(\frac{y}{b}\right)^q\right]^{\frac{1}{k}}} z^{n-1}dz.$$

To effect this formidable-looking integration we change the variables in this manner:

$$x = au^{\frac{1}{p}}, \quad y = bv^{\frac{1}{q}}, \quad z = cw^{\frac{1}{k}}$$

$$dx = a \cdot \frac{1}{p} u^{\frac{1}{p}-1} du, \quad dy = b \cdot \frac{1}{q} v^{\frac{1}{q}-1} dv, \quad dz = c \cdot \frac{1}{k} w^{\frac{1}{k}-1} dw.$$

This simplifies our iterated integral to

$$H = \frac{a^h b^m c^n}{pqk} \int_0^1 u^{\frac{h}{p}-1}\, du \int_0^{1-u} v^{\frac{m}{q}-1}\, dv \int_0^{1-u-v} w^{\frac{n}{k}-1}\, dw,$$

which immediately becomes upon integrating with respect to w

$$H = \frac{a^h b^m c^n}{pqk} \frac{k}{n} \int_0^1 u^{\frac{h}{p}-1}\, du \int_0^{1-u} v^{\frac{m}{q}-1} (1 - u - v)^{\frac{n}{k}}\, dv.$$

Here we can get a further simplification by taking advantage of the fact that u is held constant while the integration with respect to v is carried out. We do so by letting $v = (1 - u)t$ with $1 - u$ considered constant for the time being. This makes $dv = (1 - u)dt$. Thus we transform our iterated integral into another iterated integral which, as in Prob. I-11 is equivalent to a *product* of integrals:

$$H = \left[\int_0^1 u^{\frac{h}{p}-1} (1 - u)^{\frac{m}{q}+\frac{n}{k}}\, du \right] \left[\int_0^1 t^{\frac{m}{q}-1} (1 - t)^{\frac{n}{k}}\, dt \right] \frac{a^h b^m c^n}{pqk} \frac{k}{n}$$

$$= \frac{a^h b^m c^n k}{pqkn} B\left(\frac{h}{p}, \frac{m}{q} + \frac{n}{k} + 1 \right) B\left(\frac{m}{q}, \frac{n}{k} + 1 \right) \qquad \text{by Eq. (I-25.1)}$$

$$= \frac{a^h b^m c^n}{pqk} \frac{\Gamma\left(\frac{h}{p}\right) \Gamma\left(\frac{m}{q}\right) \Gamma\left(\frac{n}{k}\right)}{\Gamma\left(\frac{h}{p} + \frac{m}{q} + \frac{n}{k} + 1\right)}, \qquad \text{by Probs. I-29 and I-4.}$$

$$\text{(II-35.1)}$$

Problems: Evaluation of Certain Geometrical Magnitudes

II-36. Letting R denote the region in the first octant bounded by the coordinate planes $x = 0$, $y = 0$, $z = 0$, and the surface $(x/5)^{1/2} + (y/6)^{2/3} + (z/7)^{3/4} = 1$, apply Prob. II-35 to compute
  (1) the volume V of R,
  (2) the mass M of a solid of density $D = \sqrt{x^3 y^5 z}$ occupying R,
  (3) the moment of inertia I_z of a homogeneous solid occupying R,
  (4) The coordinates of the centroid $(\bar{x}, \bar{y}, \bar{z})$ of R.   ★

The constants a, b, c, p, q, k of Prob. II-35 are the same for all the triple integrals involved in the four parts of the present problem:

$$a = 5, \quad b = 6, \quad c = 7, \quad p = \frac{1}{2}, \quad q = \frac{2}{3}, \quad k = \frac{3}{4}.$$

For **(1)** we have

$$V = \iiint_R dV = \iiint_R x^0 y^0 z^0 \, dV,$$

which means that we take $h = 1$, $m = 1$, $n = 1$. Thus, using Eq. (II-35.1), namely

$$H = \frac{a^h b^m c^n}{pqk} \frac{\Gamma\left(\dfrac{h}{p}\right)\Gamma\left(\dfrac{m}{q}\right)\Gamma\left(\dfrac{n}{k}\right)}{\Gamma\left(\dfrac{h}{p} + \dfrac{m}{q} + \dfrac{n}{k} + 1\right)},$$

we get

$$V = \frac{(5)(6)(7)}{(1/2)(2/3)(3/4)} \frac{\Gamma(2)\Gamma(3/2)\Gamma(4/3)}{\Gamma(2 + 3/2 + 4/3 + 1)}$$

$$= 840 \frac{(1)(\sqrt{\pi}/2)\Gamma(4/3)}{(29/6)(23/6)(17/6)(11/6)\Gamma(11/6)} \cong 7.4.$$

For **(2)** we have

$$M = \iiint_R x^{\frac{3}{2}} y^{\frac{5}{2}} z^{\frac{1}{2}} \, dV.$$

Here $h = 5/2$, $m = 7/2$, $n = 3/2$. So for the mass we get

$$M = \frac{5^{\frac{5}{2}} 6^{\frac{7}{2}} 7^{\frac{3}{2}}}{(1/2)(2/3)(3/4)} \frac{\Gamma(5)\Gamma(21/4)\Gamma(2)}{\Gamma(5 + 21/4 + 2 + 1)}.$$

For **(3)** we have, with $D = \sqrt{x^3 y^5 z}$,

$$I_z = \iiint_R (x^2 + y^2) D \, dV = \iiint_R x^{\frac{7}{2}} y^{\frac{5}{2}} z^{\frac{1}{2}} \, dV + \iiint_R x^{\frac{3}{2}} y^{\frac{9}{2}} z^{\frac{1}{2}} \, dV.$$

We recall from Prob. II-35:

$$H = \iiint\limits_R x^{h-1}y^{m-1}z^{n-1}dV.$$

Hence,

$$h - 1 = \frac{7}{2} \text{ and } h - 1 = \frac{3}{2}, \quad m - 1 = \frac{5}{2} \text{ and } m - 1 = \frac{9}{2}, \quad n - 1 = \frac{1}{2}.$$

Thus

$$I_z = \frac{5^{\frac{9}{2}}6^{\frac{7}{2}}7^{\frac{3}{2}}}{(1/4)} \frac{\Gamma(9)\Gamma(21/4)\Gamma(2)}{\Gamma(69/4)} + \frac{5^{\frac{5}{2}}6^{\frac{11}{2}}7^{\frac{3}{2}}}{(1/4)} \frac{\Gamma(5)\Gamma(33/4)\Gamma(2)}{\Gamma(65/4)}.$$

For **(4)** we have

$$V\bar{x} = \iiint\limits_R x \, dV, \quad V\bar{y} = \iiint\limits_R y \, dV, \quad V\bar{z} = \iiint\limits_R z \, dV.$$

The volume V has already been computed in part **(1)**. For $\bar{x}$ we have

$$\bar{x} = \frac{1}{V} \iiint\limits_R xy^0z^0 dV.$$

Here $h - 1 = 1$; hence, $h = 2$, $m - 1 = 0$, and $n - 1 = 0$. Thus $m = 1$, $n = 1$.
Then

$$\bar{x} = \frac{1}{V} \frac{(5^2)(6)(7)}{(1/4)} \frac{\Gamma(4)\Gamma(3/2)\Gamma(4/3)}{\Gamma(47/6)}.$$

Similarly we find $\bar{y}$ except that in this case $h = 1$, $m = 2$, $n = 1$.
For $\bar{z}$, $h = 1$, $m = 1$, $n = 2$.

Reduction of the results for parts **(2)**, **(3)**, and **(4)** to an approximate numerical value may be carried out, as was done in part **(1)**, via Prob. I-4 and Table I-1.

II-37. Compute the area A enclosed between the x-axis and one arch of the curve $y = \sin^8 x$. ★

We have

$$A = \int_0^{\pi} \sin^8 x \, dx = 2 \int_0^{\pi/2} \sin^8 x \, dx$$

$$= \int_0^{\pi/2} 2 \sin^8 x \cos^0 x \, dx$$

$$= \int_0^{\pi/2} 2 \sin^8 \theta \cos^0 \theta \, d\theta$$

$$= B\left(\frac{9}{2}, \frac{1}{2}\right) \qquad \text{by Prob. I-27}$$

$$= \frac{\Gamma\left(\frac{9}{2}\right)\Gamma\left(\frac{1}{2}\right)}{\Gamma(5)} \qquad \text{by Prob. I-29}$$

$$= \frac{35\pi}{128} \qquad \text{by Probs. I-4, I-10, I-11.}$$

II-38. Compute the area A enclosed by the oval

$$y^2 = \frac{1 - x^2}{1 + x^2}. \qquad ★$$

We observe that this curve is bounded by the lines $x = \pm 1$ because y^2 would be negative for $|x| > 1$. Similarly, by solving for x^2, one finds the curve to be bounded by the lines $y = \pm 1$. The restriction on x suggests the following change of variable.

Let $x^2 = \sin t$. Then $dx = \frac{1}{2} \frac{\cos t}{\sqrt{\sin t}} \, dt$. Because the curve is symmetric to both coordinate axes we may compute the area in the first quadrant and multiply by 4. Then we have

$$A = 4 \int_0^1 y \, dx = 4 \int_0^{\pi/2} \sqrt{\frac{1 - \sin t}{1 + \sin t}} \cdot \frac{1}{2} \frac{\cos t}{\sqrt{\sin t}} \, dt.$$

In the radical of the integrand we multiply the numerator and denominator by $1 - \sin t$. Then

$$A = 2 \int_0^{\pi/2} \sqrt{\frac{(1 - \sin t)^2}{\cos^2 t}} \cdot \frac{\cos t}{\sqrt{\sin t}}\, dt$$

$$= 2 \int_0^{\pi/2} (\sin^{-1/2} t - \sin^{1/2} t)\, dt$$

$$= 2 \int_0^{\pi/2} \frac{dt}{\sin^{1/2} t} - 2 \int_0^{\pi/2} \sin^{1/2} t\, dt.$$

The second of these two integrals is proper, but the first is improper. Now we know that the improper integral must be convergent because we know that the area A is finite, since A is entirely contained within a square of side 2. However, it is not without interest to demonstrate here, on its own merits, the convergence of the improper integral. We write

$$\int_0^{\pi/2} \frac{dt}{\sin^{1/2} t} = \int_0^{\pi/2} \left(\frac{t}{\sin t}\right)^{1/2} \frac{1}{t^{1/2}} dt.$$

The ratio $t/\sin t$ approaches the limit unity as $t \to 0$. This makes the integrand to be "like" $1/t^{1/2}$ as $t \to 0$ with the consequence that the question of convergence of $\int_0^{\pi/2} dt/\sin^{1/2} t$ hinges upon the convergence or divergence of $\int_0^{\pi/2} dt/t^{1/2}$. But we know that this last integral is convergent because the exponent on t is less than unity. Therefore, the integral in question is also convergent.

Now we insert the factor $\cos^0 t$ which enables us to recognize the integrals as Beta integrals:

$$A = \int_0^{\pi/2} 2 \sin^{-1/2} t \cos^0 t\, dt - \int_0^{\pi/2} 2 \sin^{1/2} t \cos^0 t\, dt$$

$$= B\left(\frac{1}{4}, \frac{1}{2}\right) - B\left(\frac{3}{4}, \frac{1}{2}\right) \qquad \text{by Prob. I-27}$$

$$= \sqrt{\pi} \left[\frac{4\Gamma\left(\frac{5}{4}\right)}{\frac{4}{3}\Gamma\left(\frac{7}{4}\right)} - \frac{\frac{4}{3}\Gamma\left(\frac{7}{4}\right)}{\Gamma\left(\frac{5}{4}\right)} \right] \qquad \text{by Probs. I-29, I-11, and I-4}$$

$$\cong 2.85 \qquad \text{by Table I-1.}$$

II-39. Find length of lemniscate $\rho^2 = a^2 \cos 2\theta$ via the Gamma function. ★

This curve is a figure-eight with the x-axis as its axis of symmetry and with its double point at the origin. Since this curve, by symmetry, has an equal length in each of the four quadrants, we can compute the length in the first quadrant only and multiply the result by 4. The points where the curve intersects the x-axis at the origin are those for which the argument of the cosine is an integral, odd multiple of $\pi/2$. For the portion in the first quadrant we can take $0 \leq \theta \leq \pi/4$.

The length of the curve in polar coordinates is

$$s = \int_\alpha^\beta \left[\rho^2 + \left(\frac{d\rho}{d\theta}\right)^2 \right]^{1/2} d\theta.$$

From $\rho^2 = a^2 \cos 2\theta$ we find

$$\left(\frac{d\rho}{d\theta}\right)^2 = \frac{a^4 \sin^2 2\theta}{a^2 \cos 2\theta}.$$

Then

$$\frac{1}{4} s = \int_0^{\pi/4} \left(a^2 \cos 2\theta + \frac{a^2 \sin^2 2\theta}{\cos 2\theta} \right)^{1/2} d\theta = \int_0^{\pi/4} a \cos^{-1/2} 2\theta \, d\theta.$$

By a change in variable we can transform this last integral into a Beta integral. Let $2\theta = t$. The interval of integration is now from 0 to $\pi/2$; hence,

$$\int_0^{\pi/4} a \cos^{-1/2} 2\theta \, d\theta = \frac{a}{2} \int_0^{\pi/2} \cos^{-1/2} t \sin^0 t \, dt.$$

The length in the first quadrant is then $a/2 \cdot 1/2B(\frac{1}{4}, \frac{1}{2})$ by Prob. I-27. The length of the entire lemniscate is 4 times this, namely,

$$s = aB\left(\frac{1}{4}, \frac{1}{2}\right) = \frac{a\Gamma\left(\frac{1}{4}\right)\Gamma\left(\frac{1}{2}\right)}{\Gamma\left(\frac{3}{4}\right)} = \frac{4a\Gamma\left(\frac{5}{4}\right)\sqrt{\pi}}{\frac{4}{3}\Gamma\left(\frac{7}{4}\right)} \cong 5.2a \qquad \text{by Probs. I-29, I-4, I-11.}$$

Problems: Evaluation of Certain Physical Quantities

II-40. A particle of mass m on the positive x-axis is attracted toward the origin by a variable force such that the product of the magnitude of the force by the distance from the origin is a constant k. The particle starts from rest at $x = L$. Determine the time required for it to reach the origin.   ★

We start from Newton's Equation $F = ma$, wherein the force F and the acceleration a are vectors having vanishing components in all directions except along the x-axis. Equating these components along the x-axis, we have

$$-\frac{k}{x} = m\frac{d^2x}{dt^2}.$$

We interchange the sides of the equation and multiply both sides by dx/dt and integrate

$$\frac{1}{2}m\left(\frac{dx}{dt}\right)^2 = -k\int_L^x \frac{\left(\dfrac{dx}{dt}\right)}{x}\, dt$$

$$= -k\log_e x + k\log_e L.$$

We note that the derivative of $(dx/dt)^2$ is $2(dx/dt)\,(d^2x/dt^2)$.

We observe that the last equation makes the velocity $dx/dt = 0$ when $x = L$ in accordance with one of our boundary conditions.

We solve the velocity equation for dx/dt, taking the negative square root because the motion is in the negative x-direction. Then we separate the variables, and integrate:

$$t = - \sqrt{\frac{m}{2k}} \int_L^0 \left(\log_e \frac{L}{x} \right)^{-1/2} dx.$$

This looks bad at the moment with respect to integration. But when we look back in Chapter II we find a close resemblance here to Prob. II-7, so much so, that all we need is a simple change of variable, namely $x = Lu$, to give us

$$t = L \sqrt{\frac{m}{2k}} \int_0^1 u^0 \left(\log_e \frac{1}{u} \right)^{-1/2} du,$$

whence by Probs. II-7 and I-11 we obtain

$$t = L \sqrt{\frac{m\pi}{2k}}.$$

To check dimensionality we may take t in seconds, L in feet, k in foot-pounds, and $m = \dfrac{\text{weight in pounds}}{32.2 \text{ ft/sec}^2}$.

II-41. Compute the period T of vibration of a simple pendulum swinging to and fro in a 180° arc.   ★

We take the coordinate axes as shown in the diagram so that, as the pendulum P swings, the polar angle θ varies between $-\pi/2$ and $\pi/2$. The polar coordinate $r = OP$ remains constant.

As usual we let g denote the acceleration caused by the earth's gravitational force and let W denote the weight of pendulum P. We naturally consider the potential energy P.E. of P to be zero when P is at its lowest position, namely, when $\theta = 0$. Then the value of P.E.

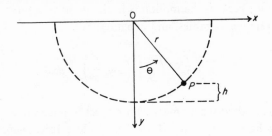

at *any* time equals the product of W by the height h reached by P at that instant:

$$\text{P.E.} = W(r - r \cos \theta).$$

The kinetic energy K.E. at each instant is given by

$$\text{K.E.} = \frac{mV^2}{2} = \frac{1}{2}\frac{W}{g}\left(r\frac{d\theta}{dt}\right)^2.$$

As P swings, the total energy remains constant: P.E. + K.E. $= C$. This fact provides us with the differential equation of the motion:

$$Wr(1 - \cos \theta) + \frac{1}{2}\frac{W}{g}r^2\left(\frac{d\theta}{dt}\right)^2 = C.$$

To determine C we take the time $t = 0$ when $\theta = \pi/2$, observing that P is then momentarily at rest, making $d\theta/dt = 0$ when $t = 0$ with $\theta = \pi/2$. This makes $C = Wr$. So our equation of motion is

$$\frac{r}{2g}\left(\frac{d\theta}{dt}\right)^2 - \cos \theta = 0.$$

Now as P swings back and forth, $d\theta/dt$ is at times negative and at times positive. But we can compute the period T by computing the time of the swing from $\theta = 0$ to $\theta = \pi/2$, and multiplying the result by 4. This allows us to use only the positive square root when we solve our last equation for $d\theta/dt$. We get

$$\sqrt{\frac{r}{2g}} \frac{d\theta}{dt} = \cos^{1/2} \theta.$$

$$dt = \sqrt{\frac{r}{2g}} \cos^{-1/2} \theta \, d\theta,$$

$$T = 4 \sqrt{\frac{r}{2g}} \int_0^{\pi/2} \cos^{-1/2} \theta \, d\theta$$

$$= 2 \sqrt{\frac{r}{2g}} \int_0^{\pi/2} 2 \sin^0\theta \cos^{-1/2} \theta \, d\theta$$

$$= \sqrt{\frac{2r}{g}} B\left(\frac{1}{2}, \frac{1}{4}\right) \qquad \text{by Prob. I-27}$$

$$= \sqrt{\frac{2r}{g}} \frac{\Gamma\left(\frac{1}{2}\right)\Gamma\left(\frac{1}{4}\right)}{\Gamma\left(\frac{3}{4}\right)} \qquad \text{by Prob. I-29}$$

$$= \sqrt{\frac{2r}{g}} \frac{\Gamma\left(\frac{1}{2}\right) 4\Gamma\left(\frac{5}{4}\right)}{\frac{4}{3}\Gamma\left(\frac{7}{4}\right)} \qquad \text{by Prob. I-4.}$$

Then, from the values in Table I-1, we find that

$$T \cong 7.416 \sqrt{\frac{r}{g}}.$$

II-42. Evaluate the radiation density integral

$$\phi = \frac{8\pi h}{c^3} \int_0^\infty \frac{\nu^3 d\nu}{e^{\frac{h\nu}{kT}} - 1}. \qquad \star$$

Let us do more here. Let us recall whence comes the formula we are to evaluate. The point of departure is the number dz of waves in a small frequency range between ν and $\nu + d\nu$ which are contained in a certain volume V as calculated by Rayleigh and Jeans:

$$dz = \frac{8\pi V}{c^3} \nu^2 d\nu, \tag{II-42.1}$$

where c is the speed of light. We take this in conjunction with Planck's quantum principle that radiation is emitted intermittently in bundles of integral multiples of a fundamental amount:

$$E = kT, \tag{II-42.2}$$

where k is the gas constant per molecule (Boltzmann's constant) and T is the temperature. Planck then showed that the average energy of an oscillator in equilibrium at temperature T is

$$E = \frac{\epsilon}{e^{\epsilon/kT} - 1}. \tag{II-42.3}$$

Planck further specified that the energy quantum ϵ be proportional to the frequency of the oscillator according to the following equation:

$$\frac{\epsilon}{\nu} = h, \tag{II-42.4}$$

where h has been found to be a universal constant (Planck's constant). Equation (II-42.3) can now be rewritten using the value of ϵ from (II-42.4).

$$E = \frac{h\nu}{e^{\frac{h\nu}{kT}} - 1}. \tag{II-42.5}$$

Now we multiply the left side and right side of (II-42.1) by the left and right side of (II-42.5) respectively and divide by the factor V.

Then we have as equation for the energy (radiation) density:

$$\frac{E\,dz}{V} = \frac{8\pi h}{c^3}\left[\frac{\nu^3 d\nu}{e^{\frac{h\nu}{kT}} - 1}\right]. \tag{II-42.6}$$

For convenience we let the left side of (II-42.6) be denoted by $d\phi$. Then the radiation density for the entire spectrum from the very small through the very large frequencies may be taken to be given by

$$\phi = \frac{8\pi h}{c^3}\int_0^\infty \frac{\nu^3 d\nu}{e^{\frac{h\nu}{kT}} - 1}. \tag{II-42.7}$$

Now how to evaluate this integral? Well, the fact that the interval of integration is from $\nu = 0$ to $\nu = \infty$ suggests the possibility of involvement with the Gamma function. This hope is strengthened by the fact that if it were not for the -1 in the denominator the integral would be $\int_0^\infty \nu^3 e^{-m\nu}d\nu$ where $m = h/kT$.

As an initial step, then, let us try the transformation $e^{-m\nu} = x$ and see what happens. We then have $-m\nu = \log_e x$, $d\nu = -(1/m)\,dx/x$. As ν runs from zero to infinity, x varies from 1 to zero. We get

$$\phi = \frac{8\pi k^4 T^4}{c^3 h^3}\int_0^1 \left(\log_e \frac{1}{x}\right)^3 \frac{1}{1-x}\,dx. \tag{II-42.8}$$

It looks as if we have not made any progress. But if we look back a bit in Chapter II we see that our integral now differs from the integral $\int_0^1 x^m (\log_e 1/x)^n\,dx$ in Prob. II-7 only in that the fraction $1/(1-x)$ takes the place of the factor x^m. This gives us the clue to our next step. We recall the expansion

$$\frac{1}{1-x} = 1 + x + x^2 + x^3 + \cdots, \qquad -1 < x < 1.$$

Denoting the constant before the integral in II-42·8 by C, we have

$$\phi = C\int_0^1 \left[\left(\log_e\frac{1}{x}\right)^3 + x\left(\log_e\frac{1}{x}\right)^3 + x^2\left(\log_e\frac{1}{x}\right)^3 + \cdots\right]dx.$$

We are confronted with this question: May we integrate this infinite series termwise, that is, term by term? If we may, our problem is solved: by Prob. II-7 with $n = 3$ we get

$$\phi = C\Gamma(4)\left[\frac{1}{1^4} + \frac{1}{2^4} + \frac{1}{3^4} + \cdots\right].$$

The series here is indeed convergent: it is a convergent case of the p-series $\frac{1}{1^p} + \frac{1}{2^p} + \frac{1}{3^p} + \cdots$, which is convergent when and only when $p > 1$.

Inasmuch as the p-series converges to a finite value for every $p > 1$, its sum for $p > 1$ defines a function of p, named after Riemann and called the Riemann Zeta function:

$$\zeta(p) = 1 + \frac{1}{2^p} + \frac{1}{3^p} + \cdots, \qquad p > 1.$$

The Riemann Zeta function, like many others including the Γ-function, has been extended into the realm of complex arguments where $p = x + iy$ with x and y real and $i = \sqrt{-1}$. In this extended realm it is found that the p-series is convergent for all complex p having real part $x > 1$.

Apparently, then, all we have to do to finish our problem is to look up the value $\zeta(4)$ in the tables of the Riemann Zeta function. There we find $\zeta(4) = \pi^4/90$, making

$$\phi = C\Gamma(4)\zeta(4),$$

that is

$$\phi = \frac{8\pi^5 k^4 T^4}{15c^3 h^3}, \text{ since } \Gamma(4) = 3! \text{ by Eq. (I-10.1)}.$$

We shall not be content, however, with our solution of this problem until we have justified that step where we integrated an infinite series of functions termwise. As a result of the termwise integration we got a convergent series of constants. But even that did not prove that the sum of the series of integrals is the same as the integral of the function $\left(\log_e \frac{1}{x}\right)^3 \frac{1}{1-x}$ which is the sum of the series of functions that we integrated termwise. To prove

$$\int_0^1 \left(\log_e \frac{1}{x}\right)^3 dx + \int_0^1 x\left(\log_e \frac{1}{x}\right)^3 dx + \int_0^1 x^2\left(\log_e \frac{1}{x}\right)^3 dx + \cdots$$
$$= \int_0^1 \left(\log_e \frac{1}{x}\right)^3 \frac{1}{1-x} dx$$

we shall express the function $\left(\log_e \frac{1}{x}\right)^3 \frac{1}{1-x}$ as a *finite* series in powers of x with a remainder term:

$$\left(\log_e \frac{1}{x}\right)^3 \frac{1}{1-x} = \left(\log_e \frac{1}{x}\right)^3 + x\left(\log_e \frac{1}{x}\right)^3 + \cdots$$
$$+ x^{n-1}\left(\log_e \frac{1}{x}\right)^3 + \frac{x^n}{1-x}\left(\log_e \frac{1}{x}\right)^3$$

and then demonstrate somehow that the integral of the remainder term approaches the limit zero as $n \to \infty$.

Our task then is to appraise the integral

$$\int_0^1 \left[\frac{\log_e^3 \frac{1}{x}}{1-x}\right] x^n \, dx$$

in terms of n. Our first thought is to compare this integral with the integral of x^n. But this does not seem feasible because the function enclosed by the brackets becomes infinite as $x \to 0$. To obviate that difficulty we write the integral of the remainder term like this:

$$\int_0^1 \left[\frac{x \log_e^3 \frac{1}{x}}{1-x}\right] x^{n-1} \, dx.$$

The situation appears now to be favorable, for we find via L'Hospital's Rule that the function now enclosed by the brackets approaches zero as limit as $x \to 0$ and as $x \to 1$. Moreover, the bracketed function is positive for $0 < x < 1$, and an examination of its derivative shows that it has a single maximum between $x = 0$ and $x = 1$, occurring where $\log_e x = 3(x - 1)$. We need not try to compute this maximum. It suffices for the argument we have in mind to know that it exists. Let us call it M.

We can now make a suitable appraisal of the integral of the remainder term:

$$\int_0^1 \left[\frac{\log_e^3 \frac{1}{x}}{1 - x} \right] x^n dx = \int_0^1 \left[\frac{x \log_e^3 \frac{1}{x}}{1 - x} \right] x^{n-1} dx < \int_0^1 M x^{n-1}\, dx = \frac{M}{n},$$

which shows at once that the limit zero is approached by the integral of the remainder term occurring when $\left(\log_e \frac{1}{x} \right)^3 \frac{1}{1 - x}$ is expanded in a finite series of powers of x, thus proving that our termwise integration was valid.

Problems: Approximation of n! for Large Integers

II-43. Show that, when n is a large positive integer, the magnitude of the difference between $\log_e (n!)$ and $(n + \frac{1}{2}) \log_e n - n$ is given with a good approximation (the larger the n, the better) by $\log_e (n! e^n / n^{n+1/2})$.   ★

This exercise and the next two, taken together, constitute a development of Stirling's formula for the factorial of a large positive integer: $\lim_{n \to \infty} [(2\pi)^{1/2} n^{n+1/2} e^{-n} / n!] = 1$.

Observing that $\log_e (n!) = \log_e 2 + \log_e 3 + \cdots + \log_e n$, where we have left out $\log_e 1$ which is zero, and comparing the area under the curve $y = \log_e x$ from $x = 1$ to $x = n$, n integral and ≥ 2, with the set of majorant rectangles each of base unity, we see that $\int_1^n \log_e x \, dx < \log_e (n!)$. Likewise, by comparing the area under $y = \log_e x$ from $x = 2$ to $x = n + 1$ with the set of underlying rectangles each of base unity, we find that

$$\text{Log}_e (n!) < \int_1^{n+1} \log_e x \, dx.$$

Evaluating these two integrals and combining the inequalities, we have

$$n \log_e n - n + 1 < \log_e (n!) < (n + 1) \log_e (n + 1) - n. \quad \text{(II-43.1)}$$

We now revamp the expression $(n + 1) \log_e (n + 1)$:

$$(n + 1) \log_e (n + 1) = (n + 1) \log_e \left[n\left(1 + \frac{1}{n}\right) \right]$$

$$= (n + 1) \left[\log_e n + \log_e \left(1 + \frac{1}{n}\right) \right]$$

$$= n \log_e n + \log_e n + \log_e \left(1 + \frac{1}{n}\right)^n$$

$$+ \log_e \left(1 + \frac{1}{n}\right).$$

We have in mind to appraise the difference between $\log_e (n!)$ and the *arithmetic average* of the two outside members of our inequality (II-43.1). This average is readily seen to be

$$n \log_e n - n + \frac{1}{2} \log_e n + \left[\frac{1}{2} + \frac{1}{2} \log_e \left(1 + \frac{1}{n}\right) \right.$$

$$\left. + \frac{1}{2} \log_e \left(1 + \frac{1}{n}\right)^n \right]. \quad \text{(II-43.2)}$$

Upon scanning this average we see that there is not much point, when n is large, to retaining all the terms thereof, because the term $\frac{1}{2}$

becomes of less and less significance as n is taken larger and larger, as does the term $\frac{1}{2} \log_e \left(1 + \frac{1}{n}\right)$, whose value is close to zero when n is large. Take a look at the last term in (II-43.2) and you will see that it approaches $\frac{1}{2} \log_e e$, namely $\frac{1}{2}$, as $n \to \infty$. So, we discard these three terms in brackets as being of less and less significance as n is taken larger. Accordingly, we consider the difference

$$E_n = \log_e (n!) - \left(n \log_e n - n + \frac{1}{2} \log_e n\right)$$

$$= \log_e (n!) - \log_e (n^n) + \log_e (e^n) - \log_e (n^{1/2})$$

$$= \log_e \left[\frac{n! e^n}{n^{n+1/2}}\right].$$

II-44. Show that

$$\lim_{n \to \infty} \left[\frac{n! e^n}{n^{n+1/2}}\right] = \sqrt{2\pi}. \qquad \star$$

QUESTIONS. How to start? What and where is the connection between π and such an expression as $n! e^n / n^{n+1/2}$? An answer: We have already found in Prob. II-28 how to express $\pi/2$ as the limit of an unending product. Let us see if we can modify that product into a form whose limit is $\sqrt{2\pi}$ and then show that the expression $n! e^n / n^{n+1/2}$ has the same limit as $n \to \infty$.

We recall that the Wallis formula

$$\prod_{n=1}^{\infty} \frac{4n^2}{4n^2 - 1} = \frac{\pi}{2} \qquad \text{(II-44.1)}$$

in Prob. II-28 can be written

$$\lim_{n \to \infty} \left\{\left[\frac{2 \cdot 4 \cdots (2n)}{1 \cdot 3 \cdots (2n - 1)}\right]^2 \frac{1}{2n + 1}\right\} = \frac{\pi}{2}.$$

It follows that

$$\lim_{n \to \infty} \left[\frac{2 \cdot 4 \cdots (2n)}{1 \cdot 3 \cdots (2n - 1)} \cdot \frac{2}{(2n + 1)^{1/2}} \right] = 2\sqrt{\pi/2} = \sqrt{2\pi}. \quad \text{(II-44.2)}$$

Our task, then, is to prove that $n!e^n/n^{n+1/2}$ has the same limit, as $n \to \infty$, as has the expression enclosed by the brackets in (II-44.2). We proceed as follows.

Let $P_n = n!e^n/n^{n+1/2}$. Then $P_n{}^2 = (n!)^2e^{2n}/n^{2n+1}$; and $P_{2n} = (2n)!e^{2n}/(2n)^{2n+1/2}$, whence

$$\frac{P_n{}^2}{P_{2n}} = \frac{(n!)^2 2^{2n}}{(2n)!} \left[\frac{(2n)^{1/2}}{n} \right]$$

$$= \frac{2 \cdot 4 \cdots (2n)}{1 \cdot 3 \cdot 5 \cdots (2n - 1)} \cdot \frac{\sqrt{2n}}{n}. \quad \text{(II-44.3)}$$

Now we begin to see the desired connection. The first fraction on the right in (II-44.3) is the same as the first fraction on the left in (II-44.2). The second fractions, although different from each other, are readily seen to be such that the limit of their ratio as $n \to \infty$ is unity. It follows at once, therefore, by (II-44.2) and (II-44.3) that

$$\lim_{n \to \infty} \frac{P_n{}^2}{P_{2n}} = \sqrt{2\pi}. \quad \text{(II-44.4)}$$

We almost have the desired result, namely, $\lim\limits_{n \to \infty} P_n = \sqrt{2\pi}$. It only takes a little adroit maneuvering with $P_n{}^2$ and P_{2n} as follows. In the first place we observe that if P_n has a limit as $n \to \infty$, then P_{2n} approaches *the same limit* because P_{2n} is merely a *later* value of P_n itself. We will proceed on the assumption that P_n *does* approach a limit as $n \to \infty$. (This can be verified, for instance, by examining the difference $\log_e P_{n+1} - \log_e P_n$ in such a way as to show that $\log_e P_n$

approaches a finite limit as $n \to \infty$, whence it follows that P_n does likewise.) Then we have

$$\lim_{n \to \infty} P_n = \frac{\lim P_n \lim P_n}{\lim P_n}$$

$$= \frac{(\lim P_n)^2}{\lim P_{2n}}$$

$$= \frac{\lim (P_n{}^2)}{\lim P_{2n}}$$

$$= \lim_{n \to \infty} \left[\frac{P_n{}^2}{P_{2n}} \right]$$

$$= \sqrt{2\pi}.$$

II-45. From Prob. II-44, where the limit (as $n \to \infty$) of the fraction $n!e^n/n^{n+1/2}$ was shown to be $\sqrt{2\pi}$, deduce the conclusion that the factorial of a large integer n is well approximated by $(n/e)^n \sqrt{2\pi n}$. ★

From the result proved in Prob. II-44 we have at once

$$\lim_{n \to \infty} \left[\frac{n!e^n}{n^{n+1/2}\sqrt{2\pi}} \right] = 1,$$

$$\lim_{n \to \infty} \left[\frac{n!}{e^{-n}n^n\sqrt{n}\sqrt{2\pi}} \right] = 1,$$

$$\lim_{n \to \infty} \left[\frac{n!}{\left(\dfrac{n}{e}\right)^n \sqrt{2\pi n}} \right] = 1.$$

Since the limit of the ratio within the brackets is unity, it follows that when n is large the value of the ratio is close to unity (the larger, the closer) so that numerator and denominator differ by only a small percentage of either. In other words, the denominator furnishes a good approximation for $n!$ when n is large:

$$\left(\frac{n}{e}\right)^n \sqrt{2\pi n} \cong n!. \qquad (\text{II-45.1})$$

The left side of (II-45.1) is known as *Stirling's (approximation) formula for n!*.

II-46. Determine $\lim\limits_{n\to\infty} \dfrac{(n!)^{1/n}}{n}$. ★

This problem clearly does not lend itself to any immediate application of L'Hospital's Rule. We shall have to do something with the expression before us in order to get it in shape for application of L'Hospital's Rule, which says that, if $f(n)$ and $g(n)$ are two differentiable functions of real variable n both of which $\to \infty$ as $n \to \infty$, then

$$\lim_{n\to\infty} \frac{f(n)}{g(n)} = \lim_{n\to\infty} \frac{f'(n)}{g'(n)}.$$

Clearly the stumbling block in proceeding lies in $n!$. What to do with it? Answer: Try Stirling. Make use of the fact that

$$\lim_{n\to\infty} \frac{n!}{\left(\dfrac{n}{e}\right)^n \sqrt{2\pi n}} = 1.$$

By multiplying the numerator and denominator of the expression given in the problem by $\left(\dfrac{n}{e}\right)^n \sqrt{2\pi n}$ we have then

$$\lim_{n\to\infty} \frac{(n!)^{1/n}}{n} = \lim_{n\to\infty} \left\{ \left[\frac{n!}{\left(\frac{n}{e}\right)^n \sqrt{2\pi n}} \right]^{1/n} \left[\frac{\left(\frac{n}{e}\right)^n \sqrt{2\pi n}}{n^n} \right]^{1/n} \right\}$$

$$= (1) \lim_{n\to\infty} \left[\frac{\left(\frac{n}{e}\right)^n \sqrt{2\pi n}}{n^n} \right]^{1/n}$$

$$= \lim_{n\to\infty} \left[\frac{1}{e} (2\pi n)^{1/2n} \right]$$

$$= \frac{1}{e} \lim_{n\to\infty} (2\pi n)^{1/2n}.$$

Now we can apply L'Hospital's Rule. Let $y = (2\pi n)^{1/2n}$. Then $\log_e y = (\log_e 2\pi n)/2n$. And

$$\lim_{n\to\infty} \log_e y = \lim_{n\to\infty} \frac{\log_e 2\pi n}{2n}$$

$$= \lim_{n\to\infty} \frac{\frac{2\pi}{2\pi n}}{2} \qquad \text{by L'Hospital's Rule}$$

$$= 0.$$

Hence, $\lim_{n\to\infty} y = e^0 = 1$. Consequently,

$$\lim_{n\to\infty} \frac{(n!)^{1/n}}{n} = \frac{1}{e}.$$

II-47. Appraise the relative error committed when Stirling's formula $\left(\frac{n}{e}\right)^n \sqrt{2\pi n}$ is taken as approximation to 10!, to 50!, to 100!. Also, compute these factorials approximately via Stirling. ⋆

We apply here the connection between Stirling's formula and the Gamma function, namely the asymptotic expansion

$$\log_e \Gamma(x) \sim \log_e \left[\sqrt{\frac{2\pi}{x}} \frac{x^x}{e^x} \right] + \frac{1}{12x} - \frac{1}{360x^3} + \cdots, \qquad x > 0,$$

in which we have used the wiggle sign $\sim$ instead of the equality sign because the alternating series indicated is divergent for all x. But it *does* have the property that any partial sum thereof for $x > 0$ differs from $\log_e \Gamma(x)$ by an amount which in absolute value is less than the last term of the partial sum. So, for factorials of integers we have the appraisal

$$\begin{aligned}
\log_e (n!) &= \log_e \Gamma(n + 1) \qquad \text{by Prob. I-10} \\
&= \log_e n\Gamma(n) \qquad \text{by Prob. I-4} \\
&= \log_e n + \log_e \Gamma(n) \\
&\cong \log_e n + \log_e \left[\sqrt{\frac{2\pi}{n}} \frac{n^n}{e^n} \right] + \frac{1}{12n} \\
&\cong \log_e \left[\sqrt{2\pi n}\left(\frac{n}{e}\right)^n \right] + \frac{1}{12n}.
\end{aligned}$$

The error in this last equation is less than $1/12n$. We observe that the expression in brackets in this last approximate equation is none other than Stirling's formula. Let us denote Stirling's formula by S:

$$S = \sqrt{2\pi n}\left(\frac{n}{e}\right)^n.$$

Thus we may write

$$\log_e (n!) - \log_e S = E, \qquad 0 < E < \frac{1}{12n}.$$

Then it follows at once that

$$\frac{S}{n!} = e^{-E}.$$

So, for the relative error R we have

$$R = \frac{n! - S}{n!} = 1 - \frac{S}{n!} = 1 - e^{-E}.$$

Replacing e^{-E} by its Maclaurin series in powers of E, we get

$$R = E - \frac{E^2}{2!} + \frac{E^3}{3!} - \cdots,$$

which shows at once that $R < E$. But we already have $E < 1/12n$. We conclude that, when $n!$ is approximated by S, the relative error committed is less than $1/12n$.

We can now quickly appraise the desired approximate errors as well as compute approximately the requested factorials. Here are the results in tabular array:

n	$S = \sqrt{2\pi n}\left(\dfrac{n}{e}\right)^n$	$R = \dfrac{n! - S}{n!} < \dfrac{1}{12n}$
10	3.63×10^6	less than 0.84%
50	3.04×10^{64}	less than 0.17%
100	9.33×10^{157}	less than 0.09%
500	1.22×10^{1134}	less than 0.002%

Two Problems in Probability

II-48. A class of 100 students line up at random in a row. Compute approximately via Stirling's formula, the probability that they will line up from left to right in the order of ranking in the class from valedictorian to the lowest.   ★

The number of different orders in which it would be possible for the class to line up from left to right in a row is 100!. The probability of falling into line in order of class standing from left to right is $1/(100!)$.

By Stirling's formula we have

$$n! \cong \left(\frac{n}{e}\right)^n \sqrt{2\pi n},$$

so that

$$100! \cong \left(\frac{100}{2.718}\right)^{100} \sqrt{(2\pi)(100)}.$$

Computing the right side by logarithms or slide rule, we get

$$100! \cong 9.3 \times 10^{157}.$$

The probability requested is approximately $\dfrac{1}{9.3 \times 10^{157}}$.

To put the matter mildly, we remark that their chances of getting in line in the order described is rather small. But it *could* happen!

II-49. Compute approximately the probability that, if a succession of 1,000 picks of one card each be made from a full bridge deck at random (that is, the picked card is replaced each time and the deck reshuffled between picks), the number of Jacks drawn will not be more than 100 nor less than 50.   ★

We observe first that this problem can be solved exactly by elementary formulas as follows. If the probability of success for every individual independent trial be p, then the probability $P(x)$ of *precisely* x successes in n trials is

$$P(x) = {}_nC_x p^x (1 - p)^{n-x} = \frac{n!}{x!(n - x)!} p^x (1 - p)^{n-x},$$

where ${}_nC_x$ denotes the number of combinations of n different things taken x at a time. It follows from this that when n trials are made the probability of $\bar{X}$ successes with $x_1 \leq \bar{X} \leq x_2$ is the *sum* of probabilities of mutually exclusive events given by

$$P(\bar{X}) = \sum_{x=x_1}^{x_2} {}_nC_x p^x (1-p)^{n-x} = \sum_{x=x_1}^{x_2} \frac{n!}{x!(n-x)!} p^x (1-p)^{n-x}.$$

In the problem at hand we would get the solution exactly by computing

$$\sum_{x=50}^{100} \frac{1000!}{x!(1000-x)!} \left(\frac{1}{13}\right)^x \left(\frac{12}{13}\right)^{1000-x}.$$

But who would want to do that for the sake of exactness? The next best thing to do might be to use Stirling's approximation $m^m e^{-m}\sqrt{2\pi m}$ for $m!$ from Prob. II-45 for each of the factorials involved in the summation. But that still leaves us 51 terms to add together after computing each term approximately by Stirling. This objection causes us to ask: If Stirling's compact formula gives a good approximation for the long product of factors involved in factorial n when n is large, might there also be obtainable a compact approximate formula for the summation

$$P(\bar{X}) = \sum_{x=x_1}^{x_2} \frac{n!}{x!(n-x)!} p^x (1-p)^{n-x},$$

such that the ratio of the approximate formula to the exact formula $P(\bar{X})$ approaches the limit unity as $n \to \infty$?

This question has been answered affirmatively by the De Moivre-Laplace theorem in the following manner. Since we are asking that the number of successes be somewhere on a relatively small spread of numbers as compared with a rather large number of trials, it is reasonable to think that the probability $P(\bar{X})$ might be given with good approximation by the area under an appropriate portion of the standard normal distribution curve

$$y = \frac{1}{\sqrt{2\pi}}\, e^{-(t^2/2)} = \phi(t)$$

as compared with the area A under the whole curve from $t = -\infty$ to $t = +\infty$. In fact we can even omit the comparison, inasmuch as the area under the whole curve is unity. For we have by Prob. II-1, with $a = 0$, $b = 1/2$, $c = 2$,

$$A = \int_{-\infty}^{\infty} \frac{1}{\sqrt{2\pi}}\, e^{-(t^2/2)} dt = 2 \int_{0}^{\infty} \frac{1}{\sqrt{2\pi}}\, e^{(-t^2/2)} dt$$

$$= \frac{2}{\sqrt{2\pi}} \frac{\Gamma\!\left(\dfrac{1}{2}\right)}{2\!\left(\dfrac{1}{2}\right)^{1/2}} = 1, \qquad \text{by Prob. I-11.}$$

Thus the whole area represents certainty in probability theory, namely unity. We should expect, then, that we might have

$$P(\bar{X}) = \int_{t_1}^{t_2} \frac{1}{\sqrt{2\pi}}\, e^{-(t^2/2)} dt,$$

where t_1 and t_2 depend somehow on n, x_1 and x_2.

That actually is the essence of the content of the De Moivre-Laplace theorem which is derived from the summation formula for $P(\bar{X})$ by replacing each factorial therein by its Stirling approximation.

The De Moivre-Laplace theorem:

$$P(\bar{X}) \cong \int_{t_1}^{t_2} \frac{1}{\sqrt{2\pi}}\, e^{-(t^2/2)} dt = \Phi(t_2) - \Phi(t_1)$$

where

$$t_1 = \frac{x_1 - np}{\sqrt{np(1-p)}}, \quad t_2 = \frac{x_2 - np}{\sqrt{np(1-p)}}, \quad \Phi(t) = \int_{0}^{t} \frac{1}{\sqrt{2\pi}} e^{-(t^2/2)} dt;$$

moreover, with t_1 and t_2 held fixed,

$$\lim_{n\to\infty} \frac{\Phi(t_2) - \Phi(t_1)}{P(\bar{\underline{X}})} = 1.$$

We observe that the function $\Phi(t)$ involved in the De Moivre-Laplace theorem is an incomplete Gamma function with a coefficient.

If we put $t^2/2 = u$ we get

$$\Phi(t) = \frac{1}{\sqrt{2\pi}} \int_0^u \frac{1}{\sqrt{2}} u^{-1/2} e^{-u} \, du = \frac{1}{\sqrt{2\pi}} \frac{1}{\sqrt{2}} \gamma\left(\frac{1}{2}, u\right) = \frac{1}{2\sqrt{\pi}} \gamma\left(\frac{1}{2}, \frac{t^2}{2}\right),$$

where $\gamma(\frac{1}{2}, u)$ is the incomplete Gamma function. (See introduction to Chap. I.)

Putting $x_1 = 50$, $x_2 = 100$, $n = 1,000$, $p = \frac{1}{13}$ in the formulas of the De Moivre-Laplace theorem and using the tabulated values of $\Phi(t)$ we get

$$t_1 = \frac{50 - (1,000)(1/13)}{\sqrt{(1,000)(1/13)(12/13)}} \cong -3.19,$$

$$t_2 = \frac{100 - (1,000)(1/13)}{\sqrt{(1,000)(1/13)(12/13)}} \cong 2.74,$$

$$\Phi(t_1) = -\Phi(-t_1) = -\Phi(3.19) \cong -0.4993,$$

$$\Phi(t_2) = \Phi(2.74) \cong 0.4969.$$

The probability of picking between 50 and 100 Jacks in 1,000 tries is approximately

$$\begin{aligned} P(\bar{\underline{X}}) &\cong \Phi(t_2) - \Phi(t_1) \\ &\cong 0.4969 - (-0.4993) \\ &= 0.9962. \end{aligned}$$

The result, which is close to unity, indicates that the chance of success is an excellent one.

REMARK.   The function $\Phi(t)$ is related to the *error function*

$$\text{erf}(t) = \frac{2}{\sqrt{\pi}} \int_0^t e^{-t^2} dt.$$

One finds that

$$\text{erf}(t) = 2\Phi(\sqrt{2}\, t).$$

A Problem in Heat Flow in a Straight Wire

II-50. A long, straight wire lies in the positive x-axis with one end at $x = 0$. Except for the end $x = 0$ the wire is kept thermally isolated after having been given initially at time $t = 0$ a continuous, positive temperature distribution $f(x)$. The end at $x = 0$ is maintained at temperature zero. If the length of the wire is so long as to be considered infinite for the purpose of the problem and if the initial temperature distribution $f(x)$ is, moreover, bounded for all x, determine the temperature distribution T as a function of position x and time t: $T = T(x, t)$.

Solve the problem also for the case where the initial temperature distribution for $x > 0$ is a positive constant C.   ★

The differential equation to be satisfied by $T = T(x, t)$ is known (see, for example, Rainville, *Elementary Differential Equations*, Macmillan, 1958) to be

$$\frac{\partial T}{\partial t} = \alpha \frac{\partial^2 T}{\partial x^2} \tag{II-50.1}$$

where α is a constant, namely the diffusivity. It has been found that in heat-flow problems a solution for the temperature T can often be obtained as a composite of particular solutions, each particular solution being a *product function* of the form FG where F is a function of time alone and G is a function of the space coordinates only. In our present problem such a particular solution of the product type for Eq. (1) is

$$T(x, t) = FG = F(t)G(x). \qquad \text{(II-50.2)}$$

For such a solution we have

$$\frac{\partial T}{\partial t} = GF', \frac{\partial^2 T}{\partial x^2} = FG''.$$

Eq. (II-50.1) then becomes

$$GF' = \alpha\, FG''.$$

Since $\alpha \neq 0$ and since the problem is such that neither G nor F will be identically zero we may divide by each of these three, obtaining

$$\frac{1}{\alpha}\frac{F'}{F} = \frac{G''}{G}. \qquad \text{(II-50.3)}$$

Since F'/F is independent of x while G''/G is independent of t, it follows that each of these two ratios must be a constant. Letting $k = (1/\alpha)(F'/F)$ and integrating, we get

$$\log_e F = \alpha k t + \log c,$$
$$F = c e^{\alpha k t}$$

where c is a constant.

The constant k cannot be zero, for this would mean that the temperature $T(x, t) = FG = cG(x)$ would hold constant at each point x of the wire with passage of time contrary to the physical set-up of the problem which is such that heat will leave the wire at the end $x = 0$ where the temperature is being maintained at zero. Nor can k be positive. A positive k would call for an increase of temperature with increasing t at each and every point x on the wire.

This could happen only if heat would come into the wire at $x = 0$. This, too, is contrary to the conditions of the problem. We conclude, therefore, that k must be taken negative.

Taking $k = -\lambda^2$ with $\lambda \neq 0$, we find from Eq. (II-50.3) that

$$G'' + \lambda^2 G = 0,$$

for which the general solution is

$$G = c_1 \sin \lambda\, x + c_2 \cos \lambda\, x,$$

where c_1 and c_2 are constants. Then the product solution assumed in Eq. (II-50.2) becomes

$$T(x,\, t) = e^{-\alpha\lambda^2 t}(A \sin \lambda x + B \cos \lambda x),$$

where $A = cc_1$ and $B = cc_2$. In order to fulfill the boundary condition $T(0,\, t) = 0$ we require

$$0 = Be^{-\alpha\lambda^2 t},$$

which means that B must be zero. We now have

$$T(x,\, t) = Ae^{-\alpha\lambda^2 t} \sin \lambda x. \qquad \text{(II-50.4)}$$

The function $T(x,\, t)$ defined in Eq. (II-50.4) will satisfy Eq. (II-50.1) for any choice of the constants A and λ. It will also meet the boundary condition $T(0,\, t) = 0$. But a single such function $T(x,\, t)$ will not satisfy the initial condition $T(x,\, 0) = f(x)$ except in the special case where the initial temperature distribution for $x > 0$ is of the form $f(x) = A \sin \lambda x$. Nor can we hope to compose a solution via an infinite series of particular solutions of the kind defined in Eq. (II-50.4) except in the case where $f(x)$ is periodic, since the wire is taken as infinitely long.

There remains the possibility of composing a solution via an integral as follows. We consider a family of solutions

$$A(\lambda)e^{-\alpha\lambda^2 t} \sin \lambda x$$

in which there shall be a member of the family for each nonnegative real number λ and in which, as indicated, there is to be determined a value for A corresponding to each λ. We then superimpose, so to speak, all such solutions in an integral

$$T(x, t) = \int_0^\infty A(\lambda)e^{-\alpha\lambda^2 t} \sin \lambda x \, d\lambda \qquad \text{(II-50.5)}$$

and see if we can satisfy the initial condition $T(x, 0) = f(x)$ as well as Eq. (II-50.1) with such a composite function.

Assuming for the moment that the function $A(\lambda)$ can be so determined as to permit differentiation of the integral in Eq. (II-50.5) with respect to the parameters x and t, we find by Leibniz's Rule for differentiation under the integral sign that $T(x, t)$ as defined in Eq. (II-50.5) satisfies Eq. (II-50.1). There remains to see if we can determine $A(\lambda)$ so that $T(x, 0) = f(x)$. This requires that

$$\int_0^\infty A(\lambda) \sin \lambda x \, d\lambda = f(x). \qquad \text{(II-50.6)}$$

We can satisfy the demand made upon $A(\lambda)$ by Eq. (II-50.6) as follows. First, we restrict $f(x)$ to be such that the integral $\int_0^\infty f(x)dx$ is convergent. Next, we define $f(x)$ for all negative x by taking $f(-x) = -f(x)$. Then we consider the Fourier integral of $f(x)$:

$$f(x) = \int_0^\infty g(\lambda) \cos x\lambda \, d\lambda + \int_0^\infty h(\lambda) \sin x\lambda \, d\lambda,$$

where

$$g(\lambda) = \frac{1}{\pi} \int_{-\infty}^\infty f(\phi) \cos \lambda\phi \, d\phi, \; h(\lambda) = \frac{1}{\pi} \int_{-\infty}^\infty f(\phi) \sin \lambda\phi \, d\phi.$$

(See, for example, Kaplan, *Advanced Calculus*, Addison-Wesley, 1952.)

Since we have built $f(x)$ into an odd function, it follows that $g(\lambda)$ vanishes and that

$$h(\lambda) = \frac{2}{\pi} \int_0^\infty f(\phi) \sin \lambda\phi \, d\phi, \tag{II-50.7}$$

making

$$f(x) = \int_0^\infty h(\lambda) \sin x\lambda \, d\lambda. \tag{II-50.8}$$

Comparing Eqs. (II-50.7) and (II-50.8) with Eq. (II-50.6), we find that Eq. (II-50.6) will hold if we take

$$A(\lambda) = h(\lambda) = \frac{2}{\pi} \int_0^\infty f(\phi) \sin \lambda\phi \, d\phi.$$

Thus, we have obtained the solution for our problem:

$$T(x, t) = \frac{2}{\pi} \int_0^\infty \left\{ \int_0^\infty e^{-\alpha\lambda^2 t} f(\phi) \sin \lambda\phi d\phi \right\} \sin \lambda x \, d\lambda. \tag{II-50.9}$$

The solution as presented in Eq. (II-50.9) may be revamped somewhat. We may write

$$T(x, t) = \frac{2}{\pi} \int\!\!\int_Q e^{-\alpha\lambda^2 t} f(\phi) \sin \lambda\phi \sin \lambda x \, dA,$$

where Q denotes the entire first quadrant of the $\phi\lambda$-plane, considering ϕ and λ as rectangular coordinates, and where dA denotes element of area. This double integral over Q is equivalent to the iterated integral

$$T(x, t) = \frac{2}{\pi} \int_0^\infty \left\{ \int_0^\infty e^{-\alpha\lambda^2 t} \sin \lambda\phi \sin \lambda x \, d\lambda \right\} f(\phi) \, d\phi. \tag{II-50.10}$$

Moreover, we have

$$\sin \lambda\phi \sin \lambda x = \frac{1}{2}\left[\cos \left\{\lambda(\phi - x)\right\} - \cos \left\{\lambda(\phi + x)\right\} \right].$$

Then by the formula obtained in Prob. II-20, namely

$$\int_0^\infty e^{-a^2\lambda^2} \cos b\lambda \, d\lambda = \frac{\sqrt{\pi}}{2a} e^{-b^2/4a^2},$$

we find that Eq. (II-50.10) becomes

$$T(x, t) = \begin{cases} \dfrac{1}{2\sqrt{\pi\alpha t}} \displaystyle\int_0^\infty f(\phi) \left[e^{-(\phi-x)^2/4\alpha t} - e^{-(\phi+x)^2/4\alpha t} \right] d\phi, & t > 0 \\[3mm] f(x), & t = 0. \end{cases}$$

$$\text{(II-50.11)}$$

Although our solution as given by Eq. (II-50.11) was obtained under the hypothesis that $f(\phi)$ is such that the integral $\int_0^\infty f(\phi)d\phi$ is convergent, we observe that the integral in Eq. (II-50.11) is convergent if $f(\phi)$ be any function which is continuous and bounded for all positive ϕ. This follows from the fact that, as $\phi \to +\infty$, the exponentials in the integral decrease like $e^{-k\phi^2}$, $k > 0$. We suspect, therefore, that our solution in Eq. (II-50.11) holds when $f(x)$ is any function which is continuous and bounded for all positive x. And this is found to be so: differentiation of $T(x, t)$ for $t > 0$ as defined in Eq. (II-50.11) partially with respect to t and twice partially with respect to x shows that $T(x, t)$ satisfies Eq. (II-50.1).

In particular then, we may apply the solution presented in Eq. (II-50.11) to the particular solution in which the initial condition is

$$T(x, 0) = \begin{cases} 0, & x = 0, \\ C > 0, & x > 0 \end{cases}$$

where C is a constant. Since in this case the factor $f(\phi) = C$ in the integrand may be taken out in front of the integral, it seems likely that the solution may be written in simple form by transforming the exponents. Writing the integral in Eq. (II-50.11) as the difference of two integrals, we transform the first integral by letting $(\phi - x)^2/4\alpha t = u^2$. This makes $\phi = x + u\sqrt{4\alpha t}$ and $d\phi = \sqrt{4\alpha t} \, du$. Similarly, we transform the second integral by letting $(\phi + x)^2/4\alpha t = v^2$. Then for $t > 0$ we have

$$T(x, t) = \frac{C}{\sqrt{\pi}} \left\{ \int_{-x/\sqrt{4\alpha t}}^{\infty} e^{-u^2} du - \int_{x/\sqrt{4\alpha t}}^{\infty} e^{-v^2} dv \right\}$$

$$= \frac{2C}{\sqrt{\pi}} \int_{0}^{x/\sqrt{4\alpha t}} e^{-u^2} du.$$

Thus, for our solution of the particular case we have

$$T(x, t) = \begin{cases} \dfrac{2C}{\sqrt{\pi}} \displaystyle\int_{0}^{x/\sqrt{4\alpha t}} e^{-u^2} du, & t > 0, \\[2mm] C, & t = 0, \end{cases} \qquad \text{(II-50.12)}$$

when C is a positive constant.

We observe that the upper right side of Eq. (II-50.12) can be expressed in terms of either the function $\Phi(t)$ or the function $\text{erf}(t)$ as defined in Prob. II-49. Thus, for $t > 0$, we have

$$T(x, t) = C \, \text{erf}(x/\sqrt{4\alpha t})$$

$$= 2C\Phi(x/\sqrt{2\alpha t}).$$

Also, as in Prob. II-49, we may express our solution of the particular case in terms of the incomplete Gamma function by letting $u^2 = w$. Doing so, we get for $t > 0$

$$T(x, t) = \frac{C}{\sqrt{\pi}} \int_{0}^{x^2/4\alpha t} w^{-1/2} e^{-w} dw$$

$$= \frac{C}{\sqrt{\pi}} \gamma(1/2, x^2/4t\alpha).$$

REMARK. The constant k to which we equated each side of Eq. (II-50.3) is sometimes referred to in textbooks as the *separation constant*. Determination of the character and/or the value of the separation constant by appropriate methods is made also in problems in Chapters IV and VI, wherever product solutions of differential equations are involved.

3 LEGENDRE

POLYNOMIALS

INTRODUCTION

Legendre's differential equation is

$$(1 - x^2)y'' - 2xy' + n(n + 1)y = 0 \qquad \text{(III-0.1)}$$

where y' and y'' denote the derivatives dy/dx and d^2y/dx^2 respectively and n is a constant. It may also be written as

$$\frac{d}{dx}\left[(1 - x^2)y'\right] + n(n + 1)y = 0. \qquad \text{(III-0.2)}$$

Legendre's differential equation is a particular case of the equation

$$(1 + Ax^H)y'' + \frac{1}{x}(B + Cx^H)y' + \frac{1}{x^2}(D + Ex^H)y = 0.$$
$$\text{(III-0.3)}$$

If in this last equation we take $A = -1$, $B = 0$, $C = -2$, $D = 0$, $E = n(n + 1)$, and $H = 2$, we get Eq. (III-0.1). Other particular cases of Eq. (III-0.3) include Bessel's equation

$$x^2y'' + xy' + (x^2 - p^2)y = 0 \qquad \text{(III-0.3a)}$$

and Gauss's equation (the hypergeometric equation)

$$x(1 - x)y'' + [\gamma - (\alpha + \beta + 1)x]y' - \alpha\beta y = 0 \quad \text{(III-0.3b)}$$

as well as equations having as particular solutions such special functions (see, for example, Rainville, *Special Functions*, Macmillan, 1960) as Hermite polynomials, Tchebicheff polynomials, Jacobi polynomials, and Laguerre polynomials.

If we let $x = \cos \phi$, then Eq. (III-0.1) becomes transformed into a trigonometric form of Legendre's equation, namely

$$y'' + (\cot \phi)y' + n(n + 1)y = 0 \quad \text{(III-0.4)}$$

where now y' and y'' denote $dy/d\phi$ and $d^2y/d\phi^2$ respectively.

The constant n in Legendre's equation we shall call the *index* and shall refer to Eq. (III-0.1) or any equivalent form thereof as Legendre's differential equation of index n. It is shown in textbooks (see, for example, Franklin, *Methods of Advanced Calculus*, McGraw-Hill, 1944) that Legendre's equation (III-0.1) has solutions of a first kind

$$y = C_0\left[1 - \frac{n(n + 1)}{2!}x^2 + \frac{n(n + 1)(n - 2)(n + 3)}{4!}x^4 - \cdots\right]$$

$$+ C_1\left[x - \frac{(n - 1)(n + 2)}{3!}x^3\right.$$

$$+ \left.\frac{(n - 1)(n + 2)(n - 3)(n + 4)}{5!}x^5 - \cdots\right] \quad \text{(III-0.5)}$$

valid for $|x| < 1$, C_0 and C_1 being arbitrary constants.

Equation (III-0.1) also has solutions of a second kind expressible in infinite series (convergent for $|x| > 1$) of negative powers of x.

Legendre's equation (III-0.1) has especial importance when the index n is a positive integer or zero. Consider the family of equations

$$(1 - x^2)y'' - 2xy' + n(n + 1)y = 0, \qquad n = 0, 1, 2, \cdots. \quad \text{(III-0.6)}$$

For each member of the family there is the general solution of the first kind given by Eq. (III-0.5). For each number n of the set 0, 1, 2, 3, $\cdots$ one of the two series in brackets in Eq. (III-0.5) reduces to a polynomial of degree n. Choosing the coefficient before the other bracket to be zero and determining the coefficient before the polynomial so that the polynomial has the value unity at $x = 1$, we get the particular solution of Legendre's equation of index n known as the *Legendre polynomial $P_n(x)$*.

The process just indicated for obtaining the family of Legendre polynomials from Eq. (III-0.5) leads to the formula

$$P_n(x) = \frac{(2n)!}{2^n(n!)^2} \left[x^n - \frac{n(n-1)}{2(2n-1)} x^{n-2} \right.$$
$$\left. + \frac{n(n-1)(n-2)(n-3)}{2 \cdot 4 \cdot (2n-1)(2n-3)} x^{n-4} - \cdots + F \right],$$

$$\text{(III-0.7)}$$

where F denotes the final term. F is a constant when n is even. F is a constant times x when n is odd. It is to be observed that $P_n(x)$ contains only even powers of x when n is even, only odd powers of x when n is odd. A compact formulation of Eq. (III-0.7) is

$$P_n(x) = \frac{1}{2^n} \sum_{k=0}^{N} \frac{(-1)^k (2n-2k)!}{k!(n-k)!(n-2k)!} x^{n-2k} \qquad \text{(III-0.8)}$$

where $N = n/2$ or $(n-1)/2$ according as n is even or odd and where, as usual, the factorial of zero is taken as unity.

A formula for $P_n(x)$ even more compact than that given by Eq. (III-0.8) is *Rodrigues's formula*

$$P_n(x) = \frac{1}{2^n n!} D^n \left[(x^2 - 1)^n \right] \qquad \text{(III-0.9)}$$

where the operator symbol D^n denotes nth derivative.

The Legendre polynomials are often called by other names. One such alternative designation is *spherical harmonics* or *zonal harmonics*. This stems from the fact that the Legendre polynomials provide the ingredients for solutions of Laplace's equation $\nabla^2 V = 0$ where the function V of space coordinates is required to take on boundary values of restricted type on zones of a spherical surface as well as be *harmonic* (have continuous second-order partial derivatives and satisfy Laplace's equation) interior to the sphere.

Yet another name for the Legendre polynomials is *Legendre coefficients*, which is used because of a characteristic property of the family—a property which brings the whole family into play at once. It is the fact that, when the function

$$W(h, x) = (1 - 2xh + h^2)^{-1/2}$$

is expanded in Maclaurin's series in powers of h, the coefficient of h^n is, for every n, the Legendre polynomial $P_n(x)$. The establishment of this property is our problem No. 1 in this chapter. $W(h, x)$ is called a *generating function* for the Legendre polynomials.

The family of Legendre polynomials enjoys a number of characteristic properties and mutual relationships. The establishment and exploitation of these properties and relationships constitute the chief concern of the exercises and problems in this chapter.

A list of the Legendre polynomials of degrees 0 through 10 as well as a table of values of $P_n(x)$, $n = 1, 2, 3, 4, 5, 6, 7$, for $0 \leq x \leq 1$ will be found at the end of this chapter. There is also a sketch of the curves $y = P_n(x)$, $n = 0, 1, 2, 3, 4$, for $-1 \leq x \leq 1$.

Problem: Coefficients in Expansion of a Generating Function

III-1. Show that, when $W(h, x) = (1 - 2xh + h^2)^{-(1/2)}$ is expanded in a series of the form

$$W(h, x) = y_0(x) + y_1(x)h + y_2(x)h^2 + \cdots + y_n(x)h^n + \cdots,$$

the following are true of the coefficients of the powers of h:

(a) $y_n(x)$ is a polynomial of degree n,

(b) $y_n(1) = 1$,

(c) $y = y_n(x)$ is a solution of the Legendre equation

$$(1 - x^2)y'' - 2xy' + n(n + 1)y = 0.$$

In other words show that for $n = 0$, 1, 2, 3, $\cdots$ we have $y_n(x)$ $= P_n(x)$ where $P_n(x)$ is the Legendre polynomial of degree n as defined in the introduction to Chapter III. ★

The expansion may be obtained via the formal binomial expansion:

$$W(h, x) = 1 + \frac{1}{2}(2xh - h^2) + \left(\frac{1}{2}\right)\left(\frac{3}{2}\right)\left(\frac{1}{2!}\right)(2xh - h^2)^2$$

$$+ \left(\frac{1}{2}\right)\left(\frac{3}{2}\right)\left(\frac{5}{2}\right)\left(\frac{1}{3!}\right)(2xh - h^2)^3 + \cdots. \quad \text{(III-1.1)}$$

An inspection of Eq. (III-1.1) shows that when every integral power of the expression $2xh - h^2$ is expanded and then like powers of h are collected, the coefficient of h^n will be a polynomial in x of degree n. This is seen to be true because the expansion of $(2xh - h^2)^m$, where m is any positive integer, will be such that the power of x in the term $(2xh)^m$ is the same of that of h, namely m, while the power of x is less than m in each of the other terms where the power of h exceeds m. So we have

$$W(h, x) = 1 + xh + \left(\frac{3}{2}x^2 - \frac{1}{2}\right)h^2 + y_3(x)h^3$$

$$+ \cdots + y_n(x)h^n + \cdots \quad \text{(III-1.2)}$$

where $y_0(x) = 1$, $y_1(x) = x$, $y_2(x) = \frac{3}{2}x^2 - \frac{1}{2}$, and $y_n(x)$ is a polynomial in x of degree n, thus demonstrating part (a) of our problem.

We should, however, be honest here and admit that we have assumed the validity of Eqs. (III-1.1) and (III-1.2) for some (as yet undetermined) interval of values of h together with an interval of values of x. On the other hand, if we do not for the moment concern ourselves with the question of convergence of the series in these equations, then we surely can say that, when the series on the right in Eq. (III-1.1) is rearranged into the series on the right in Eq. (III-1.2), the coefficient $y_n(x)$ is a polynomial in x of degree n.

But in order to effect a demonstration of part **(b)** we find it advisable to show that the series expansion of $W(h, x)$ in Eq. (III-1.1) is valid, when $x = 1$, for an interval of values of h. When $x = 1$ we have

$$W(h, 1) = (1 - 2h + h^2)^{-(1/2)} = (1 - h)^{-1}$$
$$= 1 + h + h^2 + h^3 + \cdots + h^n + \cdots$$
(III-1.3)

valid for $-1 < h < 1$. Each coefficient in the right side of Eq. (III-1.3) is unity. So, if we can show that Eq. (III-1.2) holds for $-1 < h < 1$ whenever x is taken as any number having $|x| \leq 1$, we will have shown that every polynomial coefficient $y_n(x)$ in Eq. (III-1.2) has the value 1 when x is 1.

First, we take $x = -1$ and find that

$$W(h, -1) = (1 + 2h + h^2)^{-(1/2)} = (1 + h)^{-1}$$
$$= 1 - h + h^2 - h^3 + \cdots$$

which is valid for $-1 < h < 1$. Next, let us take an h such that $0 < h < 1$ and hold it fixed for the moment while we examine the character of $W(h, x)$ for this fixed h as x varies from -1 to 1. We see at once that $W(h, x)$ varies from $(1 + h)^{-1}$ to $(1 - h)^{-1}$. Moreover, $W(h, x)$ varies monotonically, always increasing from $(1 + h)^{-1}$ to $(1 - h)^{-1}$, because the partial derivative $\partial W/\partial x = h/(1 - 2xh + h^2)^{3/2}$ has the same sign as h. Consequently, when x is taken between -1 and 1, we have $(1 + h)^{-1} < W(h, x) < (1 - h)^{-1}$, which we may write as

$$W(h, x) = (1 - \theta h)^{-1}, \qquad |\theta| < 1. \qquad \text{(III-1.4)}$$

Similarly, we find this appraisal, Eq. (III-1.4), to be true when $-1 < h < 0$. Thus, for $0 < |h| < 1$, $-1 \leq x \leq 1$ we have

$$W(h, x) = (1 - \theta h)^{-1}, \qquad |\theta| \leq 1$$
$$= 1 + \theta h + \theta^2 h^2 + \cdots + \theta^n h^n + \cdots.$$

All in all, we have found that our Eq. (III-1.2) holds for all pairs h and x such that $-1 < h < 1, -1 \leqq x \leqq 1$. (We have not actually considered the case $h = 0$ because when $h = 0$ we see at once that Eq. (III-1.2) holds, since it reduces to $1 = 1$.) We may conclude, then, that every polynomial coefficient $y_n(x)$ in Eq. (2) has the value 1 when $x = 1$.

It remains to demonstrate part (c). This will follow from the fact that $W(h, x)$ satisfies the equation

$$h^2\frac{\partial^2 W}{\partial h^2} + (1 - x^2)\frac{\partial^2 W}{\partial x^2} + 2h\frac{\partial W}{\partial h} - 2x\frac{\partial W}{\partial x} = 0, \quad \text{(III-1.5)}$$

which can be verified by direct substitution or can be shown to hold by virtue of the following considerations. If h be taken as denoting distance r from the origin of a point in space and x be taken as denoting $\cos \phi$ where ϕ is the spherical co-latitude coordinate, then $W(h, x)$, by the law of cosines, represents the reciprocal of the distance of an arbitrary point from the point where $\phi = 0, \quad r = 1$. Such a reciprocal distance is known to be a harmonic function of the variable coordinates involved and so satisfies Laplace's equation, which in terms of r and x (where $x = \cos \phi$) reduces to Eq. (III-1.5). (See, for example, Franklin, *Methods of Advanced Calculus*, McGraw-Hill, 1944.)

Let us represent Eq. (III-1.5) in operator notation:

$$\left[h^2\frac{\partial^2}{\partial h^2} + (1 - x^2)\frac{\partial^2}{\partial x^2} + 2h\frac{\partial}{\partial h} - 2x\frac{\partial}{\partial x} \right] W = 0, \quad \text{(III-I.6)}$$

and apply the operator shown in Eq. (III-1.6) to both sides of Eq. (III-1.2). By virtue of the convergence properties which we established for the series on the right in Eq. (III-1.2) we may apply the operator shown in Eq. (III-1.6) termwise to the series in Eq. (III-1.2). We observe that when this operator is applied to a term $y_n(x)h^n$, it yields h^n multiplied by a polynomial in x, namely

$$\left[(n)(n - 1)y_n(x) + (1 - x^2)y_n''(x) + 2ny_n(x) - 2xy_n'(x) \right] h^n.$$

But when the operator is applied to both sides of Eq. (III-1.2), the left side of the resulting equation is identically zero by virtue of Eq. (III-1.6). Consequently, the coefficient of each power of h must vanish in the resulting power series on the right:

$$n(n-1)y_n + (1-x^2)y_n'' + 2ny_n - 2xy_n' = 0,$$

that is,

$$(1-x^2)y_n'' - 2xy_n' + n(n+1)y_n = 0.$$

But this last equation is none other than the Legendre equation of index n with $y = y_n$. Thus the polynomial $y_n(x)$ is a solution of the Legendre equation

$$(1-x^2)y'' - 2xy' + n(n+1)y = 0.$$

Recapitulating, we have

$$W(h, x) = (1 - 2hx + h^2)^{-1/2} = \sum_{n=0}^{\infty} P_n(x)h^n, \qquad |h| < 1, |x| \leq 1. \tag{III-1.7}$$

Problem: Recurrence Relations

III-2. Show that the Legendre polynomials $P_0(x)$, $P_1(x)$, $P_2(x)$, $\cdots$, $P_n(x)$, $\cdots$ are such that for $m \geq 2$ every $P_m(x)$ is related to its two immediate predecessors by the formula

$$mP_m(x) = (2m-1)xP_{m-1}(x) - (m-1)P_{m-2}(x). \qquad \star$$

The result we are looking for can be obtained from the expansion obtained in Prob. III-1, namely

$$W = P_0(x) + P_1(x)h + P_2(x)h^2 + \cdots \tag{III-2.1}$$

where

$$W = \frac{1}{\sqrt{1 + h^2 - 2hx}}.$$

First, we observe that

$$\frac{\partial W}{\partial h} = -\frac{1}{2}(1 + h^2 - 2hx)^{-(3/2)}(2h - 2x)$$

$$= \frac{(x - h)W}{1 + h^2 - 2hx},$$

whence

$$(1 + h^2 - 2hx)\frac{\partial W}{\partial h} - (x - h)W = 0. \qquad \text{(III-2.2)}$$

Next, we differentiate termwise with respect to h in the series in Eq. (III-2.1), obtaining a series which converges to $\partial W/\partial h$ for $|h| < 1$, $|x| \leq 1$ by the well-known theorem on termwise differentiation on its interval of convergence of a power series in nonnegative integral powers of h:

$$\frac{\partial W}{\partial h} = P_1(x) + 2P_2(x)h + 3P_3(x)h^2 + \cdots + (n + 1)P_{n+1}(x)h^n + \cdots.$$

$$\text{(III-2.3)}$$

Substituting in Eq. (III-2.2) from Eqs. (III-2.1) and (III-2.3), we get

$$(1 + h^2 - 2hx)\Big[P_1(x) + 2P_2(x)h + \cdots + (n - 1)P_{n-1}(x)h^{n-2}$$

$$+ nP_n(x)h^{n-1} + (n + 1)P_{n+1}(x)h^n + \cdots\Big]$$

$$- (x - h)\Big[P_0(x) + P_1(x)h + \cdots + P_{n-1}(x)h^{n-1} + P_n(x)h^n + \cdots\Big]$$

$$= 0,$$

which upon multiplication followed by collection of terms in like powers of h becomes

$$\Big[P_1(x) - xP_0(x)\Big] + \Big[2P_2(x) - 2xP_1(x) + P_0(x) - xP_1(x)\Big]h + \cdots$$

$$+ \Big[(n + 1)P_{n+1}(x) - (2n + 1)xP_n(x) + nP_{n-1}(x)\Big]h^n + \cdots = 0.$$

$$\text{(III-2.4)}$$

Since the series in Eq. (III-2.4) converges everywhere on the interval $|h| < 1$ when x is taken as any number on the closed interval $|x| \leq 1$, the bracketed coefficients in Eq. (III-2.4) must vanish for each x on the interval $|x| \leq 1$:

$$P_1(x) - xP_0(x) = 0,$$

$$(n + 1)P_{n+1}(x) - (2n + 1)xP_n(x) + nP_{n-1}(x) = 0, \qquad n \geq 1.$$

If in this last equation we let $n + 1 = m$, we get

$$mP_m(x) = (2m - 1)xP_{m-1}(x) - (m - 1)P_{m-2}(x), \qquad m \geq 2$$

$$(\text{III-2.5})$$

Eq. (III-2.5) expresses equality of two polynomials of degree m for all x on the interval $|x| \leq 1$. It follows that Eq. (III-2.5) must hold for all x, since it is a well known theorem of algebra that, if two polynomials of degree m are equal for more than m distinct values of x, they must be equal for all values of x.

Problem: Laplace's Integral Expression of $P_n(x)$

III-3. Verify Laplace's integral formula for the Legendre polynomials:

$$P_n(x) = \frac{1}{\pi}\int_0^\pi (x + \sqrt{x^2 - 1}\, \cos \theta)^n d\theta, \qquad n = 0, 1, 2, 3, \cdots . \quad \star$$

$$(\text{III-3.1})$$

In contrast to the transparency of Rodrigues's n^{th} derivative formula it does not seem to us immediately apparent how Laplace's integral formula came to be discovered. However, if we grant knowledge of the Legendre polynomials as coefficients in the expansion given in Eq. (III-1.7) then perhaps it does not seem too far-fetched to arrive at Laplace's formula via inquisitive experimentation with the expansion (III-1.7) as follows.

The expression $1 - 2xh + h^2$ lends itself quite readily to being written as sum or difference of two squares:

$$1 - 2xh + h^2 = (1 - xh)^2 + (h\sqrt{1 - x^2})^2 = (1 - xh)^2 -$$
$$(h\sqrt{x^2 - 1})^2. \qquad \text{(III-3.2)}$$

If we choose the latter, namely the difference of two squares, we may write

$$(1 - 2xh + h^2)^{-1/2} = \frac{1}{\sqrt{a^2 - b^2}}, \qquad a = 1 - xh, \quad b = h\sqrt{x^2 - 1}.$$

But what have we gained? And why the difference of two squares rather than the sum? Well, if we happen to be astute enough at this point to recall the integral formula

$$\int_0^\pi \frac{d\theta}{a - b\cos\theta} = \frac{\pi}{\sqrt{a^2 - b^2}}, \qquad a > |b|,$$

then we see that we may write

$$(1 - 2xh + h^2)^{-1/2} = \frac{1}{\pi}\int_0^\pi \frac{d\theta}{(1 - xh) - h\sqrt{x^2 - 1}\cos\theta}.$$
$$\text{(III-3.3)}$$

In Prob. III-1 we found that the left side of Eq. (III-3.3) could be expressed as a power series of the form

$$b_0 + b_1 h + b_2 h^2 + \cdots + b_n h^n + \cdots$$

where the coefficients b_n are the Legendre polynomials. We found the series to be uniformly convergent for $-1 \leq x \leq 1$, $|h| \leq H < 1$; and we made use of a consequence thereof, namely termwise differentiability, to establish recurrence formulas for the Legendre polynomials. But now in the present problem we seek an expansion valid for some interval of values of x having $|x| > 1$ because of the radical $\sqrt{x^2 - 1}$. We also require sufficient restriction on h, when we take values of $|x| > 1$, so that we may expand the integrand $[(1 - xh)$ $- (h\sqrt{x^2 - 1}) \cos \theta]^{-1}$ on the right in Eq. (III-3.3) in an infinite series which will be termwise integrable over $0 \leqq \theta \leqq \pi$. If, then, we can validate our previous expansion Eq. (III-1.7) to meet these new requirements, it begins to look as if we can arrive at Laplace's formula via expansion of both sides of Eq. (III-3.3) in power series of the form $b_0 + b_1 h + b_2 h^2 + \cdots$, both convergent on an interval $|h| < h_0$ when x is taken to be any number such that $1 \leqq |x| \leqq x_0$. Let us see if we can establish such an interval $|h| < h_0$ after making a choice of x_0, say $x_0 = 2$.

In order to expand $[1 - (2xh - h^2)]^{-1/2}$ in an infinite series, we require $|2xh - h^2| < 1$, which will be the case if $|h| \, |2x - h| < 1$. This latter inequality will certainly be met for all x having $|x| \leqq 2$ if

$$|h| (4 + |h|) < 1,$$

$$h^2 + 4 |h| + 4 < 5,$$

$$|h| < \sqrt{5} - 2.$$

Thus we may say that the expansion in Eq. (III-1.7) converges for $|h| < \sqrt{5} - 2$ when $|x| \leqq 2$.

So much for the function on the left in Eq. (III-3.3). Now we consider the integral on the right in that equation and observe that its integrand may be written as $(1 - u)^{-1} = [1 - (xh + h\sqrt{x^2 - 1}$ $\cos \theta)]^{-1}$. Recalling that

$$(1 - u)^{-1} = 1 + u + u^2 + u^3 + \cdots + u^n + \cdots, \qquad |u| < 1, \tag{III-3.4}$$

we see that our integrand in Eq. (III-3.3) may be expanded in a series as in Eq. (III-3.4) if we have

$$|xh + h\sqrt{x^2 - 1} \cos \theta| < 1,$$

$$|h| \, |x + \sqrt{x^2 - 1} \cos \theta| < 1.$$

This last inequality will be fulfilled provided

$$|h|(|x| + \sqrt{x^2 - 1}) < 1,$$

since $|\cos \theta| < 1$ and since the absolute of a sum cannot exceed the sum of the absolutes. And now if we take any x such that $1 \leqq |x| \leqq 2$, we will have $|xh + h\sqrt{x^2 - 1} \cos \theta| < 1$ for all values of h such that

$$|h| \, (2 + \sqrt{3}) < 1,$$

$$|h| < \frac{1}{2 + \sqrt{3}} = 2 - \sqrt{3}.$$

Thus the series expansion in Eq. (III-3.4) where $u = xh + h\sqrt{x^2 - 1} \cos \theta$, will converge uniformly for $0 \leqq \theta \leqq \pi$ when we take any x and h such that $1 \leqq |x| \leqq 2$, $|h| < 2 - \sqrt{3}$. It may, therefore, be integrated termwise over $0 \leqq \theta \leqq \pi$. We note here that $\sqrt{5} - 2 < 2 - \sqrt{3}$. In place of Eq. (III-3.3) we may now write

$$\sum_{n=0}^{\infty} P_n(x)h^n = \frac{1}{\pi} \sum_{n=0}^{\infty} \int_0^{\pi} (xh + h\sqrt{x^2 - 1} \cos \theta)^n d\theta,$$

or

$$\sum_{n=0}^{\infty} P_n(x)h^n = \sum_{n=0}^{\infty} \left[\frac{1}{\pi} \int_0^{\pi} (x + \sqrt{x^2 - 1} \cos \theta)^n d\theta \right] h^n, \qquad \text{(III-3.5)}$$

valid for $1 \leqq |x| \leqq 2$, $|h| < h_0 = \sqrt{5} - 2$.

Since the series on the two sides of Eq. (III-3.5), regarded as Maclaurin series in powers of h, converge to the same value (for each choice of x such that $1 \leqq |x| \leqq 2$) on the common interval $|h| < \sqrt{5} - 2$, it follows that coefficients of like powers of h must be equal to each other for all x such that $1 \leqq |x| \leqq 2$:

$$P_n(x) = \frac{1}{\pi} \int_0^{\pi} (x + \sqrt{x^2 - 1} \cos \theta)^n d\theta, \qquad 1 \leqq |x| \leqq 2,$$

$$n = 0, 1, 2, 3, \cdots. \qquad \text{(III-3.6)}$$

Inasmuch as the function of x, call it $H_n(x)$ defined by the integral on the right in Eq. (III-3.6) equals the Legendre polynomial $P_n(x)$ for every x having $1 \leq |x| \leq 2$, we may presume that $H_n(x)$ is none other than $P_n(x)$ itself. However, let us give the argument to validate this assumption. It is as follows. We observe that when the integrand $(x + \sqrt{x^2 - 1} \cos \theta)^n$ is expanded:

$$x^n + nx^{n-1} \sqrt{x^2 - 1} \cos \theta + \frac{n(n - 1)}{2!} x^{n-2}(\sqrt{x^2 - 1})^2 \cos^2 \theta$$

$$+ \frac{n(n - 1)(n - 2)}{3!} x^{n-3}(\sqrt{x^2 - 1})^3 \cos^3 \theta$$

$$+ \cdots + (\sqrt{x^2 - 1})^n \cos^n \theta,$$

each term having an odd power of $\sqrt{x^2 - 1}$ also has an odd power of $\cos \theta$. The integral of each odd power term from $\theta = 0$ to $\theta = \pi$ will therefore vanish. On the other hand, each term having an even power of $\cos \theta$, whose integral consequently does not vanish, is multiplied by a polynomial in x. Thus the function

$$H_n(x) = \frac{1}{\pi} \int_0^\pi (x + \sqrt{x^2 - 1} \cos \theta)^n d\theta$$

is a polynomial of degree n. And since it equals $P_n(x)$ for more than n distinct values of x, namely for the infinitely many values of x having $1 \leq |x| \leq 2$, we conclude that $H_n(x)$ is $P_n(x)$. Laplace's integral formula Eq. (III-3.1) is thus established.

Problems: Determination of Specific Legendre Polynomials

III-4. Obtain the Legendre polynomials $P_0(x)$, $P_1(x)$, $P_2(x)$, $P_3(x)$ from the series solution (III-0.5) of Legendre's equation. ★

To obtain $P_0(x)$ we first take $n = 0$ in (III-0.5). Then we choose $C_1 = 0$. This gives us a family of constant solutions, namely $y = C_0$, of Legendre's equation of index $n = 0$. Finally, we choose $C_0 = 1$ and we have $P_0(x) = 1$.

To get $P_1(x)$ we take $n = 1$ and $C_0 = 0$. This gives us $y = C_1 x$.

The polynomial of this family which equals 1 at $x = 1$ is obtained by choosing $C_1 = 1$. Thus $P_1(x) = x$.

To get $P_2(x)$ we take $n = 2$ and $C_1 = 0$ and then choose C_0 so that $y = 1$ at $x = 1$. We first get

$$y = C_0\left[1 - \frac{(2)(3)}{2!}x^2\right] = C_0(1 - 3x^2).$$

Then to make $y = 1$ at $x = 1$, we require

$$1 = C_0(1 - 3)$$

$$C_0 = -\frac{1}{2}.$$

Then we have $P_2(x) = -(1/2)(1 - 3x^2) = 3x^2/2 - 1/2$.

Similarly, $P_3(x)$ is found by taking $n = 3$ with $C_0 = 0$:

$$y = C_1\left[x - \frac{(2)(5)}{3!}x^3\right] = C_1\left(x - \frac{5}{3}x^3\right).$$

To have $y = 1$ at $x = 1$ then requires $C_1 = -3/2$ with the result that $P_3(x) = 5x^3/2 - 3/2$.

III-5. Obtain the Legendre polynomial $P_4(x)$ from Rodrigues's formula

$$P_n(x) = \frac{1}{2^n n!}\frac{d^n}{dx^n}[(x^2 - 1)^n]. \qquad \star$$

We have

$$P_4(x) = \frac{1}{2^4 4!}D^4\left[(x^2 - 1)^4\right]$$

$$= \frac{1}{384}D^4(x^8 - 4x^6 + 6x^4 - 4x^2 + 1)$$

$$= \frac{1}{384}(1680x^4 - 1440x^2 + 144)$$

$$= \frac{1}{8}(35x^4 - 30x^2 + 3).$$

III-6. Obtain the Legendre polynomial $P_5(x)$ from the formula (III-0.8), namely

$$P_n(x) = \frac{1}{2^n} \sum_{k=0}^{N} \frac{(-1)^k (2n - 2k)! x^{n-2k}}{k!(n-k)!(n-2k)!},$$

where $N = n/2$ or $(n-1)/2$ according as n is even or odd.   ★

Since n is odd, namely 5, we take $N = 2$ and get

$$P_5(x) = \frac{1}{2^5} \sum_{k=0}^{2} \frac{(-1)^k (10 - 2k)! x^{5-2k}}{k!(5-k)!(5-2k)!}.$$

Then taking $k = 0$, 1 and 2 we have

$$P_5(x) = \frac{1}{32} \left[\frac{10! x^5}{0!5!5!} - \frac{8! x^3}{1!4!3!} + \frac{6! x}{2!3!1!} \right]$$

$$= \frac{1}{8}(63x^5 - 70x^3 + 15x).$$

III-7. Obtain the Legendre polynomial $P_4(x)$ directly from Legendre's equation of index 4 by assuming a polynomial solution of degree 4.   ★

We assume a solution of the form

$$y = ax^4 + bx^3 + cx^2 + dx + g \qquad \text{(III-7.1)}$$

for the equation

$$(1 - x^2)y'' - 2xy' + 20y = 0. \qquad \text{(III-7.2)}$$

This requires that

$$(1 - x^2)(12ax^2 + 6bx + 2c) - 2x(4ax^3 + 3bx^2 + 2cx + d)$$
$$+ 20(ax^4 + bx^3 + cx^2 + dx + g)$$
$$= 0, \qquad \text{(III-7.3)}$$

that is,

$$(20a - 20a)x^4 + (8b)x^3 + (12a + 14c)x^2$$
$$+ (6b + 18d)x + (2c + 20g) = 0. \qquad \text{(III-7.4)}$$

The polynomial on the left in Eq. (III-7.4) will vanish identically if and only if every coefficient vanishes. This requires $b = d = 0$, $c = -6a/7$, $g = -c/10$; but leaves a arbitrary. We thus find that Eq. (III-7.1) has a family of polynomial solutions of degree 4:

$$y = a\left(x^4 - \frac{6}{7}x^2 + \frac{6}{70}\right). \tag{III-7.5}$$

We require now that y shall equal 1 at $x = 1$, which means that we require $1 = 8a/35$, that is $a = 35/8$. This makes $c = -15/4$ and $g = 3/8$. Thus we have

$$P_4(x) = \frac{1}{8}(35x^4 - 30x^2 + 3).$$

III-8. Obtain the Legendre polynomial $P_6(x)$ by application of the recurrence formula

$$nP_n(x) = (2n - 1)xP_{n-1}(x) - (n - 1)P_{n-2}(x)$$

assuming that $P_4(x)$ and $P_5(x)$ are known.　　★

We have

$$6P_6(x) = 11xP_5(x) - 5P_4(x),$$

$$= 11x\left(\frac{63}{8}x^5 - \frac{70}{8}x^3 + \frac{15}{8}x\right) - 5\left(\frac{35}{8}x^4 - \frac{30}{8}x^2 + \frac{3}{8}\right),$$

whence

$$P_6(x) = \frac{1}{16}(231x^6 - 315x^4 + 105x^2 - 5).$$

III-9. Obtain the Legendre polynomial $P_2(x)$ from Laplace's integral formula

$$P_n(x) = \frac{1}{\pi}\int_0^\pi (x + \sqrt{x^2 - 1}\cos\phi)^n d\phi.　　★$$

We have

$$P_2(x) = \frac{1}{\pi}\int_0^\pi (x + \sqrt{x^2 - 1} \cos \phi)^2 d\phi$$

$$= \frac{1}{\pi}\int_0^\pi (x^2 + 2x\sqrt{x^2 - 1} \cos \phi + [x^2 - 1] \cos^2 \phi) d\phi$$

$$= \frac{1}{\pi}\left[x^2\phi + 2x\sqrt{x^2 - 1} \sin \phi + (x^2 - 1)\left(\frac{\phi}{2} + \frac{\sin 2\phi}{4}\right)\right]_{\phi=0}^{\phi=\pi}$$

$$= \frac{1}{\pi}\left[x^2\pi + (x^2 - 1)\pi/2 \right]$$

$$= \frac{1}{2}(3x^2 - 1).$$

Problem: Rodrigues's Expression for $P_n(x)$

III-10. Verify Rodrigues's formula, namely

$$P_n(x) = \frac{1}{2^n n!} D^n\left[(x^2 - 1)^n \right], \qquad n = 0, 1, 2, 3, \cdots, \qquad \text{(III-10.1)}$$

where $P_n(x)$ denotes the Legendre polynomial of degree n and $D^n[(x^2 - 1)^n]$ denotes the nth order derivative $d^n[(x^2 - 1)^n]/dx^n$, it being understood that the derivative of order zero $D^0 f(x) = f(x)$ and that 0! is taken to be unity.   ★

The case of $n = 0$ is verified at once, for we have

$$\frac{1}{2^0 0!} D^0\left[(x^2 - 1)^0 \right] = 1 = P_0(x) \qquad \text{by Prob. III-4.}$$

For $n > 0$ we apply the binomial theorem and get

$$\frac{1}{2^n n!} D^n\left[(x^2 - 1)^n \right] = \frac{1}{2^n n!} D^n\left[\sum_{k=0}^{n} \frac{(-1)^k n! x^{2n-2k}}{k!(n - k)!} \right].$$

$$\text{(III-10.2)}$$

Since the powers in the summation are all even, when we differentiate the summation n times we find that the resulting summation yields vanishing terms for $k > n/2$ when n is even and for $k > (n - 1)/2$ when n is odd. Thus we have

$$\frac{1}{2^n n!} D^n \left[(x^2 - 1)^n \right]$$

$$= \frac{1}{2^n} \sum_{k=0}^{N} \frac{(-1)^k (2n - 2k)(2n - 2k - 1) \cdots (n - 2k + 1) x^{n-2k}}{k!(n - k)!},$$

$$\text{(III-10.3)}$$

where $N = n/2$ or $(n - 1)/2$ according as n is even or odd.

The collection of factors in the numerator of each term of the summation in Eq. (III-10.3) suggests that we multiply numerator and denominator of the kth term by $(n - 2k)!$. Doing so to each term of the summation, we get

$$\frac{1}{2^n n!} D^n \left[(x^2 - 1)^n \right] = \frac{1}{2^n} \sum_{k=0}^{N} \frac{(-1)^k (2n - 2k)! x^{n-2k}}{k!(n - k)!(n - 2k)!}$$

$$= P_n(x) \qquad \text{by Eq. (III-0.8).} \quad \text{(III-10.4)}$$

This formula is named after Olinde Rodrigues, French mathematician, 1791–1854.

Problem: Expansion of a Given Function in Legendre Polynomials

III-11. Prove that any given polynomial

$$H(x) = C_n x^n + C_{n-1} x^{n-1} + C_{n-2} x^{n-2} + \cdots + C_1 x + C_0$$

can be expressed as

$$H(x) = A_n P_n(x) + A_{n-1} P_{n-1}(x) + A_{n-2} P_{n-2}(x)$$
$$+ \cdots + A_1 P_1(x) + A_0 P_0(x),$$

where the coefficients A_n, A_{n-1}, $\cdots$, A_0 are constants and the functions $P_n(x)$, $P_{n-1}(x)$, $\cdots$, $P_0(x)$ are the Legendre polynomials of respective degrees n, $n-1$, $\cdots$, 0. ★

First we take $P_n(x)$ and multiply it by such a constant A_n so that the coefficient of x^n in the resulting polynomial $A_n P_n(x)$ is C_n. The coefficient of x^{n-1} in $A_n P_n(x)$ is zero, since by Eq. (III-0.7) the powers of x appearing in $P_n(x)$ are all even when n is even, all odd when n is odd. Next we take $P_{n-1}(x)$ and multiply it by such a constant A_{n-1} so that the coefficient of x^{n-1} in $A_{n-1} P_{n-1}(x)$ is C_{n-1}. Thus the polynomial $T_n(x) = A_n P_n(x) + A_{n-1} P_{n-1}(x)$ is such that the coefficients of x^n and x^{n-1} are respectively C_n and C_{n-1}.

Now we add to $T_n(x)$ such a multiple of $P_{n-2}(x)$, call it $A_{n-2} P_{n-2}(x)$, that the coefficient of x^{n-2} in the polynomial $T_n(x) + A_{n-2} P_{n-2}(x)$ is C_{n-2}. Thus the coefficients of x^n, x^{n-1}, x^{n-2} in the polynomial

$$T_n(x) + A_{n-2} P_{n-2}(x) = A_n P_n(x) + A_{n-1} P_{n-1}(x) + A_{n-2} P_{n-2}(x)$$

are respectively C_n, C_{n-1}, C_{n-2}. Continuing in this manner by successive additions of appropriately multiplied Legendre polynomials of lesser degrees, we finally obtain a polynomial of degree n whose coefficients of the respective powers of x are precisely those of $H(x)$, thus obtaining

$$H(x) = A_n P_n(x) + A_{n-1} P_{n-1}(x) + A_{n-2} P_{n-2}(x) + \cdots$$
$$+ A_1 P_1(x) + A_0 P_0(x).$$

EXAMPLE.   We find that when the steps set forth in the foregoing demonstration are carried out with respect to the polynomial

$$H(x) = 2x^4 - 6x^3 + 5x^2 + 10x - 4,$$

we obtain

$$H(x) = \frac{16}{35} P_4(x) - \frac{12}{5} P_3(x) + \frac{94}{21} P_2(x) + \frac{32}{5} P_1(x) - \frac{29}{15} P_0(x),$$

taking the Legendre polynomials $P_4(x)$, $P_3(x)$, $P_2(x)$, $P_1(x)$, $P_0(x)$ from the list at the end of this chapter.

Problem: $|P_n(x)| \leq 1$ *for* $-1 \leq x \leq 1$

III-12. Show that on the interval $-1 \leq x \leq 1$ the absolute value of every Legendre polynomial does not exceed unity:

$$-1 \leq P_n(x) \leq 1, \qquad |x| \leq 1, \qquad n = 0, 1, 2, 3, \cdots. \qquad \star$$

$$\text{(III-12.1)}$$

The property of the Legendre polynomials indicated by the inequality (III-12.1) is suggested by a glance at Fig. III-1, where the property is seen to hold for the first few Legendre polynomials. In this connection we remark also that, once (III-12.1) is established, it will follow that max $[P_n(x), \; -1 \leq x \leq 1]$, namely unity, is attained by every $P_n(x)$ at $x = 1$, since $P_n(1) = 1$ for every n by the definition of the Legendre polynomials given in the introduction to Chapter III.

The question is: How to proceed with the proof? One possibility for procedure that comes to mind is the following. Since we hope to prove that, for $-1 \leq x \leq 1$, the values taken on by $P_n(x)$ are never in excess of unity nor less than -1, this thought taken in conjunction with the fact that $P_n(x)$ is an even function when n is even suggests that we try taking $x = \cos \theta$ and find some convenient means of expressing $P_n(x)$ in terms of $\cos \theta$ in a way which will reveal the truth of (III-12.1). Such an expression does not seem easy to find. One way of surmounting the difficulty here is a way that is often fruitful: it is to have recourse to the *complex* variable, even though we are dealing with functions of a real variable. It turns out that we can take advantage of the Euler expression for $\cos \theta$, namely $\cos \theta = (e^{i\theta} + e^{-i\theta})/2$, where $i = \sqrt{-1}$.

In the expansion (III-1.7) we take $x = \cos \theta$ and $h = z$, where z denotes the complex variable and θ is real. We have

$$(1 - 2z \cos \theta + z^2)^{-1/2} = [1 - z(e^{i\theta} + e^{-i\theta}) + z^2]^{-1/2}$$

$$= (1 - ze^{i\theta})^{-1/2}(1 - ze^{-i\theta})^{-1/2}.$$

For, $|z| < 1$ each of the factors on the right may be expanded in a series:

$$(1 - ze^{i\theta})^{-1/2} = 1 + \frac{1}{2}ze^{i\theta} + \frac{1 \cdot 3}{2 \cdot 4}z^2 e^{2i\theta} + \frac{1 \cdot 3 \cdot 5}{2 \cdot 4 \cdot 6}z^3 e^{3i\theta} + \cdots,$$

$$(1 - ze^{-i\theta})^{-1/2} = 1 + \frac{1}{2}ze^{-i\theta} + \frac{1 \cdot 3}{2 \cdot 4}z^2 e^{-2i\theta}$$
$$+ \frac{1 \cdot 3 \cdot 5}{2 \cdot 4 \cdot 6}z^3 e^{-3i\theta} + \cdots.$$

For $|z| < 1$ the formal product of these two series gives a series expansion for $(1 - 2z\cos\theta + z^2)^{-1/2}$. It follows then by the expansion in Eq. (III-1.7) that the coefficient of z^n therein is $P_n(\cos\theta)$:

$$P_n(\cos\theta) = \frac{1 \cdot 3 \cdot 5 \cdots (2n - 1)}{2 \cdot 4 \cdot 6 \cdots 2n}\left[\left(e^{ni\theta} + e^{-ni\theta}\right)\right.$$
$$+ \frac{1}{2}\frac{2n}{2n - 1}(e^{(n-2)i\theta} + e^{-(n-2)i\theta})$$
$$+ \frac{1 \cdot 3}{2 \cdot 4}\frac{2n(2n - 2)}{(2n - 1)(2n - 3)}(e^{(n-4)i\theta} + e^{-(n-4)i\theta}) + \cdots$$
$$\left. + C(e^{bi\theta} + e^{-bi\theta})\right]$$

where the final term in the brackets is such that $b = 0$ when n is even, $b = 1$ when n is odd; and the coefficient C is as follows.

$$C = \frac{1 \cdot 3 \cdot 5 \cdots (n - 1)}{2 \cdot 4 \cdot 6 \cdots (n)}\frac{(2n)(2n - 2)(2n - 4)\cdots(n + 2)}{(2n - 1)(2n - 3)(2n - 5)\cdots(n + 1)},$$
$$n \text{ even}, \quad \text{(III-12.2a)}$$

$$C = \frac{1 \cdot 3 \cdot 5 \cdots (n - 2)}{2 \cdot 4 \cdot 6 \cdots (n - 1)}\frac{(2n)(2n - 2)(2n - 4)\cdots(n + 3)}{(2n - 1)(2n - 3)(2n - 5)\cdots(n + 2)}$$
$$n \text{ odd}. \quad \text{(III-12.2b)}$$

Since $e^{i\theta} + e^{-i\theta} = 2\cos\theta$, we find that

$$P_n(\cos\theta) = \frac{1 \cdot 3 \cdot 5 \cdots (2n-1)}{2 \cdot 4 \cdot 6 \cdots (2n)} 2\Bigg[\cos n\theta$$

$$+ \frac{1}{2}\frac{2n}{2n-1}\cos\Big\{(n-2)\theta\Big\}$$

$$+ \frac{1 \cdot 3}{2 \cdot 4}\frac{(2n)(2n-2)}{(2n-1)(2n-3)}\cos\Big\{(n-4)\theta\Big\}$$

$$+ \cdots + C\cos b\theta\,\Bigg], \qquad \text{(III-12.3)}$$

where b and C are the same constants as in the previous paragraph. Every cosine in this last equation has a positive coefficient. It follows, therefore, that the maximum possible value for $P_n(\cos\theta)$ can occur when and only when all the cosines are unity. For the range $0 \leqq \theta \leqq \pi$ which we selected to correspond to the range $-1 \leqq x \leqq 1$, the maximum of $P_n(\cos\theta)$ will occur at $\theta = 0$, which means that max $[P_n(x), \ -1 \leqq x \leqq 1]$ occurs at $x = 1$. Since $P_n(1) = 1$ for every Legendre polynomial, we have

$$\max[P_n(x), \ -1 \leqq x \leqq 1] = 1, \qquad n = 0, 1, 2, 3, \cdots.$$

The minimum possible value for $P_n(\cos\theta)$ can occur when and only when all the cosines in Eq. (III-12.3) are -1. When n is odd, every cosine in Eq. (3) equals -1 at $\theta = \pi$. Thus, we find that $\min[P_n(x), \ -1 \leqq x \leqq 1]$, n odd, occurs at $x = -1$ and must equal -1, because $P_n(-1) = -P_n(1)$ by Eq. (III-0.7) when n is odd. Since -1 is thus the minimum value possible that can be taken by any $P_n(x)$ on the interval $-1 \leqq x \leqq 1$, appraisal (III-12.1) is established.

REMARKS. 1. When n is even, a study of the structure of the right side of Eq. (III-12.3) shows that not all the cosines therein can equal -1 simultaneously. The minimum of $P_n(x)$ for $-1 \leqq x \leqq 1$

is therefore not -1 when n is even. The minimum value on $-1 \leqq x \leqq 1$ for $P_2(x)$ and for $P_4(x)$, for example, are as follows.

$$\min[P_2(x), \quad -1 \leqq x \leqq 1] = -1/2 \quad \text{at } x = 0,$$

$$\min[P_4(x), \quad -1 \leqq x \leqq 1] = -3/7 \quad \text{at } x = \pm\sqrt{3/7}.$$

2. An alternative way of writing Eq. (III-12.3) is

$$P_n(\cos \theta) = \frac{(2n)!}{2^{2n}(n!)^2}\left[2 \cos n\theta + \frac{1 \cdot n}{1 \cdot (2n - 1)} 2 \cos (n - 2)\theta \right.$$

$$\left. + \frac{1 \cdot 3 \cdot n(n - 1)}{1 \cdot 2 \cdot (2n - 1)(2n - 3)} 2 \cos(n - 4)\theta + \cdots + F \right]$$

$$\text{(III-12.4)}$$

where the terminating term F is the same as in (III-12.3).

Problems: Recurrence Relations

III-13. For the Legendre polynomials $P_n(x)$, $n = 1, 2, 3, \cdots$ establish the recurrence relation

$$xP_n'(x) - P_{n-1}'(x) = nP_n(x). \quad \star \quad \text{(III-13.1)}$$

We follow the method of procedure used in Prob. III-2. The corresponding initial step in the present problem is to observe that

$$h\frac{\partial W}{\partial h} - (x - h) \frac{\partial W}{\partial x} = 0, \quad \text{(III-13.2)}$$

which is readily seen to be so; for we have

$$\frac{\partial W}{\partial h} = \frac{x - h}{(1 - 2xh + h^2)^{3/2}}, \quad \text{(III-13.3)}$$

and

$$\frac{\partial W}{\partial x} = \frac{h}{(1 - 2xh + h^2)^{3/2}}. \quad \text{(III-13.4)}$$

It is then apparent that Eq. (III-13.2) follows from Eqs. (III-13.3) and (III-13.4).

Differentiating the series on the right in Eq. (III-1.7) with respect to h and with respect to x, then putting the resulting series in Eq. (III-13.2), we obtain

$$h\sum_{n=0}^{\infty} P_n(x)nh^{n-1} - (x - h)\sum_{n=0}^{\infty} P_n'(x)h^n = 0. \qquad \text{(III-13.5)}$$

We observe that in Eq. (III-13.5) both summations may be taken starting with $n = 1$, since the term given by $n = 0$ vanishes in both summations. This is at once evident in the first summation and is so in the second summation because $P_0(x) = 1$, which makes $P_0'(x) = 0$. Thus, Eq. (III-13.5) may be written

$$\sum_{n=1}^{\infty}\Big[nP_n(x) - (x - h)P_n'(x)\Big] h^n = 0, \qquad \text{(III-13.6)}$$

valid for $|h| < 1$, $-1 \leqq x \leqq 1$. An equivalent expression for Eq. (III-13.6) is

$$\sum_{n=1}^{\infty}\Big[nP_n(x) - xP_n'(x) + P_{n-1}'(x)\Big] h^n = 0.$$

It follows as in Prob. III-2 that the coefficients of the respective powers of h must all vanish for $-1 \leqq x \leqq 1$, thus yielding Eq. (III-13.1). Then by the same argument as in the last paragraph of Prob. III-2, we realize that Eq. (III-13.1) is valid for all values of x.

III-14. Show that

$$P_n'(x) - xP_{n-1}'(x) = nP_{n-1}(x), \qquad n = 1, 2, 3, \cdots,$$
$$\text{(III-14.1)}$$

where $P_n(x)$ and $P_{n-1}(x)$ denote the Legendre polynomials of degree n and $n - 1$ respectively.   ★

Equation (III-14.1) can be made to follow from Eqs. (III-2.5) and (III-13.1) as follows. Differentiation of Eq. (III-2.5) yields

$$mP'_m(x) = (2m - 1)[xP'_{m-1}(x) + P_{m-1}(x)] - (m - 1)P'_{m-2}(x).$$
$$\text{(III-14.2)}$$

Equation (III-13.1) when written with $m - 1$ in place of n becomes

$$P'_{m-2}(x) = xP'_{m-1}(x) - (m - 1)P_{m-1}(x). \qquad \text{(III-14.3)}$$

Replacing P'_{m-2} in Eq. (III-14.2) by its equivalent as given by the right side of Eq. (III-14.3) and collecting terms in the resulting equation, we get

$$mP'_m(x) = mxP'_{m-1}(x) + m^2P_{m-1}(x), \qquad m \geqq 2. \quad \text{(III-14.4)}$$

Dividing both sides of Eq. (III-14.4) by m and transposing the first term on the right side, we have

$$P'_m(x) - xP'_{m-1}(x) = mP_{m-1}(x), \qquad m \geqq 2. \quad \text{(III-14.5)}$$

Although Eq. (III-14.5) has thus been established for $m \geqq 2$, we can surmise that, since the subscript $m - 2$ does not appear therein, Eq. (III-14.5) should also hold for $m = 1$. And this is seen at once to be so, for if we take $m = 1$ in Eq. (III-14.5) we have

$$P'_1(x) - xP'_0(x) = 1 \cdot P_0(x),$$

which is true because $P_1(x) = x$ and $P_0(x) = 1$. Thus, writing n for m in Eq. (III-14.5) and changing the qualification $n \geqq 2$ to $n \geqq 1$, we have Eq. (III-14.1).

III-15. Show that the Legendre polynomials $P_n(x)$, $n = 1, 2, 3, \cdots$ satisfy the recurrence relation

$$(1 - x^2)P'_n(x) = \frac{n(n + 1)}{2n + 1}\left[P_{n-1}(x) - P_{n+1}(x)\right]. \qquad \star$$
$$\text{(III-15.1)}$$

Since the relationship to be established involves only one derivative $P'_n(x)$, we first eliminate the derivative $P'_{n-1}(x)$ from Eqs. (III-13.1) and (III-14.1). This is done by multiplying both sides of Eq. (III-13.1) by $(-x)$ and adding Eq. (III-14.1). Thus we obtain

$$(1 - x^2)P'_n(x) = n\left[P_{n-1}(x) - xP_n(x)\right]. \quad \text{(III-15.2)}$$

Comparing this result with the equation (III-15.1) to be established, we see that we wish to retain $P_{n-1}(x)$ but need to replace $xP_n(x)$ by something involving $P_{n+1}(x)$. This we can do by appropriate application of Eq. (III-2.5) as follows. In Eq. (III-2.5) we let $m - 1 = n$. Then Eq. (III-2.5) becomes

$$xP_n(x) = \frac{1}{2n + 1}\left[(n + 1)P_{n+1}(x) + nP_{n-1}(x)\right].$$

Upon substituting the right side of this last equation for $xP_n(x)$ in Eq. (III-15.2) and collecting terms we get Eq. (III-15.1).

Problems: Orthogonality Property and Related Property

III-16. From the fact that the Legendre polynomial $P_n(x)$ of degree n satisfies the equation (III-0.2) namely,

$$\frac{d}{dx}\left[(1 - x^2)P'_n(x)\right] + n(n + 1)P_n(x) = 0,$$

deduce the orthogonality property of the Legendre polynomials:

$$\int_{-1}^{1} P_m(x)P_n(x)dx = 0. \qquad m \neq n. \qquad \star \qquad \text{(III-16.1)}$$

How do we make a start toward solving this problem? We get a clue from the problem itself, which is to demonstrate the vanishing of the integral of the product $P_m(x)P_n(x)$. So, if we transpose the second term of Legendre's equation to the other side of the equation, we will have $P_n(x)$ there all by itself, except for a mere constant factor, waiting to be multiplied by $P_m(x)$ so that we can integrate the product $P_m(x)P_n(x)$. To be sure, we shall have to multiply the left side also by $P_m(x)$ and then integrate. And this looks promising, because the left side will then lend itself to integration by parts. Accordingly, we multiply both sides of the equation by $P_m(x)$ and we have

$$P_m(x)\frac{d}{dx}\left[(1-x^2)P_n'(x)\right] = -n(n+1)P_m(x)P_n(x).$$

Integrating over the interval $-1 \leq x \leq 1$ and applying integration by parts on the left, we get

$$\left[P_m(x)(1-x^2)P_n'(x)\right]_{x=-1}^{x=1}$$

$$-\int_{-1}^1 (1-x^2)P_n'(x)P_m'(x)dx = -n(n+1)\int_{-1}^1 P_m(x)P_n(x)dx,$$

which reduces to

$$\int_{-1}^1 (1-x^2)P_n'(x)P_m'(x)dx = n(n+1)\int_{-1}^1 P_m(x)P_n(x)dx,$$

$$\text{(III-16.2)}$$

because the factor $1-x^2$ vanishes at $x=1$ and at $x=-1$.

What to do next? There is one thing in our last equation which suggests a possibility: the equation would be unchanged if m and n were interchanged. Let us make use of this fact and see what happens. We start all over again, writing the Legendre equation for $P_m(x)$ instead of $P_n(x)$:

$$\frac{d}{dx}\left[(1-x^2)P_m'(x)\right] = -m(m+1)P_m(x).$$

Then we multiply both sides by $P_n(x)$ and integrate over the interval $-1 \leqq x \leqq 1$. Applying integration by parts to the left side, we get

$$\int_{-1}^{1} (1 - x^2)P_m'(x)P_n'(x)dx = m(m + 1)\int_{-1}^{1} P_n(x)P_m(x)dx.$$
(III-16.3)

The integral on the left in Eq. (III-16.3) is the same as the integral on the left in Eq. (III-16.2). And the right sides differ only in the coefficients of the integrals. Thus, we can eliminate the unwanted integrals on the left by subtracting Eq. (III-16.3) from Eq. (III-16.2):

$$\left[(m)(m + 1) - (n)(n + 1)\right]\int_{-1}^{1} P_m(x)P_n(x)dx = 0.$$
(III-16.4)

Upon examination of Eq. (III-16.4) we are confronted with three possibilities: the bracketed expression may equal zero, the integral may equal zero, or both may equal zero. If $m \neq n$ the bracketed expression does *not* vanish, for we have

$$m(m + 1) - n(n + 1) = (m^2 - n^2) + (m - n)$$
$$= (m - n)(m + n + 1).$$

Neither factor on the right in this last equation vanishes when $m \neq n$ because both m and n are nonnegative inasmuch as they are degrees of two Legendre polynomials. Thus, since the bracketed expression in Eq. (III-16.4) does not vanish, it follows that the integral must vanish:

$$\int_{-1}^{1} P_m(x)P_n(x)dx = 0, \qquad m \neq n.$$

REMARKS. Our demonstration gives no clue as to the value of the integral when $m = n$, for when $m = n$ the bracketed expression in Eq. (III-16.4) *does* vanish and we learn nothing about the integral

$$\int_{-1}^{1} P_n(x)P_n(x)dx.$$

But we do know two things about the integral of $[P_n(x)]^2$: (1) it has a positive value, (2) its value is a function of the degree n of $P_n(x)$. So our next problem will be to find a formula for the integral of $[P_n(x)]^2$.

One immediate consequence of the orthogonality property is

$$\int_{-1}^{1} P_m(x)dx = 0, \qquad m \neq 0. \qquad \text{(III-16.5)}$$

This is readily seen to be true because

$$\int_{-1}^{1} P_m(x)dx = \int_{-1}^{1} P_m(x)[1]dx$$

$$= \int_{-1}^{1} P_m(x)P_0(x)dx$$

$$= 0, \qquad m \neq 0 \qquad \text{by Eq. (III-16.1)}.$$

For $m = 0$ we have

$$\int_{-1}^{1} P_0(x)dx = \int_{-1}^{1} dx = 2.$$

III-17. Develop a formula for the value of the integral

$$\int_{-1}^{1} \left[P_n(x) \right]^2 dx$$

where $P_n(x)$ denotes an arbitrary Legendre polynomial.   ★

Since the integrand is never negative and is not identically zero we see that the value of the integral must be positive for every n. Moreover, Fig. III-1 suggests that the relative maxima and minima of $P_n(x)$ decrease in size with increasing n. This makes us suspect that the value of the integral of $[P_n(x)]^2$ will decrease as n increases, perhaps toward the limit zero as $n \to \infty$. But these observations do not give much clue as to the actual value of the integral. We must search further if we are to deduce the desired formula for the actual value of the integral.

Looking through the solution of Prob. III-16, we notice that it was integration by parts which played an important role in the solution there. Perhaps it can be turned to account in the present problem if we can somehow write the integrand $[P_n(x)]^2$ as a product which involves a derivative. This thought brings to mind Rodrigues's formula for $P_n(x)$ in Prob. III-10, which allows us to write

$$\int_{-1}^{1}\left[P_n(x)\right]^2 dx = \int_{-1}^{1}\left[\frac{1}{2^n n!}\, D^n(x^2-1)^n\right]^2 dx$$

$$= \frac{1}{2^{2n}(n!)^2}\int_{-1}^{1} D^n(x^2-1)^n \cdot D^n(x^2-1)^n dx.$$

$$\text{(III-17.1)}$$

Now we can apply integration by parts to the integral on the right in Eq. (III-17.1). This integral, apart from its coefficient, thus equals

$$\left[D^n(x^2-1)^n \cdot D^{n-1}(x^2-1)^n\right]_{x=-1}^{x=1}$$

$$-\int_{-1}^{1} D^{n-1}(x^2-1)^n \cdot D^{n+1}(x^2-1)^n dx.$$

The integrated part in brackets in the last expression vanishes because the $(n-1)^{\text{th}}$ derivative of $(x^2-1)^n$ contains the factor x^2-1 and therefore vanishes at $x=1$ and $x=-1$.

Applying integration by parts to the remaining integral, we get

$$-\left[D^{n+1}(x^2-1)^n \cdot D^{n-2}(x^2-1)^n\right]_{x=-1}^{x=1}$$

$$+(-1)^2\int_{-1}^{1} D^{n-2}(x^2-1)^n \cdot D^{n+2}(x^2-1)^n dx.$$

Again the integrated part vanishes. Continuing thus n times, we get

$$\int_{-1}^{1}\left[P_n(x)\right]^2 dx = \frac{(-1)^n}{2^{2n}(n!)^2}\int_{-1}^{1} (x^2-1)^n \cdot D^{2n}(x^2-1)^n dx. \quad \text{(III-17.2)}$$

Let us examine the second factor of the integrand on the right in Eq. (III-17.2):

$$D^{2n}(x^2 - 1)^n = D^{2n}[x^{2n} + c_1 x^{2n-2} + c_2 x^{2n-4} + \cdots + c_{2n}],$$

where c_1, c_2, $\cdots$, c_{2n} are constant coefficients. By the time we differentiate the bracketed polynomial $2n$ times, the $2n^{\text{th}}$ derivative of every term will have become zero except that of the first term, whose $2n^{\text{th}}$ derivative is $(2n)!$. Thus Eq. (III-17.2) becomes

$$\int_{-1}^{1} \left[P_n(x) \right]^2 dx = \frac{(-1)^n (2n)!}{2^{2n}(n!)^2} \int_{-1}^{1} (x^2 - 1)^n dx. \quad \text{(III-17.3)}$$

But $(-1)^n (x^2 - 1)^n = (1 - x^2)^n$. If now we place x^0 in the integrand we have

$$\int_{-1}^{1} \left[P_n(x) \right]^2 dx = \frac{(2n)!}{2^{2n}(n!)^2} 2 \int_{0}^{1} x^0 (1 - x^2)^n dx.$$

This last integral is a Beta integral, as in Prob. II-17 with $a = 1$, $b = 0$, $c = 2$, and $d = n$. Then we have

$$\int_{-1}^{1} \left[P_n(x) \right]^2 dx = \frac{2(2n)!}{2^{2n}(n!)^2} \frac{\Gamma(1/2)\Gamma(n+1)}{(2n+1)\Gamma(n+[1/2])}. \quad \text{(III-17.4)}$$

For $(2n)!$ we write $2n\Gamma(2n)$ by Eqs. (I-10.1) and (I-4.1); and then from the Legendre duplication formula in Prob. I-18 we have

$$\Gamma(2n) = \frac{\Gamma(n)\Gamma[n + (1/2)]}{2^{1-2n}\Gamma(1/2)}.$$

Substituting this value of $\Gamma(2n)$ in Eq. (III-17.4) we have

$$\int_{-1}^{1} \left[P_n(x) \right]^2 dx = \frac{2}{2n+1}. \quad \text{(III-17.5)}$$

REMARK. We may normalize the Legendre polynomials with respect to the interval $-1 \leqq x \leqq 1$ by taking

$$L_n(x) = \sqrt{\frac{2n+1}{2}} P_n(x). \tag{III-17.6}$$

Thus, the polynomials $L_n(x)$, $n = 0, 1, 2, \cdots$ not only retain the orthogonal property (III-16.1) but are now such that

$$\int_{-1}^{1} \left[L_n(x) \right]^2 dx = 1 \tag{III-17.7}$$

for all n.

Problems: Expansion of a Given Function in Legendre Polynomials

III-18. If $f(x)$ is bounded on the interval $H: -1 \leqq x \leqq 1$, is continuous on H except for a finite number of discontinuities, and such that for each subinterval of H on which $f(x)$ is continuous the curve $y = f(x)$ is rectifiable, then there exists (see, for example, Whittaker and Watson, *Modern Analysis*, Cambridge, 1927) a series of Legendre polynomials with constant coefficients

$$A_0 P_0(x) + A_1 P_1(x) + \cdots + A_n P_n(x) + \cdots \tag{III-18.1}$$

(a) which converges everywhere on H,

(b) converges to $f(x)$ at each point of continuity of $f(x)$ on H,

(c) is such that the series after multiplication by an arbitrary $P_k(x)$ is termwise integrable on H to the integral of $f(x)P_k(x)$ on H.

Show, then, that the coefficients in Eq. (III-18.1) are given by the formula

$$A_k = \frac{2k+1}{2} \int_{-1}^{1} f(x) P_k(x) dx, \qquad k = 0, 1, 2, 3, \cdots. \qquad \star \tag{III-18.2}$$

By part **(c)** of the given expansion we have

$$\int_{-1}^{1} f(x)P_k(x)dx = \int_{-1}^{1} A_0 P_0(x)P_k(x)dx + \int_{-1}^{1} A_1 P_1(x)P_k(x)dx$$

$$+ \cdots + \int_{-1}^{1} A_k P_k(x)P_k(x)dx + \cdots. \qquad \text{(III-18.3)}$$

By the orthogonality property established in Prob. III-16 every term on the right in Eq. (III-18.3) vanishes except the integral

$$\int_{-1}^{1} A_k P_k(x)P_k(x)dx,$$

which except for its constant factor A_k equals $2/(2k + 1)$ by Prob. III-17. Formula (III-18.2) follows at once by solving Eq. (III-18.3) for A_k.

REMARKS. 1. If $f(x)$ is continuous and has continuous second derivative on H, then the series

$$\sum_{n=0}^{\infty} A_n P_n(x)dx, \qquad A_n = \frac{2n + 1}{2} \int_{-1}^{1} f(x)P_n(x)dx \qquad \text{(III-18.4)}$$

converges to $f(x)$ uniformly on H (see, for instance, Kaplan, *Advanced Calculus*, Addison-Wesley, 1952).

2. By the theory of functions of a real variable a consequence of the hypothesis of the finiteness of the number of discontinuities of $f(x)$ on H is that the discontinuities are all of the first kind (simple saltus). See, for example, Hobson, *Functions of a Real Variable*, Cambridge, 1926.

Eq. (III-18.1) converges at each point x_0 of discontinuity to the average of the two functional limits as $x \to x_0$. This behavior of (III-18.1) is thus like that of a Fourier series expansion at a point of simple saltus discontinuity.

3. When $f(x)$ on $-1 \leq x \leq 1$ is such that it possesses a series expansion in terms of normalized Legendre polynomials defined in Eq. (III-17.6), namely

$$f(x) = \sum_{n=0}^{\infty} B_n L_n(x) \qquad \text{(III-18.5)}$$

satisfying conditions (a), (b) and (c) of Prob. III-18, the coefficients B_n are given by

$$B_n = \int_{-1}^{1} f(x) L_n(x) dx, \qquad n = 0, 1, 2, \cdots. \qquad \text{(III-18.6)}$$

III-19. Let $f(x)$ be so defined on the interval $H: -1 \leq x \leq 1$ so that its expansion thereon in series of normalized Legendre polynomials

$$f(x) = \sum_{n=0}^{\infty} B_n L_n(x), \qquad \text{(III-19.1)}$$

where $L_n(x)$ is defined by Eq. (III-17.6), converges uniformly on H. Show that the coefficients B_0, B_1, B_2, $\cdots$ are such that

$$\int_{-1}^{1} \Big[f(x) \Big]^2 dx = \sum_{n=0}^{\infty} B_n^2. \qquad \star \qquad \text{(III-19.2)}$$

It is immediately apparent what we have to do to establish Eq. (III-19.2): multiply both sides of Eq. (III-19.1) by $f(x)$ and integrate over H. We get

$$\int_{-1}^{1} \Big[f(x) \Big]^2 dx = \sum_{n=0}^{\infty} \Big[B_n \int_{-1}^{1} f(x) L_n(x) dx \Big]. \qquad \text{(III-19.3)}$$

Termwise integration is valid here because of the assumed uniform convergence in Eq. (III-19.1). For every n the value of the nth integral on the right in Eq. (III-19.3) is B_n by Eq. (III-18.6). Thus, Eq. (III-19.3) reduces at once to Eq. (III-19.2).

Problem: Evaluation of Integrals Involving Legendre Polynomials

III-20. Evaluate $\int_{-1}^{1} x^m P_n(x) dx$ where $P_n(x)$ denotes the Legendre polynomial of degree n and m is a positive integer or zero. ★

It would seem that, in order to evaluate this integral, we should take advantage of some distinguishing characteristic feature of the Legendre polynomials. Moreover, the fact that the integrand is a product leads us to think of the possibility of applying integration by parts. And in this connection it looks as if Rodrigues's formula for $P_n(x)$ established in Prob. III-10 might be just the thing to use, since it expresses $P_n(x)$ as a derivative. True, the derivative is of order n. But successive integration by parts will lower the order of the derivative so that we may be able to reach a final integral yielding the value we are seeking. Let us proceed and see how it comes out.

Using Rodrigues's formula for $P_n(x)$ we have

$$\int_{-1}^{1} x^m P_n(x) dx = \frac{1}{2^n n!} \int_{-1}^{1} x^m D^n \left[(x^2 - 1)^n \right] dx,$$

where D^n is operator notation for d^n/dx^n.

Integrating by parts, we take

$$u = x^m, \quad du = mx^{m-1} dx,$$

$$dv = \frac{1}{2^n n!} D^n \left[(x^2 - 1)^n \right] dx, \quad v = \frac{1}{2^n n!} D^{n-1} \left[(x^2 - 1)^n \right].$$

We take note of the handling of the operator notation D^n. Upon integrating the order of the derivative becomes lower; thus we write D^{n-1}. If we were to differentiate, the order of the derivative would become higher and we would write D^{n+1}. We have now

$$\int_{-1}^{1} x^m P_n(x) dx = \frac{1}{2^n n!} \left\{ \left[x^m D^{n-1}(x^2 - 1)^n \right]_{x=-1}^{x=1} \right.$$
$$\left. - m \int_{-1}^{1} x^{m-1} D^{n-1} \left[(x^2 - 1)^n \right] dx \right\}.$$

The first expression on the right of the last equation vanishes because the function $(x^2 - 1)^n$ has the property that any derivative thereof of order *less* than n contains $x^2 - 1$ as a factor. This is seen by observing that the first derivative $D[(x^2 - 1)^n]$ contains the factor $(x^2 - 1)^{n-1}$, whence the second derivative $D^2[(x^2 - 1)^n]$ contains the factor $(x^2 - 1)^{n-2}$. In general for $m < n$ we see thus that the mth derivative $D^m[(x^2 - 1)^n]$ contains the factor $(x^2 - 1)^{n-m}$.

Applying integration by parts to the remaining integral and then repeating the process, we see that, by the observation made in the preceding paragraph, the integrated part will vanish each time the process is applied as long as the order of the derivative of $(x^2 - 1)^n$ in the integrated part is less than n. This observation leads us to distinguish two cases as follows:

(a) $m < n$. Applying integration by parts m times, we get

$$\int_{-1}^{1} x^m P_n(x)dx = \frac{(-1)^m m!}{2^n n!} \int_{-1}^{1} D^{n-m}\left[(x^2 - 1)^n\right] dx$$

$$= \frac{(-1)^m m!}{2^n n!} \left[D^{n-m-1}(x^2 - 1)^n \right]_{x=-1}^{x=1}$$

$$= 0, \quad m < n, \tag{III-20.1}$$

because $D^{n-m-1}[(x^2 - 1)^n]$ is either $(x^2 - 1)^n$ itself (in case $m = n - 1$), or is a derivative of $(x^2 - 1)^n$ of order less than n (in case $m < n - 1$) and so contains $x^2 - 1$ as a factor.

(b) $m \geqq n$. In this case we apply integration by parts n times, obtaining

$$\int_{-1}^{1} x^m P_n(x)dx = C_{mn} \int_{-1}^{1} x^{m-n}(x^2 - 1)^n dx$$

where

$$C_{mn} = \frac{(-1)^n(m)(m - 1)(m - 2) \cdots (m - [n - 1])}{2^n n!}.$$

Let us multiply numerator and denominator of the coefficient before the integral on the right by $(m - n)!$. In case $m = n$ we take

0! to be unity as remarked in Prob. I-10. Let us also put the factor $(-1)^n$ into the integrand. Thus we get

$$\int_{-1}^{1} x^m P_n(x)dx = \frac{m!}{2^n n!(m-n)!} \int_{-1}^{1} x^{m-n}(1-x^2)^n dx, \qquad m \geqq n.$$

Inspection of the integrand in the integral on the right tells us that when $m - n$ is odd (that is, when m is even and n odd or when m is odd and n even) the integrand is a polynomial $Q(x)$ consisting entirely of odd powers of x. Thus, $Q(-x) = -Q(x)$; and it follows that the integral $Q(x)$ from $x = -1$ to $x = 1$ vanishes. Accordingly, we have

$$\int_{-1}^{1} x^m P_n(x)dx = 0, \qquad m > n, \quad m-n \text{ odd}. \qquad \text{(III-20.2)}$$

But if $m - n$ is even (that is, when m and n are both odd or both even), then the integrand is a polynomial consisting entirely of even powers of x, so that the integral from $x = -1$ to $x = 1$ is equal to twice the integral of the same integrand from $x = 0$ to $x = 1$:

$$\int_{-1}^{1} x^m P_n(x)dx$$
$$= \frac{m!}{2^n n!(m-n)!} 2\int_{0}^{1} x^{m-n}(1-x^2)^n dx, \qquad m \geqq n, \quad m-\text{n even}.$$

The integral on the right can now be evaluated by Eq. (II-17.1), taking $a = 1$, $b = m - n$, $c = 2$, $d = n$. We obtain

$$\int_{-1}^{1} x^m P_n(x)dx$$
$$= \frac{m!\Gamma\left(\dfrac{m-n+1}{2}\right)}{2^{n-1}(m-n)!(m+n+1)\Gamma\left(\dfrac{m+n+1}{2}\right)},$$
$$m \geqq n, \quad m-n \text{ even}. \qquad \text{(III-20.3)}$$

It is interesting to note that when $m = n$, this formula reduces by successive applications of Eq. (I-4.1) to a simpler expression as follows.

$$\int_{-1}^{1} x^n P_n(x)dx$$

$$= \frac{n!\Gamma\left(\frac{1}{2}\right)}{2^{n-1}(2n + 1)\left(\frac{2n - 1}{2}\right)\left(\frac{2n - 3}{2}\right) \cdots \left(\frac{3}{2}\right)\left(\frac{1}{2}\right)\Gamma\left(\frac{1}{2}\right)}$$

$$= \frac{n!2^n}{2^{n-1}(2n + 1)(2n - 1)(2n - 3) \cdots (3)(1)}$$

$$= \frac{n!2^n 2^n n!}{2^{n-1}(2n + 1)(2n)(2n - 1)(2n - 2)(2n - 3) \cdots (4)(3)(2)(1)}$$

$$= \frac{2^{n+1}(n!)^2}{(2n + 1)!}. \tag{III-20.4}$$

Recapitulating, we have

$$\int_{-1}^{1} x^m P_n(x)dx = \begin{cases} 0, & m < n \\ 0, & m > n, \quad m - n \text{ odd} \\ \dfrac{m!\Gamma\left(\dfrac{m - n + 1}{2}\right)}{2^{n-1}(m - n)!(m + n + 1)\Gamma\left(\dfrac{m + n + 1}{2}\right)}, \\ \qquad\qquad m \geqq n, \quad m - n \text{ even.} \end{cases}$$

$$\tag{III-20.5}$$

Problem: Character and Location of Zeros of $P_n(x)$

III-21. Prove that the zeros of each Legendre polynomial $P_n(x)$ of positive degree are (**a**) all real, (**b**) distinct, (**c**) all between $x = -1$ and $x = 1$. ★

We recall that a zero of a function $f(x)$ is a value of x where $f(x) = 0$. We note that the phrase "of positive degree" was put in the statement of this problem because the Legendre polynomial of degree 0, namely, $P_0(x) = 1$, does not have any zeros.

First of all, let us see how one might be led to surmise the properties (a), (b), (c) which we are to establish. This surmise is a natural one to make upon observing the entries in the last column of this table:

n	Equation $P_n(x) = 0$	Solutions
1	$x = 0$	$x = 0$
2	$\dfrac{3}{2}x^2 - \dfrac{1}{2} = 0$	$x = \pm\sqrt{1/3}$
3	$\dfrac{5}{2}x^3 - \dfrac{3}{2}x = 0$	$x = 0,\ \pm\sqrt{3/5}$
4	$\dfrac{35}{8}x^4 - \dfrac{15}{4}x^2 + \dfrac{3}{8} = 0$	$x = \pm\sqrt{\dfrac{-3 + \sqrt{4.8}}{7}},\ \pm\sqrt{\dfrac{-3 - \sqrt{4.8}}{7}}$

The table shows that properties (a), (b), and (c) hold for $n = 1$, 2, 3, 4.

To surmise properties (a), (b), and (c) is one thing; to establish them is another. How to proceed? Perhaps we can show that the properties (a), (b), (c) follow as a consequence of properties already established in earlier problems. Or, to put it the other way around, perhaps we can show that assumption of the falsity of properties (a), (b), (c) will lead to an absurd conclusion, namely, the contradiction of results already established. Let us make an initial attempt with property (b).

If we suppose that $P_n(x)$ has a multiple zero x_0, then both $P_n(x_0) = 0$ and $P_n'(x_0) = 0$. Can $P_n(x)$ and its derivative $P_n'(x)$ both vanish at the same x_0? If that could happen, then by Eq. (III-15.2) $P_{n-1}(x)$ would vanish there. Then, by Prob. III-2, $P_{n-2}(x)$ would also vanish there. And by continuation of the same argument, we could eventually conclude that $P_0(x_0) = 0$, *contrary* to the fact that $P_0(x) \equiv 1$, as found in Prob. III-4. The (false) premise that $P_n(x)$ has a multiple zero is to be rejected. Hence, property (b) must be true.

Let us try this same procedure of indirect logic with properties

(a) and (c). Let us assume that the zeros of $P_n(x)$ lying between $x = -1$ and $x = 1$ are *less* in number than n. If this assumption leads to an absurd conclusion, namely, the contradiction of a property already established, then we reject that assumption. First we ask ourselves what property shown in earlier problems might be related to the nature and number of zeros of $P_n(x)$. Perhaps the integral property of Eq. (III-20.5), whereby

$$\int_{-1}^{+1} P_n(x)Q_h(x)dx = 0$$

when $Q_h(x)$ is any polynomial of degree $h < n$, will be of help.

Assuming that the zeros of $P_n(x)$ which lie between $x = -1$ and $x = 1$ are $r_1, r_2, \cdots, r_h$ where $1 \leq h < n$, we can formulate a polynomial $Q_h(x)$ of degree h having its zeros at $r_1, r_2, \cdots, r_h$ like this:

$$Q_h(x) = (x - r_1)(x - r_2) \cdots (x - r_h).$$

We see that $Q_h(1) > 0$, because each linear factor in $Q_h(x)$ is positive when $x = 1$. Moreover, $P_n(1) > 0$ because $P_n(1) = 1$ by the definition of Legendre polynomials given in the introduction to Chapter III. It follows, then, that $Q_h(x)$ and $P_n(x)$ always have the same sign on the interval $-1 \leq x \leq 1$, because as x varies from $+1$ to -1, $Q_h(x)$ and $P_n(x)$ both start out with a positive value and both cross the x-axis always at the same x. Thus the product $P_n(x)Q_h(x)$ is never negative and is positive at every x on the interval $-1 \leq x \leq 1$ except at the h points where both vanish. Consequently, we have

$$\int_{-1}^{1} P_n(x)Q_h(x)dx > 0,$$

in contradiction to the property established in Eq. (III-20.5).

We conclude that $P_n(x)$ must have at least n zeros between $x = 1$ and $x = -1$. Since $P_n(x)$ cannot have more than n zeros, it follows that all the zeros of $P_n(x)$ must be real and must lie on the open interval $-1 < x < 1$.

It will be observed that we assumed $P_n(x)$ to have at least one zero between $x = 1$ and $x = -1$. This assumption is valid for every $n \geq 1$ by virtue of Eq. (III-16.5) because the integral of $P_n(x)$ over the interval $-1 \leq x \leq 1$ would not vanish if $P_n(x)$ were of constant sign on the interval.

Problem: Evaluation of Derivatives

III-22. Letting P_n denote the Legendre polynomial $P_n(x)$ of degree n, show that

$$\frac{dP_n}{dx} = P_0 + 5P_2 + 9P_4 + \cdots$$
$$+ (2n - 5)P_{n-3} + (2n - 1)P_{n-1}, \qquad n \text{ odd}; \qquad \text{(III-22.1)}$$

and that

$$\frac{dP_n}{dx} = 3P_1 + 7P_3 + 11P_5 + \cdots + (2n - 5)P_{n-3}$$
$$+ (2n - 1)P_{n-1}, \qquad n \text{ even}, \quad n \neq 0. \qquad *$$
$$\text{(III-22.2)}$$

Since the derivative dP_n/dx is a polynomial of degree $n - 1$, we know by Prob. III-11 that dP_n/dx can be expressed as a finite sum of the form $A_0 + A_1P_1 + A_2P_2 + \cdots + A_{n-1}P_{n-1}$, where the coefficients $A_0, A_1, \cdots, A_{n-1}$ are constants. But the demonstration there of the *existence* of such an expansion $A_0 + A_1P_1 + \cdots + A_{n-1}P_{n-1}$, although usable in any given case under the assumption that the polynomials $P_0, P_1, P_2, \cdots, P_{n-1}$ are all at hand, does not appear to be feasible for determining the actual coefficients in a general formula of the kind to be established in the present problem. However, we do have in Prob. III-18 a general formula for determination of the A_k in the expansion of a given function (of properly restricted type). If we attempt to find the A_k for the present problem via Prob. III-18 we should expect to find (by Prob. III-20) that all A_k having subscript greater than $n - 1$ vanish; also we should expect to be able to determine the value for each A_k having subscript less than or equal to $n - 1$. Let us proceed and see if this turns out to be so.

By Prob. III-18 we have

$$\frac{dP_n}{dx} = \sum_{k=0}^{\infty} A_k P_k, \qquad A_k = \frac{2k + 1}{2} \int_{-1}^{1} \frac{dP_n}{dx} P_k dx. \tag{III-22.3}$$

In the integral formula for A_k in Eq. (III-22.3) we may replace dP_n/dx by the known-to-exist (by Prob. III-11) linear combination

$$C_0 P_0 + C_1 P_1 + C_2 P_2 + \cdots + C_{n-1} P_{n-1}, \tag{III-22.4}$$

where the coefficients $C_0, C_1, \cdots, C_n$ are constants. Then the integral for A_k becomes a finite sum of integrals:

$$A_k = \frac{2k + 1}{2}\left[\int_{-1}^{1} C_0 P_0 P_k dx + \int_{-1}^{1} C_1 P_1 P_k dx \right.$$
$$\left. + \int_{-1}^{1} C_2 P_2 P_k dx + \cdots + \int_{-1}^{1} C_{n-1} P_{n-1} P_k dx \right]. \tag{III-22.5}$$

(a)   For $k \geqq n$, each of the integrals in Eq. (III-22.5) vanishes by Prob. III-16, making $A_k = 0$. When $k < n$ we return to Eq. (III-22.3) and distinguish two categories:

(b) $k < n$, $n + k$ even. When $n + k$ is even, then n and k are both odd or both even. In either event we see by Eq. (III-0.7) that the integrand $(dP_n/dx)P_k$ will be the product of two polynomials, one of which is made up of odd powers of x while the other one comprises only even powers. The resulting polynomial $Q(x)$, therefore, consists only of odd powers of x. This makes $Q(-x) = -Q(x)$, so that the integral $Q(x)$ from $x = -1$ to $x = 1$ is zero:

$$\int_{-1}^{1} \frac{dP_n}{dx} P_k dx = 0, \qquad k < n, \quad n + k \text{ even,}$$

making $A_k = 0$.

(c) $k < n$, $n + k$ odd. Now n and k are such that one is even while the other is odd. This makes the product $(dP_n/dx)P_k$ a product of two polynomials such that both consist only of odd powers or both consist only of even powers of x. In either event their product is a polynomial $H(x)$ consisting only of even powers of x, so that $H(-x) = H(x)$, and the only conclusion we can draw at the moment is that

$$\int_{-1}^{1} \frac{dP_n}{dx} P_k dx = 2 \int_{0}^{1} \frac{dP_n}{dx} P_k dx.$$

This is true but not revealing. There is, however, one clue as to possible fruitful procedure in the very nature of our integrand, which suggests trying the (often useful) technique of integration by parts. We get

$$\int_{-1}^{1} P_k \frac{dP_n}{dx} dx = P_k(1)P_n(1) - P_k(-1)P_n(-1) - \int_{-1}^{1} P_n \frac{dP_k}{dx} dx.$$

Since n and k are such that one is even while the other is odd, we see by Eq. (III-0.7) that $P_k(-1)P_n(-1) = -1$. Moreover, we have $P_k(1)P_n(1) = (1)(1) = 1$. Accordingly, we are left with

$$\int_{-1}^{1} P_k \frac{dP_n}{dx} dx = 2 - \int_{-1}^{1} P_n \frac{dP_k}{dx} dx.$$

It looks like we are not getting anywhere. But a careful look at the right side of this last equation restores our hope, because, since the subscripts n and k have become interchanged and since we are dealing with the case $k < n$, then by the argument used in case (a) we have

$$\int_{-1}^{1} P_n \frac{dP_k}{dx} dx = 0.$$

Thus, we find that

$$\int_{-1}^{1} P_k \frac{dP_n}{dx} dx = 2, \qquad k < n, \quad n + k \text{ odd},$$

making $A_k = 2k + 1$ for such k.

Recapitulating, we have

$$A_k = \begin{cases} 0, & k \geqq n, \\ 0, & k < n, \quad k + n \text{ even}, \\ 2k + 1, & k < n, \quad k + n \text{ odd}. \end{cases} \quad \text{(III-22.6)}$$

When these values for the coefficients are placed in Eq. (III-22.3) we have at once the required formulas Eqs. (III-22.1) and (III-22.2) as given in the statement of the problem.

As an example we will expand $dP_7(x)/dx$ in a series of Legendre polynomials. The respective coefficients turn out to be as follows.

A_0: here $k = 0$ and $n = 7$. $0 + 7$ is odd and k is less than n. Thus, by Eq. (6) the coefficient is $2k + 1$ or 1.

A_1: $k = 1, n = 7, n + k$ is even. Hence $A_1 = 0$ by Eq. (III-22.6).

A_2: $k = 2, n = 7$. $n + k$ is odd. A_2 is then $2k + 1$ or 5.

A_3, A_5, A_7 are zero by the same reasoning as was used for A_1.

A_4 and A_6 are found to be 9 and 13.

A_8: here $k = 8$. Since $k > n$, we see by Eq. (III-22.6) that $A_8 = 0$.

Then we have

$$\frac{dP_7}{dx} = A_0 P_0(x) + A_2 P_2(x) + A_4 P_4(x) + A_6 P_6(x)$$

$$= P_0(x) + 5P_2(x) + 9P_4(x) + 13P_6(x).$$

Similarly, $dP_8(x)/dx$ is expanded as follows.

A_0: here $k = 0$ and $n = 8$. The sum of k and n is even. Therefore, $A_0 = 0$.

A_1: $k + n = 1 + 8$, which is odd. $A_1 = 2(1) + 1 = 3$.

All even subscripts yield coefficients equal to zero. Then we have

$$\frac{dP_8}{dx} = 3P_1(x) + 7P_3(x) + 11P_5(x) + 15P_7(x).$$

Problems: Recurrence Relations

III-23. Show that the Legendre polynomials $P_1(x)$, $P_2(x)$, $P_3(x)$, $\cdots$ satisfy the relation

$$\frac{dP_{n+1}}{dx} - \frac{dP_{n-1}}{dx} = (2n+1)P_n. \quad \star \qquad \text{(III-23.1)}$$

We notice that the polynomials whose derivatives are on the left side of Eq. (III-23.1) differ by 2 in degree. Now the polynomials occurring on the right in Eqs. (III-22.1) and (III-22.2) advance in degree by 2. So we can solve the present problem via Prob. III-22 as follows. We write Eq. (III-22.1) both for dP_m/dx and dP_{m-2}/dx, where m is any odd integer ≥ 3:

$$\frac{dP_m}{dx} = P_0 + 5P_2 + 9P_4 + \cdots + (2m-5)P_{m-3} + (2m-1)P_{m-1},$$
$$\text{(III-23.2)}$$

$$\frac{dP_{m-2}}{dx} = P_0 + 5P_2 + \cdots + (2m-9)P_{m-5} + (2m-5)P_{m-3},$$
$$\text{(III-23.2a)}$$

and subtract, obtaining

$$\frac{dP_m}{dx} - \frac{dP_{m-2}}{dx} = (2m-1)P_{m-1}. \qquad \text{(III-23.3)}$$

This same procedure applied to Eq. (III-22.2) shows that Eq. (III-23.3) of the present problem as obtained for m odd and ≥ 3 also holds when m is any even integer ≥ 2. Since Eq. (III-23.3) is thus true for every integer $m \geq 2$, we may take $n = m - 1$ and write Eq. (III-23.3) in the form

$$\frac{dP_{n+1}}{dx} - \frac{dP_{n-1}}{dx} = (2n+1)P_n, \qquad n = 1, 2, 3, \cdots,$$

as required in the statement of our problem.

III-24. If $P_{n-1}(x), P_n(x), P_{n+1}(x)$ be any three successive Legendre polynomials, show that

$$\int_t^1 P_n(x)dx = \frac{1}{2n+1}\left[P_{n-1}(t) - P_{n+1}(t)\right]. \quad \star$$

(III-24.1)

Since (III-24.1) involves three successive Legendre polynomials, we should be able to obtain it from the relation established in Prob. III-23. Since Eq. (III-23.1) is true for all values of x, the integrals of its right and left sides over any chosen interval will be equal. Thus, we get

$$(2n+1)\int_t^1 P_n(x)dx = \int_t^1 \frac{d}{dx}\left[P_{n+1}(x) - P_{n-1}(x)\right]dx,$$

$$\int_t^1 P_n(x)dx = \frac{1}{2n+1}\left\{\left[P_{n+1}(1) - P_{n-1}(1)\right] - \left[P_{n+1}(t) - P_{n-1}(t)\right]\right\}$$

$$= \frac{1}{2n+1}\left[P_{n-1}(t) - P_{n+1}(t)\right],$$

because $P_{n+1}(1) = P_{n-1}(1) = 1$ by the definition of Legendre polynomials as given in the introduction to Chapter III.

Problem: Evaluation of $P_{2n}(0)$

III-25. Evaluate $P_{2n}(0)$ and $P'_{2n+1}(0)$; also show that

$$P'_{2n+1}(0) = (2n+1)P_{2n}(0). \quad \star$$

To evaluate $P_{2n}(0)$ we have only to determine the constant term of $P_{2n}(x)$. We can use the summation formula for $P_n(x)$ given in Eq. (III-0.8) and write it for $P_{2n}(x)$:

$$P_{2n}(x) = \sum_{k=0}^n \frac{(-1)^k(4n-2k)!x^{2n-2k}}{2^{2n}k!(2n-k)!(2n-2k)!}.$$

The constant term is immediately seen to be the final term of the summation where $k = n$. Thus, we have

$$P_{2n}(0) = \frac{(-1)^n(4n-2n)!}{2^{2n}n!(2n-n)!(2n-2n)!} = \frac{(-1)^n(2n)!}{2^{2n}(n!)^2}.$$

$$(\text{III-25.1})$$

It will be observed that we have taken 0! to be unity. This convention is implied in Eq. (III-0.8).

For every Legendre polynomial of odd degree we have $P_{2n+1}(0) = 0$ because a Legendre polynomial of odd degree is comprised entirely of odd powers of x. But the derivative of $P_{2n+1}(x)$ at $x = 0$ does not equal zero, since the derivative will contain a constant. We can compute the value of this derivative by differentiating Eq. (III-0.8) written for $P_{2n+1}(x)$ and evaluating the result at $x = 0$. By Eq. (III-0.8) we have

$$P_{2n+1}(x) = \sum_{k=0}^{n} \frac{(-1)^k(4n+2-2k)!x^{2n+1-2k}}{2^{2n+1}k!(2n+1-k)!(2n+1-2k)!},$$

whence

$$P'_{2n+1}(x) = \sum_{k=0}^{n} \frac{(-1)^k(4n+2-2k)!(2n+1-2k)x^{2n-2k}}{2^{2n+1}k!(2n+1-k)!(2n+1-2k)!}.$$

The value for $P'_{2n+1}(0)$ is the constant term of this last summation, namely, the term for which $k = n$:

$$P'_{2n+1}(0) = \frac{(-1)^n(2n+2)!}{2^{2n+1}n!(n+1)!}.$$

$$(\text{III-25.2})$$

A comparison of this formula for $P'_{2n+1}(0)$ with the formula for $P_{2n}(0)$ suggests a possibility of a relationship between the two. Let us see if we can show such a relationship. Starting with the formula for $P'_{2n+1}(0)$ we have

$$P'_{2n+1}(0) = \frac{(-1)^n(2n+2)(2n+1)(2n)!}{2^{2n+1}n!(n+1)n!}$$

$$= (2n+1)\left[\frac{(-1)^n(2n)!}{2^{2n}(n!)^2}\right]$$

$$= (2n+1)P_{2n}(0).$$

Problem: Evaluation of Integrals Involving Legendre Polynomials

III-26. Evaluate $\int_0^1 P_m(x)dx$ where $P_m(x)$ denotes the Legendre polynomial of degree m, $m \neq 0$. $\star$

Since no particular Legendre polynomial is specified, our problem is to obtain a formula for the value of the integral in terms of m. Recalling that $P_m(x)$ is made up only of even powers of x when m is even and only of odd powers when m is odd (see Eq. (III-0.7)), we become aware that our problem divides itself naturally into two cases.

(a) m even. By Eq. (III-0.7) we have $P_m(-x) = P_m(x)$, so that

$$\int_0^1 P_m(x)dx = \frac{1}{2}\int_{-1}^1 P_m(x)dx. \tag{III-26.1}$$

But the integral on the right in Eq. (III-26.1) equals zero by Eq. (III-16.5). Consequently, we have

$$\int_0^1 P_m(x)dx = 0, \qquad m \text{ even}, \quad m \neq 0. \tag{III-26.2}$$

(b) m odd. It appears likely in this case that we can evaluate our integral by taking $t = 0$ in Eq. (III-24.1), which we may do because that equation is an identity in t. We get

$$\int_0^1 P_m(x)dx = \frac{1}{2m+1}\left[P_{m-1}(0) - P_{m+1}(0)\right], \qquad m \text{ odd}.$$
$$\tag{III-26.3}$$

But when m is odd, then $m - 1$ and $m + 1$ are both even; and we may apply Prob. III-25, taking first $n = m - 1$ and then $n = m + 1$. Thus, Eq. (III-26.3) becomes

$$\int_0^1 P_m(x)dx = \frac{1}{2m + 1}\left[\frac{(-1)^{(m-1)/2}(m - 1)!}{2^{m-1}\left[\left(\frac{m - 1}{2}\right)!\right]^2}\right.$$

$$\left. - \frac{(-1)^{(m+1)/2}(m + 1)!}{2^{m+1}\left[\left(\frac{m + 1}{2}\right)!\right]^2}\right]$$

$$= \frac{(-1)^{(m-1)/2}(m - 1)!(2m + 1)(m + 1)}{(2m + 1)2^{m+1}\left(\frac{m + 1}{2}\right)!\left(\frac{m + 1}{2}\right)\left(\frac{m - 1}{2}\right)!}.$$

Finally, then

$$\int_0^1 P_m(x)dx = \frac{(-1)^{(m-1)/2}(m - 1)!}{2^m\left(\frac{m + 1}{2}\right)!\left(\frac{m - 1}{2}\right)!}, \qquad m \text{ odd}.$$

$$(\text{III-26.4})$$

REMARK. $\displaystyle\int_0^1 P_0(x)dx = \int_0^1 1dx = 1.$

Problem: Evaluation of Derivatives

III-27. Determine the value of $P_n{}^{(r)}(1)$ where $P_n(x)$ denotes the Legendre polynomial of degree n and where $P_n{}^{(r)}(1)$ denotes, as usual, the value at $x = 1$ of the rth derivative $D^r[P_n(x)]$.   ★

If $r > n$ the problem is trivial. Since $P_n(x)$ is a polynomial of degree n, all its derivatives of order greater than n vanish identically.

When $r \leq n$ the problem is not trivial and we need to use some property of the Legendre polynomials by which we can obtain a formula for the rth derivative of an arbitrary $P_n(x)$. One possibility that comes to mind is to try and exploit the characteristic expansion established in Prob. III-1, namely,

$$(1 - 2xh + h^2)^{-1/2} = \sum_{n=0}^{\infty} P_n(x)h^n. \qquad \text{(III-27.1)}$$

If we may differentiate this expansion termwise r times with respect to x and find the resulting expansion valid at $x = 1$, then we should be able to determine $P_n^{(r)}(1)$ by equating coefficients of like powers of h in the two sides of the resulting expansion. Let us proceed in this manner and see if we can validate termwise differentiation of the expansion.

Let us first see what we get for the rth derivative with respect to x of the function on the left of Eq. (III-27.1). Holding h constant, we get the following for the first few derivatives.

$$D_x[(1 - 2xh + h^2)^{-1/2}] = h(1 - 2xh + h^2)^{-3/2},$$

$$D_x^2[(1 - 2xh + h^2)^{-1/2}] = 1 \cdot 3h^2(1 - 2xh + h^2)^{-5/2},$$

$$D_x^3[(1 - 2xh + h^2)^{-1/2}] = 1 \cdot 3 \cdot 5h^3(1 - 2xh + h^2)^{-7/2}.$$

It is now apparent that for the rth derivative we will have

$$D_x^r[(1 - 2xh + h^2)^{-1/2}]$$

$$= 1 \cdot 3 \cdot 5 \cdots (2r - 1)h^r(1 - 2xh + h^2)^{-(2r+1)/2}$$

$$= \frac{2^{1-r}\Gamma(2r)}{\Gamma(r)} h^r(1 - 2xh + h^2)^{-(2r+1)/2}, \qquad \text{(III-27.2)}$$

by Prob. I-17.

Let us assume for the moment that the maximum value taken by $P_n^{(r)}(x)$ on the interval $-1 \leqq x \leqq 1$ is taken at $x = 1$. Then, if the series

$$P_0^{(r)}(1) + P_1^{(r)}(1)h + P_2^{(r)}(1)h^2 + \cdots$$

converges for a value of h, the series

$$P_0^{(r)}(x) + P_1^{(r)}(x)h + P_2^{(r)}(x)h^2 + \cdots$$

will converge *uniformly* for $-1 \leqq x \leqq 1$ with that value for h. It will then follow that we may differentiate Eq. (III-27.1) r times with respect to x and set $x = 1$ in the result, applying Eq. (III-27.2) with $x = 1$:

$$\frac{2^{1-r}\Gamma(2r)}{\Gamma(r)} h^r (1-h)^{-2r-1} = \sum_{n=0}^{\infty} P_n^{(r)}(1) h^n, \qquad |h| < 1.$$

$$\text{(III-27.3)}$$

We may now expand the function on the left in powers of h valid for $|h| < 1$ via the Maclaurin series for $(1-h)^{-2r-1}$. This is easily done for it is readily seen that the nth derivative of $f(h) = (1-h)^{-2r-1}$ is

$$f^{(n)}(h) = (2r+1)(2r+2) \cdots (2r+n)(1-h)^{-2r-1-n},$$

so that we have

$$\frac{f^{(n)}(0)}{n!} = \frac{(2r+1)(2r+2) \cdots (2r+n)}{n!}$$

$$= \frac{(1)(2)(3) \cdots (2r)(2r+1)(2r+2) \cdots (2r+n)}{(2r)!n!}$$

$$= \frac{(2r+n)!}{(2r)!n!}.$$

Accordingly, Eq. (III-27.3) becomes

$$\frac{2^{1-r}\Gamma(2r)}{(2r)!\Gamma(r)} \sum_{n=0}^{\infty} \frac{(2r+n)!}{n!} h^{r+n} = \sum_{n=0}^{\infty} P_n^{(r)}(1) h^n, \qquad |h| < 1.$$

$$\text{(III-27.4)}$$

In Eq. (III-27.4) we change the index of the right-hand side from n to $n+r$. Since the two power series in h are equal in value for every h on the common interval $-1 < h < 1$, the coefficients of like powers of h must be equal:

$$\frac{2^{1-r}\Gamma(2r)(2r + n)!}{(2r)!\Gamma(r)n!} = P_{n+r}{}^{(r)}(1).$$

Letting $n + r = m$, we can rewrite our last equation as

$$P_m{}^{(r)}(1) = \frac{2^{1-r}\Gamma(2r)(m + r)!}{\Gamma(r)(2r)!(m - r)!}, \qquad r \leqq m. \qquad \text{(III-27.5)}$$

Eq. (III-27.5) together with the fact that $P_m{}^{(r)}(1) = 0$ when $r > m$ is the solution of our problem. An inspection of Eq. (III-27.5), however, indicates that we can give our solution an alternative formulation by use of the Legendre duplication formula from Prob. I-18. Applying it together with Eqs. (I-4.3) and (I-10.1) and (I-11.1) we get, with m replaced by n,

$$P_n{}^{(r)}(1) = \frac{\Gamma(n + r + 1)}{2^r r!\Gamma(n - r + 1)}, \qquad r \leqq n. \qquad \text{(III-27.6)}$$

In particular, when $r = n$, we get

$$P_n{}^{(n)}(1) = \frac{(2n)!}{2^n n!} \quad \text{by Eq. (I-10.1).}$$

This checks with the value for $P_n{}^{(n)}(1)$ obtained by observing that $P_n{}^{(n)}(x)$ is a constant, namely, the constant obtained by taking the nth derivative of Rodrigues's formula in Prob. III-10:

$$P_n{}^{(n)}(x) = D^n\left\{D^n\left[\frac{1}{2^n n!}(x^2 - 1)^n\right]\right\}$$

$$= \frac{1}{2^n n!}D^{2n}\left[(x^2 - 1)^n\right] = \frac{(2n)!}{2^n n!}.$$

$$\text{(III-27.7)}$$

We assumed that the maximum value taken by $P_n{}^{(r)}(x)$ on the interval $-1 \leqq x \leqq 1$ is taken at $x = 1$. This assumption can be validated by appropriate modification of the arguments employed in Prob. III-12 to show that the maximum of $P_n(x)$ on the interval $-1 \leqq x \leqq 1$ is taken at $x = 1$.

Problems: Evaluation of Integrals Involving Legendre Polynomials

III-28. Evaluate $\int_0^1 P_n(y^p)dy$, where p is a constant and positive, and $P_n(y)$ denotes the Legendre polynomial of degree n.   ★

We first let $y^p = x$. Then we have $dy = (1/p)x^{(1/p)-1} dx$. Letting I denote the integral to be evaluated and taking $s = 1/p$, we have

$$\int_0^1 P_n(y^p)dy = I = \int_0^1 sx^{s-1}P_n(x)dx. \tag{III-28.1}$$

We observe that the new integral for I in Eq. (III-28.1) will be improper in case $p > 1$ and n is even. But the improper integral is convergent, because every power of x in the integrand is greater than -1. In all other cases the transformed integral is proper.

The transformed integral suggests integration by parts: $\int u\,dv = uv - \int v\,du$. The question is: How shall we pick the parts? Shall we, as in previous problems, replace $P_n(x)$ by Rodrigues's formula and then take $dv = P_n(x)dx$? Or, shall we take $dv = sx^{s-1}dx$? Both look promising at first glance. But let us look ahead. If we take $u = sx^{s-1}$ and $dv == P_n(x)dx$ and continue in this vein, the exponent on x will decrease with each successive application of integration by parts. When s is an integer, this procedure would lead us to an integral without a factor of the form x^q and this might be amenable to evaluation. But when s is not an integer, such a termination would not occur.

If, on the other hand, we take $dv = sx^{s-1}dx$ with $u = P_n(x)$, the exponent on x would increase with each successive application of integration by parts. This looks unfavorable. But the order of derivative of $P_n(x) = (1/2^n n!)D^n[(x^2 - 1)^n]$ would increase, so that after n successive applications of integration by parts we would have then an integrand of the form

$$(\text{a constant})x^q D^{2n}[(x^2 - 1)^n] = (\text{a constant})x^q,$$

because $D^{2n}[(x^2 - 1)^n]$ is a constant, namely $(2n)!$. An integral with such an integrand is elementary and readily evaluated. So, we take $u = P_n(x) = (1/2^n n!) D^n[(x^2 - 1)^n]$ with $dv = sx^{s-1}dx$ and get

$$I = \left[P_n(x)x^s \right]_{x=0}^{x=1} - \int_0^1 x^s \frac{1}{2^n n!} D^{n+1}\left[(x^2 - 1)^n \right]dx.$$

$$\text{(III-28.2)}$$

The integrated part equals unity because $P_n(1) = 1$ for every n by the definition of Legendre polynomials in the introduction to Chapter III.

We apply integration by parts to the remaining integral, taking $u = (1/2^n n!) D^{n+1}[(x^2 - 1)^n] = P_n'(x)$ with $dv = x^s ds$. Then we have

$$I = 1 - \left[P_n'(x)\frac{x^{s+1}}{s+1} \right]_{x=0}^{x=1} + \int_0^1 \frac{x^{s+1}}{s+1} \frac{1}{2^n n!} D^{n+2}\left[(x^2 - 1)^n \right] dx.$$

$$\text{(III-28.3)}$$

Continuing in this manner, we obtain, after n steps of the procedure, the following:

$$I = 1 - \frac{P_n'(1)}{s+1} + \frac{P_n''(1)}{(s+1)(s+2)} - \cdots$$

$$+ \frac{(-1)^{n-1}P_n^{(n-1)}(1)}{(s+1)(s+2)\cdots(s+n-1)}$$

$$+ \frac{(-1)^n}{(s+1)(s+2)\cdots(s+n-1)}$$

$$\frac{1}{2^n n!} \int_0^1 x^{s+n-1} D^{2n}\left[(x^2 - 1)^n \right] dx.$$

$$\text{(III-28.4)}$$

The integral remaining is now easily evaluated, for $D^{2n}[(x^2 - 1)^n] = (2n)!$. Thus, except for this constant factor the value of the integral remaining is $1/(s + n)$.

Our integral I is now evaluated. It seems, however, that we should be able to formulate our result directly in terms of n and s. And this we can do by applying Eqs. (III-27.6) and (III-27.7), namely

$$P_n{}^{(k)}(1) = \frac{\Gamma(n + k + 1)}{2^k k! \Gamma(n - k + 1)}, \quad k \leq n \text{ and } P_n{}^{(n)}(1) = \frac{(2n)!}{2^n n!},$$

to evaluate the derivatives of $P_n(x)$ at $x = 1$. Thus we get

$$I = 1 + \sum_{k=1}^{n} \frac{(-1)^k \Gamma(n + k + 1)}{2^k k! [\Gamma(n - k + 1)](s + 1)(s + 2) \cdots (s + k)}.$$

$$\text{(III-28.5)}$$

It seems, further, that there should be some way of incorporating the lone term unity within the summation. This we can do. At the same time we can put our result in more compact and more elegant form by replacing unity with $\Gamma(s + 1)/\Gamma(s + 1)$ and at the same time multiplying each term in the summation by $\Gamma(s + 1)/\Gamma(s + 1)$. This allows us, by Eq. (I-4.1), to replace $[\Gamma(s + 1)](s + 1)(s + 2) \cdots (s + k)$ with $\Gamma(s + k + 1)$. Finally, then, we have, replacing s with $1/p$,

$$I = \int_0^1 P_n(y^p)dy = \sum_{k=0}^{n} \frac{(-1)^k \Gamma(n + k + 1)\Gamma\left(\dfrac{1}{p} + 1\right)}{2^k k! \Gamma(n - k + 1)\Gamma\left(\dfrac{1}{p} + k + 1\right)}, \quad p > 0.$$

$$\text{(III-28.6)}$$

REMARKS. 1. If we recall that formula (III-28.6) was obtained by starting with the change of variable $y^p = x$, we see that we can formulate a corollary result as follows. We have, as in Eq. (III-28.1)

$$\int_0^1 P_n(y^p)dy = \int_0^1 \frac{1}{p} x^{(1/p)-1} P_n(x)dx, \quad p > 0.$$

If, now, we let $q = (1/p) - 1$, we have

$$\int_0^1 x^q P_n(x) dx$$

$$= \Gamma(q + 1) \sum_{k=0}^{n} \frac{(-1)^k \Gamma(n + k + 1)}{2^k k! \Gamma(n - k + 1)\Gamma(q + k + 2)}, \qquad q > -1.$$

$$\text{(III-28.7)}$$

Notice that the exponent on x need not be an integer. Compare with Eq. (III-20.5) where the exponent on x is required to be an integer.

2. The particular cases of (III-28.7) in which $q = -1/2$ and $q = 1/2$ yield quite simple results. For, if we compute these two cases for the first several values of n, either by (III-28.7) or by actual integration (making use of Table III-1), we get the results exhibited in the following table.

	n	0	1	2	3	4	5	6
$\int_0^1 x^{-1/2} P_n(x) dx$		2	$\frac{2}{3}$	$-\frac{2}{5}$	$-\frac{2}{7}$	$\frac{2}{9}$	$\frac{2}{11}$	$-\frac{2}{13}$
$\int_0^1 x^{1/2} P_n(x) dx$		$\frac{2}{3}$	$\frac{2}{5}$	$\frac{2}{3 \cdot 7}$	$-\frac{2}{5 \cdot 9}$	$-\frac{2}{7 \cdot 11}$	$\frac{2}{9 \cdot 13}$	$\frac{2}{11 \cdot 15}$

The results here tabulated lead us to surmise the following:

$$\int_0^1 x^{-\frac{1}{2}} P_n(x) dx = \begin{cases} \dfrac{2(-1)^{n/2}}{2n + 1}, & n \text{ even} \qquad \text{(III-28.8a)} \\[3mm] \dfrac{2(-1)^{(n-1)/2}}{2n + 1}, & n \text{ odd}; \qquad \text{(III-28.8b)} \end{cases}$$

$$\int_0^1 x^{\frac{1}{2}} P_n(x) dx = \begin{cases} \dfrac{2(-1)^{(n+2)/2}}{(2n - 1)(2n + 3)}, & n \text{ even} \qquad \text{(III-28.9a)} \\[3mm] \dfrac{2(-1)^{(n+3)/2}}{(2n - 1)(2n + 3)}, & n \text{ odd}. \qquad \text{(III-28.9b)} \end{cases}$$

The surmises expressed in (III-28.8a), (III-28.8b), (III-28.9a), (III-28.9b) are correct, as we will now demonstrate. We carry out the details of proof only for (III-28.8a). The other three are demonstrated in the same manner with minor modifications.

Let us assume that all four formulas hold for two Legendre polynomials $P_{n-2}(x)$ and $P_{n-1}(x)$. Then, by the recursion formula (III-2.5), namely

$$nP_n(x) = (2n - 1)xP_{n-1}(x) - (n - 1)P_{n-2}(x), \qquad n \geqq 2, \quad \text{(III-28.10)}$$

we will show that (III-28.8a) holds also for $P_n(x)$. By (III-28.10) we have

$$\int_0^1 x^{-1/2}P_n(x)dx = \frac{1}{n}\left[(2n - 1)\int_0^1 x^{1/2}P_{n-1}(x)dx \right.$$
$$\left. - (n - 1)\int_0^1 x^{-1/2}P_{n-2}(x)dx \right]. \quad \text{(III-28.11)}$$

Applying the assumed formulas (III-28.9b) and (III-28.8a) respectively to the first and second integrals on the right in (III-28.11), we get, remembering that n and $n - 2$ are even while $n - 1$ is odd,

$$\int_0^1 x^{-1/2}P_n(x)dx$$

$$= \frac{2}{n}\left[\frac{(2n - 1)(- 1)^{(n-1+3)/2}}{(2[n - 1] - 1)(2[n - 1] + 3)} - \frac{(n - 1)(- 1)^{(n-2)/2}}{2(n - 2) + 1} \right]$$

$$= \frac{2}{n}\left[\frac{(- 1)^{(n+2)/2}(2n - 1)}{(2n - 3)(2n + 1)} - \frac{(- 1)^{(n-2)/2}(n - 1)}{2n - 3} \right]$$

$$= \frac{2(- 1)^{n/2}(2n^2 - 3n)}{n(2n - 3)(2n + 1)}$$

$$= \frac{2(- 1)^{n/2}}{2n + 1}.$$

Thus, when (III-28.8a), (III-28.8b), (III-28.9a), (III-28.9b) all hold for $n - 2$ and $n - 1$, then (III-28.8a) holds also for n. But all four formulas hold for $n = 0, 1, 2, 3, 4, 5, 6$. Therefore, (III-28.8a) holds also for $n = 7, 8, 9, 10, \cdots$. Thus, (III-28.8a) holds for every Legendre polynomial $P_n(x)$.

The other three formulas are established in similar manner with appropriate modifications. If we let $y^2 = x$, then

$$\int_0^1 P_n(y^2)dy = \frac{1}{2} \int_0^1 x^{-1/2} P_n(x)dx, \tag{III-28.12}$$

$$\int_0^1 y^2 P_n(y^2)dy = \frac{1}{2} \int_0^1 x^{1/2} P_n(x)dx. \tag{III-28.13}$$

Thus, by virtue of (III-28.12) and (III-28.13) together with (III-28.8a) (III-28.8b), (III-28.9a), (III-28.9b), we have

$$\int_0^1 P_n(y^2)dy = \frac{(-1)^N}{2n+1}$$

where $N = n/2$ when n is even and $N = (n-1)/2$ when n is odd;

$$\int_0^1 y^2 P_n(y^2)dy = \frac{(-1)^M}{(2n-1)(2n+3)},$$

where $M = (n+2)/2$ when n is even and $M = (n+3)/2$ when n is odd.

III-29. Evaluate $\int_0^1 (1 - x^2)[P_n{}'(x)]^2 dx$. $\star$

Let us denote the integral to be evaluated by A. Since the integrand in A is an even function we may take

$$A = \frac{1}{2} \int_{-1}^1 (1 - x^2) \left[P_n{}'(x) \right]^2 dx.$$

The reason for changing the interval of integration to the interval $-1 \leqq x \leqq 1$ is that we hope somehow to take advantage of the properties established for the Legendre polynomials on this interval in Probs. III-16 and III-17. An initial step is suggested by the integrand. It is that we try making use of Eq. (III-15.1). We get

$$A = \frac{1}{2} \int_{-1}^{1} \left\{ (1 - x^2) P_n{}'(x) \right\} P_n{}'(x) dx$$

$$= \frac{1}{2} \int_{-1}^{1} \left\{ \frac{n(n + 1)}{2n + 1} \left[P_{n-1}(x) - P_{n+1}(x) \right] \right\} P_n{}'(x) dx.$$

But we still have a derivative factor in the integrand. Do we have any formulas for the derivative $P_n{}'(x)$ in terms of the Legendre polynomials themselves? Yes; we find such formulas established in Prob. III-22. So, when n is even ($n \neq 0$) we have

$$A = \frac{n(n + 1)}{2(2n + 1)} \int_{-1}^{1} \left[P_{n-1}(x) - P_{n+1}(x) \right] \left[3P_1 + 7P_3 \right. $$
$$\left. + \cdots + (2n - 1)P_{n-1} \right] dx.$$

And now we have our integrand in a form where we can readily evaluate the integrals via Probs. III-16 and III-17. When we form all possible products of pairs of Legendre polynomials indicated in the integrand and then integrate the individual products over $-1 \leq x \leq 1$ we find by Prob. III-16 that all such integrals vanish by the orthogonality property except one, namely the integral of $[P_{n-1}(x)]^2$ which by Prob. III-17 equals $2/[2(n - 1) + 1] = 2/(2n - 1)$. Thus, for n even ($n \neq 0$) we get

$$A = \frac{n(n + 1)}{2(2n + 1)} \frac{(2n - 1)2}{2n - 1} = \frac{n(n + 1)}{2n + 1}.$$

Similarly, we get the same value for A when n is odd.

We have had to exclude so far the case $n = 0$ because the formulas we used involved $P_{n-1}(x)$. However, it is evident that our result $A = n(n + 1)/(2n + 1)$ holds also when $n = 0$ inasmuch as

$P_0{}'(x)$ is identically zero because $P_0(x) \equiv 1$ by Prob. (III-4). Thus, for every Legendre polynomial $P_n(x)$ without exception we have,

$$\int_0^1 (1 - x^2)\bigg[P_n{}'(x)\bigg]^2 dx = \frac{n(n + 1)}{2n + 1}.$$

III-30. Evaluate $\int_0^1 \bigg[xP_n(x)\bigg]^2 dx.$ ∗

This integral, like the other integrals in the several problems immediately preceding, will most likely be amenable to integration by parts. Moreover, since the integrand is an even function, we have

$$\int_0^1 \bigg[xP_n(x)\bigg]^2 dx = \frac{1}{2} \int_{-1}^1 \bigg[xP_n(x)\bigg]^2 dx. \qquad \text{(III-30.1)}$$

By thus changing the interval of integration to the interval $-1 \leq x \leq 1$ we can, as in preceding problems, take advantage of Rodrigues's formula for $P_n(x)$ so that integration by parts with $dv = P_n(x)dx = (1/2^n n!)D^n[(x^2 - 1)^n]\, dx$ will yield (as, for example, in Prob. III-17) an integrated part which vanishes at both $x = 1$ and $x = -1$. Taking dv as just indicated together with $u = x^2 P_n(x)$ and letting I denote the integral to be evaluated, we get

$$I = \frac{1}{2} \frac{(-1)}{2^n n!} \int_{-1}^1 D\bigg[x^2 P_n(x)\bigg] D^{n-1}\bigg[(x^2 - 1)^n\bigg] dx.$$

Again we integrate by parts, this time taking $dv = D^{n-1}[(x^2 - 1)^n]dx$ and continue thus in succession n times in all. The result is seen to be

$$I = \frac{1}{2^{n+1}n!}\int_{-1}^1 \bigg\{ D^n\bigg[x^2 P_n(x)\bigg]\bigg\} (1 - x^2)^n dx. \qquad \text{(III-30.2)}$$

At this point scrutiny of the integrand indicates that further integration by parts would not yield anything making for progress toward the solution. In fact our integrand looks more complicated now than it was to start with. But a closer look at it reveals something. If we visualize the polynomial $x^2 P_n(x)$ written out termwise via Eq. (III-0.7) as

$$x^2 P_n(x) = \frac{(2n)!}{2^n (n!)^2} \left[x^{n+2} - \frac{n(n-1)}{2(2n-1)} x^n \right.$$

$$+ \frac{n(n-1)(n-2)(n-3)}{2 \cdot 4(2n-1)(2n-3)} x^{n-2} - \cdots + x^2 F \right], \quad \text{(III-30.3)}$$

we see at once that every term of its nth derivative will vanish except the first two:

$$D^n \left[x^2 P_n(x) \right] = \frac{(2n)!}{2^n (n!)^2} \left[(n+2)(n+1) \cdots (3) x^2 - \frac{n(n-1)n!}{2(2n-1)} \right]$$

$$= \frac{(2n)!}{2^n (n!)^2} \left[\frac{(n+2)!}{2} x^2 - \frac{n(n-1)n!}{2(2n-1)} \right]. \quad \text{(III-30.4)}$$

So the integral in Eq. (III-30.2), except for the constant coefficients involved, becomes a difference of two integrals:

$$I = \frac{1}{2^{n+1} n!} \left[A \int_{-1}^{1} x^2 (1 - x^2)^n dx - B \int_{-1}^{1} (1 - x^2)^n dx \right] \quad \text{(III-30.5)}$$

$$= \frac{1}{2^{n+1} n!} \left[2A \int_{0}^{1} x^2 (1 - x^2)^n dx - 2B \int_{0}^{1} (1 - x^2)^n dx \right]. \quad \text{(III-30.6)}$$

The coefficients A and B in Eqs. (III-30.5) and (III-30.6) are to be picked up from Eq. (III-30.4). It is clear that Eq. (III-30.6) follows at once from Eq. (III-30.5) because the integrands are even functions. And now the integrals in Eq. (III-30.6) can be calculated via Prob. II-17. Moreover, their values can be simplified by the Legendre duplication formula. From Prob. II-17 we have

$$\int_{0}^{1} x^b (1 - x^c)^d dx = \frac{\Gamma\left(\dfrac{b+1}{c}\right) \Gamma(d+1)}{(b + cd + 1) \Gamma\left(\dfrac{b + cd + 1}{c}\right)}.$$

From Prob. I-18 we have the duplication formula

$$\frac{\Gamma(1/2)}{\Gamma[n + (1/2)]} = \frac{\Gamma(n)}{2^{1-2n}\Gamma(2n)}.$$

The result turns out to be

$$I = \frac{1}{2}\left[\frac{(n + 2)(n + 1)}{(2n + 3)(2n + 1)} - \frac{n(n - 1)}{(2n + 1)(2n - 1)}\right],$$

which we write finally as

$$\int_0^1 \left[xP_n(x)\right]^2 dx = \frac{2n^2 + 2n - 1}{(2n + 3)(2n + 1)(2n - 1)}.$$

III-31. Evaluate $\int_{-1}^1 (1 - x)^{-1/2}P_n(x)dx$ where $P_n(x)$ denotes the Legendre polynomial of degree n. ★

The integral to be evaluated is improper, since the factor $(1 - x)^{-1/2}$ in the integrand becomes infinite as $x \to 1$. The integral is, however, convergent. This is seen to be so by comparison with the integral $\int_{-1}^1 (1 - x)^{-1/2}dx$, which is readily recognized to be convergent because the exponent on $1 - x$ is greater than -1. The factor $P_n(x)$ is everywhere continuous; and it equals unity at $x = 1$ by the definition of Legendre polynomials. Consequently, the given integral is convergent, since the ratio of its integrand to $(1 - x)^{-1/2}$ approaches unity as limit as $x \to 1$.

The technique to be used in evaluating the given integral is apparently going to be integration by parts, which has already been found to be the means of evaluating a number of integrals in the previous problems of the present chapter. The technique has been seen to be fruitful of results because, in most cases, in the succession of integration by parts the integrated part vanishes each time, if dv has been chosen as $P_n(x)dx$, by virtue of the Rodrigues formula for $P_n(x)$ as, for example in Prob. III-16 or III-30.

As in the above problems we replace $P_n(x)$ by its Rodrigues formula, namely $P_n(x) = (1/2^n n!)D^n[(x^2 - 1)^n]$. Then in applying the formula for integration by parts, namely $\int u dv = uv - \int v du$, we first take $dv = P_n(x)dx$ with corresponding choice of dv in the later stages. Compare Prob. III-20 or III-30. After n such successive applications of integration by parts we apply (to the coefficient of the remaining integral) Prob. I-17, whereby

$$\frac{1 \cdot 3 \cdot 5 \cdots (2n - 1)}{2^n} = \frac{2^{1-2n}\Gamma(2n)}{\Gamma(n)}.$$

Then we have

$$\int_{-1}^{1} (1 - x)^{-1/2} P_n(x)dx$$

$$= \frac{2^{1-2n}\Gamma(2n)}{\Gamma(n)} \frac{(-1)^n}{2^n n!} \int_{-1}^{1} (1 - x)^{-n-(1/2)}(x^2 - 1)^n dx$$

$$= \frac{2^{1-3n}\Gamma(2n)}{n!\Gamma(n)} \int_{-1}^{1} (1 - x)^{-1/2}(1 + x)^n dx. \quad \text{(III-31.1)}$$

In Chapter II we found that an integral of the kind appearing here on the right in Eq. (III-31.1) could be evaluated in terms of the Beta function and thence by Eq. (I-29.1) in terms of the Gamma function. We let $1 + x = 2\cos^2 \theta$. Then $dx = -4\cos \theta \sin \theta d\theta$. Also $(1 - x)^{-1/2} = 1/(\sqrt{2} \sin \theta)$. Then we find that

$$\int_{-1}^{1} (1 - x)^{-1/2}(1 + x)^n dx = 2^{n+1/2} \int_{0}^{\pi/2} 2 (\sin \theta)^0 (\cos \theta)^{2n+1} d\theta$$

$$= 2^{n+1/2} B[n + 1, 1/2] \quad \text{by Prob. I-27}$$
$$\text{and I-26}$$

$$= \frac{2^{n+1/2}\Gamma(n + 1)\Gamma(1/2)}{\Gamma(n + [3/2])} \quad \text{by Eq. (I-29.1)}$$

$$= \frac{2^{n+1/2}(n!)\sqrt{\pi}}{[n + (1/2)]\Gamma[n + (1/2)]} \quad \begin{array}{l} \text{by Eqs.} \\ \text{(I-4.1),} \\ \text{(I-10.1),} \\ \text{(I-11.1).} \end{array}$$

Substituting this value for the integral on the right in Eq. (III-31.1) and applying the Legendre duplication formula established in Prob. I-18, we get

$$\int_{-1}^{1} (1-x)^{-1/2} P_n(x) dx = \frac{2\sqrt{2}}{2n+1}.$$

III-32. Show that

$$\int_{-1}^{1} P_n(x) \log_e (1-x) dx = \frac{-2}{n(n+1)}, \qquad n = 1, 2, \cdots. \qquad \star$$

Before we commence calculating we note that the integrals in question are improper because the factor $\log_e (1-x)$ in the integrand causes the integrand to become negatively infinite as $x \to 1$. The integrals in question are all convergent, however. This may be seen as follows. The factor $P_n(x)$ is continuous throughout; and $P_n(1) = 1$ by the definition of Legendre polynomials as mentioned in the introduction to Chapter III. Consequently, we can neglect this factor in examining for convergence. It is then a question of the convergence of $\int_{-1}^{1} \log_e (1-x) dx$. Now, it is readily seen that $\int_{-1}^{1} (1-x)^{-p} dx$ is convergent if $0 < p < 1$. But, by L'Hospital's Rule, we find that

$$\lim_{x \to 1} \frac{\log_e (1-x)}{(1-x)^{-p}} = 0, \qquad 0 < p.$$

Thus, the integral $\int_{-1}^{1} \log_e (1-x) dx$ is also convergent.

The actual calculation of the integral in question is a bit tedious but really not difficult. The procedure is quite the same as in several previous problems in this chapter, for example, Prob. III-20 or Prob. III-30. We shall, therefore, indicate the procedure but omit most of the details. We integrate by parts n times in succession, starting with $u = \log_e (1-x)$ and $dv = P_n(x) dx$, replacing $P_n(x)$ by Rodrigues's formula. Each time the integrated part vanishes, because

$D^{n-r}(x^2 - 1)^n$ contains $(x^2 - 1)^r$ as a factor for every r such that $0 \leqq r < n$. This was pointed out in Prob. III-20. We get finally

$$\int_{-1}^{1} P_n(x) \log_e (1 - x)dx = \frac{(-1)^{n-1}(n-1)!}{2^n n!} \int_{-1}^{1} (1-x)^{-n}(x^2 - 1)^n dx$$

$$= \frac{(n-1)!}{(-1)2^n n!} \int_{-1}^{1} (1-x)^{-n}(1-x^2)^n dx$$

$$= \frac{-1}{2^n n} \int_{-1}^{1} (1+x)^n dx$$

$$= \frac{-1}{2^n n} \left[\frac{(1+x)^{n+1}}{n+1} \right]_{x=-1}^{x=1}$$

$$= \frac{-2}{n(n+1)}.$$

REMARK. It will be observed in the statement of the problem that the formula to be established holds for all the Legendre polynomials except $P_0(x)$. The formula for the value of the integral, namely $-2/n(n+1)$, cannot, of course, hold for $n = 0$ because of the factor n in the denominator, since the integral of $\log_e (1-x)$ taken over the interval $-1 \leqq x \leqq 1$ is convergent and thus has a finite value. Moreover, if one takes Rodrigues's formula for $P_0(x)$, namely $P_0(x) = D^0[(x^2 - 1)^0]$, and attempts to carry out the steps of the demonstration given above where n was taken implicitly to be greater than zero, one finds that the procedure yields only the indeterminate form $-\infty + \infty$ for the integral of $P_0(x) \log_e (1-x)$. By direct integration, however, one finds that

$$\int_{-1}^{1} P_0(x) \log_e (1-x)dx = \int_{-1}^{1} \log_e (1-x)dx = 2(\log_e 2 - 1).$$

III-33. Evaluate $\int_{-1}^{1} (1 + hx^2)^{-n-(3/2)} P_{2n}(x)dx$, $|h| < 1$, where n is a positive integer or zero and $P_{2n}(x)$ is a Legendre polynomial of even degree. $\quad \star$

Wondering how to get started on this problem, we said to ourselves, "Now, if only the coefficient of $P_{2n}(x)$ were a positive integral power of x itself, then we could easily evaluate the integral via Prob. III-20." And such reflection led to the thought, "Well, we could get a series of such coefficients for $P_{2n}(x)$ by expansion of $(1 + hx^2)^{-n-(3/2)}$ into an infinite series via formal binomial expansion." Indeed, this is the very line of attack we shall take.

For $|h| < 1$, $|x| \leq 1$ we have

$$(1 + hx^2)^{-n-3/2}$$

$$= 1 + \sum_{j=1}^{\infty} \frac{(-1)^j (n + \frac{3}{2})(n + \frac{3}{2} + 1) \cdots (n + \frac{3}{2} + j - 1)}{j!} (hx^2)^j.$$
$$\text{(III-33.1)}$$

For each h on the interval $|h| < 1$ the expansion Eq. (III-33.1) is uniformly convergent for $|x| \leq 1$ and will remain uniformly convergent for $|x| \leq 1$ when both sides are multiplied by $P_{2n}(x)$, since by Prob. III-12 we have $|P_{2n}(x)| \leq 1$ on the closed interval $|x| \leq 1$. Because of its uniform convergence on $|x| \leq 1$ the expansion

$$(1 + hx^2)^{-n-(3/2)}P_{2n}(x) =$$

$$\left[1 + \sum_{j=1}^{\infty} \frac{(n + \frac{3}{2})(n + \frac{3}{2} + 1) \cdots (n + \frac{3}{2} + j - 1)}{j!} (-h)^j x^{2j} \right] P_{2n}(x)$$
$$\text{(III-33.2)}$$

may be integrated termwise over the interval $-1 \leq x \leq 1$:

$$\int_{-1}^{1} (1 + hx^2)^{-n-3/2} P_{2n}(x) dx =$$

$$\sum_{j=1}^{\infty} \int_{-1}^{1} \frac{(n + \frac{3}{2})(n + \frac{3}{2} + 1) \cdots (n + \frac{3}{2} + j - 1)}{j!} (-h)^j x^{2j} P_{2n}(x) dx,$$
$$\text{(III-33.3)}$$

$n \neq 0$. We have used here the fact that $\int_{-1}^{1} P_{2n}(x)dx = 0, \quad n \neq 0$ by Eq. (III-16.5).

And now we can evaluate each and every integral on the right in Eq. (III-33.3) via the results obtained in Prob. III-20, which tell us first of all that every integral on the right in Eq. (III-33.3) having $j < n$ has the value zero. For $j \geqq n$ none of the integrals vanishes. Since we have to compute the sum of their individual values via Eq. (III-20.5), it looks as if we can make the task a bit less cumbersome by taking as index of summation a letter which represents the excess of j over n when $j \geqq n$: we let $s = j - n$. Thus, in place of Eq. (III-33.3) we now have

$$\int_{-1}^{1} (1 + hx^2)^{-n-3/2} P_{2n}(x)dx$$

$$= \sum_{s=0}^{\infty} \left[\frac{(-h)^{n+s}(n + \tfrac{3}{2})(n + \tfrac{3}{2} + 1) \cdots (n + \tfrac{3}{2} + n + s - 1)}{(n + s)!} \right.$$

$$\left. \int_{-1}^{1} x^{2n+2s} P_{2n}(x)dx \right]. \quad \text{(III-33.4)}$$

We note that the summation is now again from zero to infinity. Now we replace each integral on the right in Eq. (III-33.4) by its value as given by Eq. (III-20.5), observing that the m of Eq. (III-20.5) is now $2n + 2s$ while the n is now $2n$. Thus for the nth term of the series on the right in Eq. (III-33.4) we get

$$\frac{(n + \tfrac{3}{2})(n + \tfrac{3}{2} + 1) \cdots (2n + s + \tfrac{1}{2})(-h)^{n+s}}{(n + s)!}$$

$$\left[\frac{(2n + 2s)!\Gamma(s + \tfrac{1}{2})}{2^{2n}(2s)!\Gamma(2n + s + \tfrac{3}{2})} \right],$$

which by application of Eq. (I-4.3), by multiplying numerator and denominator by $n + \tfrac{1}{2}$, by cancelling out common parts of factorials where possible, by using up the powers of 2 from 2^{2n} to make such cancellation possible, reduces to

$$\frac{2(-h)^{n+s}(n + \tfrac{1}{2})(n + \tfrac{3}{2}) \cdots (n + s - \tfrac{1}{2})}{(2n + 1)s!} \quad \text{when } s \geqq 1,$$

and reduces to

$$\frac{2(-h)^n}{2n + 1} \quad \text{when } s = 0.$$

Thus we get

$$\int_{-1}^{1} (1 + hx^2)^{-n-(3/2)} P_{2n}(x)dx$$

$$= \frac{2(-h)^n}{2n + 1}\left[1 + \sum_{s=1}^{\infty} \frac{(-1)^s(n + \tfrac{1}{2})(n + \tfrac{3}{2}) \cdots (n + s - \tfrac{1}{2})h^s}{s!} \right].$$

$$\text{(III-33.5)}$$

The integral of our problem is now evaluated. The result, however, may be simplified considerably by scrutinizing the infinite series in brackets in Eq. (III-33.5), which is none other than the Maclaurin series in powers of h for the function $(1 + h)^{-n-(1/2)}$. And so finally, for $n = 1, 2, 3, \cdots$ we have

$$\int_{-1}^{1} (1 + hx^2)^{-n-(3/2)} P_{2n}(x)dx = \frac{2(-h)^n}{(2n + 1)(1 + h)^{n+(1/2)}}, \quad |h| < 1.$$

$$\text{(III-33.6)}$$

There is yet a loose thread to be trimmed off. It is always a good idea to be thorough. We have established Eq. (III-33.6) for $n \neq 0$. However, we can show by simple integration that Eq. (III-33.6) holds also for $n = 0$. When this is done we can, in good conscience, write finis.

When $n = 0$ we have $P_{2n}(x) = P_0(x) = 1$ by Prob. III-4. Then the integral to be evaluated reduces simply to $\int_{-1}^{1} (1 + hx^2)^{-3/2}dx$, which is readily evaluated via the transformation $\sqrt{h}x = \tan \phi$; and we get

$$\int_{-1}^{1} (1 + hx^2)^{-3/2}dx = 2\int_{0}^{1} (1 + hx^2)^{-3/2}dx$$

$$= \frac{2}{\sqrt{h}}\int_{0}^{\text{arc tan }\sqrt{h}} \cos \phi \, d\phi$$

$$= \frac{2}{(1 + h)^{1/2}},$$

which is what the right side of Eq. (III-33.6) equals when $n = 0$. Thus, we may say that Eq. (III-33.6) holds for all Legendre polynomials of even degree including $P_0(x)$:

$$\int_{-1}^{1} (1 + hx^2)^{-n-(3/2)}P_{2n}(x)dx$$

$$= \frac{2(-h)^n}{(2n + 1)(1 + h)^{n+(1/2)}}, \quad |h| < 1, \quad n = 0, 1, 2, \cdots.$$

$$(\text{III-33.7})$$

III-34. Evaluate $\int_{-1}^{1}(1 - 2xh + h^2)^{-1/2}P_n(x)dx$, where $P_n(x)$ denotes the Legendre polynomial of degree n and h is a real constant such that $|h| < 1$. ★

For each value of h such that $|h| < 1$ we may regard the expansion (III-1.7) as an expansion of the function $(1 - 2xh + h^2)^{-1/2}$ in a series of functions of x uniformly convergent on the closed interval $-1 \leq x \leq 1$, because on that closed interval we have $|P_n(x)| \leq 1$ for all n by Prob. III-12. It follows that when every term of the expansion (III-1.7) is multiplied by one and the same $P_n(x)$, the resulting series will converge to $(1 - 2xh + h^2)^{-1/2}P_n(x)$ uniformly for $-1 \leqq x \leqq 1$, and may therefore be integrated termwise over that interval:

$$\int_{-1}^{1} (1 - 2xh + h^2)^{-1/2} P_n(x) dx$$

$$= \int_{-1}^{1} \sum_{m=0}^{\infty} \left[P_m(x) h^m \right] P_n(x) dx, \qquad |h| < 1. \quad \text{(III-34.1)}$$

Upon carrying out the termwise integration indicated on the right in Eq. (III-34.1) we find that every term vanishes by the orthogonality property established in Prob. III-16 *except* the term where $m = n$, which by Eq. (III-17.5) yields $[2/(2n + 1)]h^n$. Thus, we get

$$\int_{-1}^{1} (1 - 2xh + h^2)^{-1/2} P_n(x) dx$$

$$= \frac{2h^n}{2n + 1}, \qquad n = 1, 2, \cdots, \quad |h| < 1. \quad \text{(III-34.2)}$$

III-35. Show that if h is a real constant such that $|h| < 1$ and if n is a positive integer or zero, then

$$\int_{-1}^{1} (1 - 2hx + h^2)^{-n-(1/2)} (1 - x^2)^n dx = \frac{2^{2n+1}(n!)^2}{(2n + 1)!}. \qquad \star$$

$$\text{(III-35.1)}$$

We remark that, in contrast to Prob. III-34, the value of the integral presented in the present problem is independent of h. Since, however, the Legendre polynomial $P_n(x) = \dfrac{1}{2^n n!} D^n \left[(x^2 - 1)^n \right]$ by Rodrigues's formula, it looks as if we may be able to reduce the present problem to an appropriately managed application of Rodrigues's formula as follows.

In Eq. (III-34.2) we replace $P_n(x)$ in the integral by Rodrigues's formula from Prob. III-10 and integrate by parts, taking $u = (1 - 2hx + h^2)^{-1/2}$ and $dv = (1/2^n n!) D^n[(x^2 - 1)^n] dx$. As in similarly handled previous problems (such as Prob. III-30, for instance) we integrate thus n times in succession, arriving at

$$\frac{(-1)^n \cdot 1 \cdot 3 \cdot 5 \cdots (2n-1)h^n}{2^n n!} \int_{-1}^{1} (1 - 2hx + h^2)^{-n-1/2}$$

$$(x^2 - 1)^n dx = \frac{2h^n}{2n+1}. \quad \text{(III-35.2)}$$

We use the factor $(-1)^n$ before the integral to change the factor $(x^2 - 1)^n$ in the integrand to $(1 - x^2)^n$. And by Prob. I-17 we replace the product of the factors $1 \cdot 3 \cdot 5 \cdots (2n-1)$ by $2^{1-n}\Gamma(2n)/\Gamma(n)$. And if $h \neq 0$, both sides of Eq. (III-35.2) may be divided by h^n. If $h = 0$, the left side of Eq. (III-35.1) reduces to $\int_{-1}^{1}(1 - x^2)^n dx$, whose value by Prob. II-17 equals the right side of Eq. (III-35.1). Continuing with $h \neq 0$, we have now reduced Eq. (III-35.2) to

$$\frac{2^{-2n+1}\Gamma(2n)}{n!\Gamma(n)} \int_{-1}^{1} (1 - 2hx + h^2)^{-n-1/2}(1 - x^2)^n dx = \frac{2}{2n+1},$$

whence by Eq. (I-10.1) we obtain Eq. (III-35.1).

In the foregoing we have considered n to be a positive integer, because we performed integration by parts n times. When $n = 0$, we have for $h \neq 0$

$$\int_{-1}^{1} (1 - 2hx + h^2)^{-1/2} dx = \frac{-1}{h}\left[(1 - h) - (1 + h)\right] = 2.$$

When $n = h = 0$ we have

$$\int_{-1}^{1} dx = 2.$$

In both cases, the result agrees with the value given by the right side of Eq. (III-35.1) when $n = 0$, provided as usual we take 0! to be unity as remarked in Prob. I-10.

III-36. Show that $\int_0^\pi P_n(\cos\theta)\cos n\theta\, d\theta = B(n + \frac{1}{2}, \frac{1}{2})$, where P_n denotes the Legendre polynomial of degree n and B denotes the Beta Function. ✶

We take the formula for $P_n(\cos\theta)$ that we obtained in Eq. (III-12.3), multiply each term thereof by $\cos n\theta$, then integrate the result term by term. For every integral *except the first one* we will have (except for a constant factor)

$$\int_0^\pi \cos n\theta \cos\left[(n - 2k)\theta\right] d\theta,$$

which by a trigonometric identity becomes

$$\int_0^\pi \frac{1}{2}\left[\cos(2n - 2k)\theta + \cos 2k\theta\right] d\theta.$$

This integral equals zero, because each of the two terms integrates (except for a constant factor) to the sine of an integral multiple of θ and hence vanishes at both $\theta = \pi$ and $\theta = 0$.

For the integral of the first term we have

$$\frac{1 \cdot 3 \cdot 5 \cdots (2n - 1)(2)}{2 \cdot 4 \cdot 6 \cdots (2n)} \int_0^\pi \cos^2 n\theta\, d\theta$$

$$= \frac{1 \cdot 3 \cdot 5 \cdots (2n - 1)}{2 \cdot 4 \cdot 6 \cdots (2n)}\pi$$

$$= \frac{\Gamma(n + \frac{1}{2})}{\sqrt{\pi}\,\Gamma(n + 1)}\pi \qquad \text{by Prob. I-19}$$

$$= \frac{\Gamma(n + \frac{1}{2})\Gamma(\frac{1}{2})}{\Gamma(n + 1)} \qquad \text{by Eq. (I-1.11)}$$

$$= B\left(n + \frac{1}{2}, \frac{1}{2}\right) \qquad \text{by Prob. I-29.}$$

Table III-1

Legendre Polynomials

$P_0(x) = 1$, $P_n(1) = 1$, $P_n(-1) = (-1)^n$

$P_1(x) = x$

$P_2(x) = \frac{1}{2}(3x^2 - 1)$

$P_3(x) = \frac{1}{2}(5x^3 - 3x)$

$P_4(x) = \frac{1}{8}(35x^4 - 30x^2 + 3)$

$P_5(x) = \frac{1}{8}(63x^5 - 70x^3 + 15x)$

$P_6(x) = \frac{1}{16}(231x^6 - 315x^4 + 105x^2 - 5)$

$P_7(x) = \frac{1}{16}(429x^7 - 693x^5 + 315x^3 - 35x)$

$P_8(x) = \frac{1}{128}(6435x^8 - 12012x^6 + 6930x^4 - 1260x^2 + 35)$

$P_9(x) = \frac{1}{128}(12155x^9 - 25740x^7 + 18018x^5 - 4620x^3 + 315x)$

$P_{10}(x) = \frac{1}{256}(46189x^{10} - 109395x^8 + 90090x^6 - 30030x^4 + 3465x^2 - 63)$

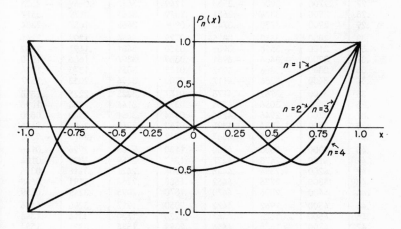

Legendre Polynomials

Figure III-1

TABLE III-2
Legendre Polynomials

x	$P_1(x)$	$P_2(x)$	$P_3(x)$	$P_4(x)$	$P_5(x)$	$P_6(x)$	$P_7(x)$
0.00	0.0000	−.5000	0.0000	0.3750	0.0000	−.3125	0.0000
.01	.0100	−.4998	−.0150	.3746	.0187	−.3118	−.0219
.02	.0200	−.4994	−.0300	.3735	.0374	−.3099	−.0436
.03	.0300	−.4986	−.0449	.3716	.0560	−.3066	−.0651
.04	.0400	−.4976	−.0598	.3690	.0744	−.3021	−.0862
.05	.0500	−.4962	−.0747	.3657	.0927	−.2962	−.1069
.06	.0600	−.4946	−.0895	.3616	.1106	−.2891	−.1270
.07	.0700	−.4926	−.1041	.3567	.1283	−.2808	−.1464
.08	.0800	−.4904	−.1187	.3512	.1455	−.2713	−.1651
.09	.0900	−.4878	−.1332	.3449	.1624	−.2606	−.1828
.10	.1000	−.4850	−.1475	.3379	.1788	−.2488	−.1995
.11	.1100	−.4818	−.1617	.3303	.1947	−.2360	−.2151
.12	.1200	−.4784	−.1757	.3219	.2101	−.2220	−.2295
.13	.1300	−.4746	−.1895	.3129	.2248	−.2071	−.2427
.14	.1400	−.4706	−.2031	.3032	.2389	−.1913	−.2545
.15	.1500	−.4662	−.2166	.2928	.2523	−.1746	−.2649
.16	.1600	−.4616	−.2298	.2819	.2650	−.1572	−.2738
.17	.1700	−.4566	−.2427	.2703	.2769	−.1389	−.2812
.18	.1800	−.4514	−.2554	.2581	.2880	−.1201	−.2870
.19	.1900	−.4458	−.2679	.2453	.2982	−.1006	−.2911
.20	.2000	−.4400	−.2800	.2320	.3075	−.0806	−.2935
.21	.2100	−.4338	−.2918	.2181	.3159	−.0601	−.2943
.22	.2200	−.4274	−.3034	.2037	.3234	−.0394	−.2933
.23	.2300	−.4206	−.3146	.1889	.3299	−.0183	−.2906
.24	.2400	−.4136	−.3254	.1735	.3353	.0029	−.2861
.25	.2500	−.4062	−.3359	.1577	.3397	.0243	−.2799
.26	.2600	−.3986	−.3461	.1415	.3431	.0456	−.2720
.27	.2700	−.3906	−.3558	.1249	.3453	.0669	−.2625
.28	.2800	−.3824	−.3651	.1079	.3465	.0879	−.2512
.29	.2900	−.3738	−.3740	.0906	.3465	.1087	−.2384
.30	.3000	−.3650	−.3825	.0729	.3454	.1292	−.2241
.31	.3100	−.3558	−.3905	.0550	.3431	.1492	−.2082
.32	.3200	−.3464	−.3981	.0369	.3397	.1686	−.1910
.33	.3300	−.3366	−.4052	.0185	.3351	.1873	−.1724
.34	.3400	−.3266	−.4117	−.0000	.3294	.2053	−.1527
.35	.3500	−.3162	−.4178	−.0187	.3225	.2225	−.1318
.36	.3600	−.3056	−.4234	−.0375	.3144	.2388	−.1098
.37	.3700	−.2946	−.4284	−.0564	.3051	.2540	−.0870
.38	.3800	−.2834	−.4328	−.0753	.2948	.2681	−.0635
.39	.3900	−.2718	−.4367	−.0942	.2833	.2810	−.0393
.40	.4000	−.2600	−.4400	−.1130	.2706	.2926	−.0146
.41	.4100	−.2478	−.4427	−.1317	.2569	.3029	.0104
.42	.4200	−.2354	−.4448	−.1504	.2421	.3118	.0356
.43	.4300	−.2226	−.4462	−.1688	.2263	.3191	.0608
.44	.4400	−.2096	−.4470	−.1870	.2095	.3249	.0859
.45	.4500	−.1962	−.4472	−.2050	.1917	.3290	.1106
.46	.4600	−.1826	−.4467	−.2226	.1730	.3314	.1348
.47	.4700	−.1686	−.4454	−.2399	.1534	.3321	.1584
.48	.4800	−.1544	−.4435	−.2568	.1330	.3310	.1811
.49	.4900	−.1398	−.4409	−.2732	.1118	.3280	.2027
.50	.5000	−.1250	−.4375	−.2891	.0898	.3232	.2231

TABLE III-2—(*Continued*)   **193**

x	$P_1(x)$	$P_2(x)$	$P_3(x)$	$P_4(x)$	$P_5(x)$	$P_6(x)$	$P_7(x)$
.50	.5000	−.1250	−.4375	−.2891	.0898	.3232	.2231
.51	.5100	−.1098	−.4334	−.3044	.0673	.3166	.2422
.52	.5200	−.0944	−.4285	−.3191	.0441	.3080	.2596
.53	.5300	−.0786	−.4228	−.3332	.0204	.2975	.2753
.54	.5400	−.0626	−.4163	−.3465	−.0037	.2851	.2891
.55	.5500	−.0462	−.4091	−.3590	−.0282	.2708	.3007
.56	.5600	−.0296	−.4010	−.3707	−.0529	.2546	.3102
.57	.5700	−.0126	−.3920	−.3815	−.0779	.2366	.3172
.58	.5800	.0046	−.3822	−.3914	−.1028	.2168	.3217
.59	.5900	.0222	−.3716	−.4002	−.1278	.1953	.3235
.60	.6000	.0400	−.3600	−.4080	−.1526	.1721	.3226
.61	.6100	.0582	−.3475	−.4146	−.1772	.1473	.3188
.62	.6200	.0766	−.3342	−.4200	−.2014	.1211	.3121
.63	.6300	.0954	−.3199	−.4242	−.2251	.0935	.3023
.64	.6400	.1144	−.3046	−.4270	−.2482	.0646	.2895
.65	.6500	.1338	−.2884	−.4284	−.2705	.0347	.2737
.66	.6600	.1534	−.2713	−.4284	−.2919	.0038	.2548
.67	.6700	.1734	−.2531	−.4268	−.3122	−.0278	.2329
.68	.6800	.1936	−.2339	−.4236	−.3313	−.0601	.2081
.69	.6900	.2142	−.2137	−.4187	−.3490	−.0926	.1805
.70	.7000	.2350	−.1925	−.4121	−.3652	−.1253	.1502
.71	.7100	.2562	−.1702	−.4036	−.3796	−.1578	.1173
.72	.7200	.2776	−.1469	−.3933	−.3922	−.1899	.0822
.73	.7300	.2994	−.1225	−.3810	−.4026	−.2214	.0450
.74	.7400	.3214	−.0969	−.3666	−.4107	−.2518	.0061
.75	.7500	.3438	−.0703	−.3501	−.4164	−.2808	−.0342
.76	.7600	.3664	−.0426	−.3314	−.4193	−.3081	−.0754
.77	.7700	.3894	−.0137	−.3104	−.4193	−.3333	−.1171
.78	.7800	.4126	.0164	−.2871	−.4162	−.3559	−.1588
.79	.7900	.4362	.0476	−.2613	−.4097	−.3756	−.1999
.80	.8000	.4600	.0800	−.2330	−.3995	−.3918	−.2397
.81	.8100	.4842	.1136	−.2021	−.3855	−.4041	−.2774
.82	.8200	.5086	.1484	−.1685	−.3674	−.4119	−.3124
.83	.8300	.5334	.1845	−.1321	−.3449	−.4147	−.3437
.84	.8400	.5584	.2218	−.0928	−.3177	−.4120	−.3703
.85	.8500	.5838	.2603	−.0506	−.2857	−.4030	−.3913
.86	.8600	.6094	.3001	−.0053	−.2484	−.3872	−.4055
.87	.8700	.6354	.3413	.0431	−.2056	−.3638	−.4116
.88	.8800	.6616	.3837	.0947	−.1570	−.3322	−.4083
.89	.8900	.6882	.4274	.1496	−.1023	−.2916	−.3942
.90	.9000	.7150	.4725	.2079	−.0411	−.2412	−.3678
.91	.9100	.7422	.5189	.2698	.0268	−.1802	−.3274
.92	.9200	.7696	.5667	.3352	.1017	−.1077	−.2713
.93	.9300	.7974	.6159	.4044	.1842	−.0229	−.1975
.94	.9400	.8254	.6665	.4773	.2744	.0751	−.1040
.95	.9500	.8538	.7184	.5541	.3727	.1875	.0112
.96	.9600	.8824	.7718	.6349	.4796	.3151	.1506
.97	.9700	.9114	.8267	.7198	.5954	.4590	.3165
.98	.9800	.9406	.8830	.8089	.7204	.6204	.5115
.99	.9900	.9702	.9407	.9022	.8552	.8003	.7384
1.00	1.0000	1.0000	1.0000	1.0000	1.0000	1.0000	1.0000

TABLE III-3

Derivatives of Legendre Polynomials at $x = 1$.

$$P_n^{(r)}(1) = \frac{d^r}{dx^r}\left[P_n(x)\right]_{x=1} = \frac{1}{2^r r!}\frac{\Gamma(n+r+1)}{\Gamma(n-r+1)}, \quad n \geqq r.$$

(From Prob. III-27)

				n			
	1	2	3	4	5	6	7
$P_n^{(1)}(1)$	1	3	6	10	15	21	28
$P_n^{(2)}(1)$		3	15	45	105	210	378
$P_n^{(3)}(1)$			15	105	420	1260	3150
$P_n^{(4)}(1)$				105	945	4725	17325
$P_n^{(5)}(1)$					945	10395	62370
$P_n^{(6)}(1)$						10395	135135
$P_n^{(7)}(1)$							135135

4 APPLICATIONS

OF LEGENDRE POLYNOMIALS

INTRODUCTION

As mentioned in the introductory paragraphs in Chapter III, Legendre polynomials have applications in problems requiring the solution of Laplace's equation $\nabla^2 V = 0$ where V must assume prescribed values of certain type on zones of a spherical surface. In this chapter we present a selection of a few typical problems in temperature distribution in steady-state heat flow, also in gravitational potential and electrostatic potential. Preparatory thereto are several exercises in the expansion of functions (continuous and discontinuous) in series of Legendre polynomials.

Other purely mathematical applications (capable, however, of being found useful in problems arising in the sciences and engineering) are to be found in application of Gauss's mechanical quadrature theorem which shows the advantage of evaluating the ordinates at the zeros of Legendre polynomials when approximating a definite integral. A pair of exercises in the application of Gauss's approximation method, together with a few miscellaneous purely mathematical applications such as the expression of $\pi/2$ as an infinite product of integrals involving the Legendre polynomials, concludes the applications selected for this chapter.

A table of the zeros of $P_n(x)$, $n = 2, 3, 4, 5, 6$, together with the associated Gaussian weight coefficients is given at the end of this chapter.

Problems: Specific Series Expansions

IV-1. Expand the following function in a series of Legendre polynomials:

$$f(x) = \begin{cases} -1, & -1 \leqq x < 0, \\ 0, & x = 0, \\ 1, & 0 < x \leqq 1. \end{cases} \quad \star$$

By Prob. III-20 we have

$$f(x) = \sum_{n=0}^{\infty} A_n P_n(x), \qquad A_n = \frac{2n+1}{2} \int_{-1}^{1} f(x) P_n(x) dx. \tag{IV-1.1}$$

Every A_n for which n is even equals zero because the integrand of the integral which defines A_n is an odd function, being the product of the given odd function $f(x)$ by the even function $P_n(x)$ which is a polynomial comprised only of even powers of x by Eq. (III-0.7). By similar reasoning we find that the integral involved in defining A_n when n is odd equals twice the integral of the same integrand taken over the right half, namely, $0 \leqq x \leqq 1$, of the given interval of definition of $f(x)$. Accordingly, for every odd n we have

$$A_n = \frac{2n+1}{2} \left[2 \int_0^1 P_n(x)[1] dx \right]$$

$$= (2n+1) \int_0^1 P_n(x) dx.$$

Applying Eq. (III-26.4), we have

$$A_n = \begin{cases} \dfrac{(2n+1)(-1)^{(n-1)/2}(n-1)!}{2^n \left(\dfrac{n+1}{2}\right)! \left(\dfrac{n-1}{2}\right)!}, & n \text{ odd}, \\ 0, & n \text{ even}. \end{cases} \tag{IV-1.2}$$

Working out the first several coefficients by Eq. (IV-1.2) and substituting in Eq. (IV-1.1) we get

$$\frac{3}{2}P_1(x) - \frac{7}{8}P_3(x) + \frac{11}{16}P_5(x) - \frac{75}{128}P_7(x) + \frac{135}{256}P_9(x) - \frac{483}{1024}P_{11}(x)$$

$$+ \cdots + A_n P_n(x) + \cdots$$

$$= \begin{cases} -1, & -1 \leqq x < 0, \\ 0, & x = 0, \\ 1, & 0 < x \leqq 1, \end{cases} \qquad \text{(IV-1.3)}$$

where every A_n is defined by Eq. (IV-1.2).

REMARKS. 1. If C be any constant, then the series obtained by multiplying the series on the left in Eq. (IV-1.3) will converge to the function $Cf(x)$:

$$C\left[\frac{3}{2}P_1(x) - \frac{7}{8}P_3(x) + \frac{11}{16}P_5(x) - \cdots + A_n P_n(x) + \cdots\right]$$

$$= \begin{cases} -C, & -1 \leqq x < 0, \\ 0, & x = 0, \\ C, & 0 < x \leqq 1, \end{cases} \qquad \text{(IV-1.4)}$$

where every A_n is defined by Eq. (IV-1.2).

2. Since only polynomials of odd degree occur in the series on the left in Eqs. (IV-1.3) and (IV-1.4), the series not only converges to zero at $x = 0$ but is such that every individual term of the series equals zero at $x = 0$. This may seem like a trite remark. But it is worthwhile in certain applications (see, for example, Prob. IV-11).

3. The expansion (IV-1.3) converges rather slowly at and near the end points of the interval. At $x = 1$ the partial sum made up of the first six nonvanishing terms equals 793/1024, which is not very close to unity. But in applied problems requiring the use of expansion

(IV-1.3) or (IV-1.4) the slowness of convergence is no drawback for computation with a modern high-speed computer.

4. If C be any constant, one finds in similar manner that the expansion of the function

$$g(x) = \begin{cases} C, & 0 < x \leqq 1, \\ \dfrac{C}{2}, & x = 0, \\ 0, & -1 \leqq x < 0 \end{cases}$$

in Legendre polynomials is

$$g(x) = C\left[\frac{1}{2}P_0(x) + \frac{3}{4}P_1(x) - \frac{7}{16}P_3(x) + \frac{11}{32}P_5(x) - \cdots \right.$$
$$\left. + B_nP_n(x) + \cdots \right],$$

where

$$B_n = \frac{2n+1}{2}\int_0^1 P_n(x)dx.$$

For every even $n \geqq 2$ we have $B_n = 0$. Every B_n for which n is odd is given by half of the upper right side of Eq. (IV-1.2).

IV-2. Expand the function

$$f(x) = \begin{cases} -1, & -1 \leqq x < 0 \\ x, & 0 < x \leqq 1 \end{cases}$$

in a series of Legendre polynomials. ⋆

The function to be expanded in this problem has not been defined at $x = 0$. Moreover, it could not be made continuous at $x = 0$ no matter what value would be assigned to it there. In that respect it is like the function given in Prob. IV-1. Since the series expansion in Prob. IV-1 converges to zero at $x = 0$ (as was to be expected by the second remark in Prob. III-18) and since we wanted the function in that problem to be zero at $x = 0$ for the sake of applications, we defined it as zero when $x = 0$.

In the present problem we might let $f(x)$ remain undefined at $x = 0$, observing, however, that the requested series expansion will converge at $x = 0$ to $- 1/2$ by the second remark in Prob. III-18. We used the word "might" in the preceding sentence because of a technicality involved. We wish to use definite integrals (Riemann integrals) over $- 1 \leqq x \leqq 1$ to evaluate the coefficients of the desired expansion. The definite integral (Riemann integral) of a bounded function over a finite interval is customarily defined for a function which has been assigned a value at every point of a *closed* interval. We should, therefore, consider the function $f(x)$ of this problem to have been assigned a value C at $x = 0$. The value chosen for C will not, however, have any effect upon the values of the integrals to be computed.

In applying Eq. (III-18.2) to obtain the coefficients of the desired expansion we naturally find it convenient to write the integral involved as the sum of two integrals, one over $- 1 \leqq x \leqq 0$ and the other over $0 \leqq x \leqq 1$. We get

$$A_n = \frac{2n + 1}{2} \left[\int_{-1}^{0} (- 1)P_n(x)dx + \int_{0}^{1} xP_n(x)dx \right]. \quad \text{(IV-2.1)}$$

Here we can make use of some of the integral formulas obtained in problems in Chapter III as follows. When n is even, $P_n(x)$ is an even function of x and is an odd function of x when n is odd. Accordingly, we have

$$\int_{-1}^{0} (- 1)P_n(x)dx = - \int_{0}^{1} P_n(x)dx = 0, \qquad n \text{ even,} \quad n > 0$$
$$\text{by Eq. (III-26.2);}$$

$$\int_{-1}^{0} (- 1)P_n(x)dx = \int_{0}^{1} P_n(x)dx$$

$$= \frac{(- 1)^{(n-1)/2}(n - 1)!}{2^n\left(\dfrac{n + 1}{2}\right)!\left(\dfrac{n - 1}{2}\right)!}, \qquad n \text{ odd}$$
$$\text{by Eq. (III-26.4).}$$

Also,

$$\int_0^1 x P_n(x)dx = \frac{1}{2}\int_{-1}^1 x P_n(x)dx, \qquad n \text{ odd}$$

$$= 0, \qquad n \text{ odd}, \quad n > 1 \text{ by Eq. (III-20.5)}.$$

For the fourth type of integral to be involved in Eq. (IV-2.1), namely, integral over $0 \leqq x \leqq 1$ of $x P_n(x)$ when n is even, we may take $P_n(x)$, $n = 2, 4, 6, 8, 10$ from Table III-1 and integrate $x P_n(x)$ term by term. For n even and greater than 10 we can get $P_n(x)$ from Eq. (III-0.7). Finally, we find by direct integration that Eq. (IV-2.1) yields $A_0 = -\frac{1}{4}$, $A_1 = \frac{5}{4}$.

Making use of the formulas and procedures developed in the preceding paragraph for evaluating the integrals involved in Eq. (IV-2.1), we obtain

$$-\frac{1}{4}P_0(x) + \frac{5}{4}P_1(x) + \frac{5}{16}P_2(x) - \frac{7}{16}P_3(x) - \frac{3}{32}P_4(x)$$

$$+ \frac{11}{32}P_5(x) + \frac{13}{256}P_6(x) - \cdots + A_n P_n(x) + \cdots$$

$$= \begin{cases} -1, & -1 \leqq x < 0, \\ -\frac{1}{2}, & x = 0, \\ x, & 0 < x \leqq 1, \end{cases}$$

where every A_n is given by Eq. (III-18.2). Here we have put in $-\frac{1}{2}$ at $x = 0$ on the right, since we know (by the second remark in Prob. III-18) that the expansion on the left converges at $x = 0$ to the arithmetic mean of -1 and zero.

At $x = 0$ we find the partial sum of the series through the term $P_6(x)$ equals (by Table III-2) to approximately $-.4572$. The convergence is thus rather slow. But slowness of convergence is no essential disadvantage when electronic computing machines are available.

IV-3. Obtain an approximate value for $\sqrt[3]{(1/2)}$ via the first several terms of a Legendre polynomial series expansion for the function $f(x) = x^{1/3}$. ★

This problem is seen to be simply an exercise in application of Prob. III-18 and Eq. (III-28.7). In order that we use Eq. (III-28.7) where the integral is taken only from $x = 0$ to $x = 1$ we take the function $f(x)$ to be expanded via Prob. III-18 as follows:

$$f(x) = \begin{cases} 0, & -1 \leqq x \leqq 0, \\ x^{1/3}, & 0 \leqq x \leqq 1. \end{cases} \tag{IV-3.1}$$

Then for this $f(x)$ we have by Prob. III-18

$$f(x) = \sum_{n=0}^{\infty} A_n P_n(x) \quad \text{where } A_n = \frac{2n+1}{2} \int_0^1 x^{1/3} P_n(x)\,dx. \tag{IV-3.2}$$

Evaluating the first few coefficients in the series expansion by way of Eq. (III-28.7), taking $q = \frac{1}{3}$, and making use of Eqs. (I-4.3) and (I-10.1), we get

$$A_0 = \left(\frac{1}{2}\right)\left(\frac{3}{4}\right)(1) = .3750,$$

$$A_1 = \left(\frac{3}{2}\right)\left(\frac{3}{4}\right)\left(1 - \frac{3}{7}\right) \cong .6429,$$

$$A_2 = \left(\frac{5}{2}\right)\left(\frac{3}{4}\right)\left(1 - \frac{9}{7} + \frac{27}{70}\right) = .1875,$$

$$A_3 = \left(\frac{7}{2}\right)\left(\frac{3}{4}\right)\left(1 - \frac{18}{7} + \frac{27}{14} - \frac{81}{182}\right) \cong .1893.$$

In similar manner we also obtain

$$A_4 = -.1056, \quad A_5 = .1522, \quad A_6 = .0761, \quad A_7 = -.1159.$$

Taking $x = .5$ in Eq. (IV-3.2) and taking the values of $P_n(.5)$, $n = 0, 1, 2, \cdots, 7$ from Table III-2, we find that

$$f\left(\frac{1}{2}\right) = \sqrt[3]{\frac{1}{2}} \cong \sum_{n=0}^{7} A_n P_n\left(\frac{1}{2}\right)$$

$$\cong .3750 + .3214 - .0234 + .0828 + .0305 + .0137$$
$$+ .0246 - .0258$$
$$\cong .799.$$

REMARK.   In tables of cube roots one finds that $\sqrt[3]{1/2} \cong .794$. Our approximation is in error by about $\frac{3}{5}$ of 1 per cent.

IV-4.   On the closed interval $L: -b \leq x \leq b$ let the function $f(x)$ be continuous and have continuous second derivative. Show that $f(x)$ can be expanded on L in a uniformly convergent series of Legendre polynomials of appropriate argument.   ★

The expansion to be established in this problem is evidently a modification of the expansion considered in Prob. III-18 in that the particular interval $H: -1 \leq x \leq 1$ is now to be replaced by the interval $L: -b \leq x \leq b$.

We first write the expansion considered in Prob. III-18 in terms of the letter t in place of x, writing also $g(t)$ in place of $f(x)$:

$$g(t) = \sum_{n=0}^{\infty} A_n P_n(t), \quad |t| \leq 1, \quad A_n = \frac{2n+1}{2} \int_{-1}^{1} g(t) P_n(t) dt. \tag{IV-4.1}$$

Next, we relate x to t by the transformation $x = bt$. Taking $g(t)$ in expansion (IV-4.1) to be $f(bt)$ and observing that $dt/dx = 1/b$, we find that the expansion (IV-4.1) is equivalent to

$$f(x) = \sum_{n=0}^{\infty} A_n P_n\left(\frac{x}{b}\right), \quad |x| \leq b, \quad A_n = \frac{2n+1}{2b} \int_{-b}^{b} f(x) P_n\left(\frac{x}{b}\right) dx. \tag{IV-4.2}$$

Since $g(t) = f(bt)$ has continuous derivative on $H: -1 \leq x \leq 1$, the curve $y = g(t)$ has finite length on H, so that all parts of the hypothesis of Prob. III-18 are fulfilled. It follows, then, that the expansion (IV-4.2) is valid everywhere on L. Moreover, the convergence to $f(x)$ is uniform on L by the first remark in Prob. III-18.

IV-5. Expand the following function in a series of Legendre polynomials of appropriate argument:

$$f(x) = \begin{cases} 0, & -3 \leq x \leq 0, \\ x^3, & 0 \leq x \leq 3. \end{cases}$$

Indicate explicitly the first several coefficients of the expansion. *

This problem is an exercise in application of the expansion obtained in Prob. IV-4. Since $f(x) = 0$ on the left half of the interval prescribed in the present problem, we have

$$f(x) = \sum_{n=0}^{\infty} A_n P_n\left(\frac{x}{3}\right) \quad \text{where } A_n = \frac{2n+1}{6} \int_0^3 x^3 P_n\left(\frac{x}{3}\right) dx.$$

We can simplify somewhat the computation of the coefficients $A_0, A_1, A_2, \cdots$ by taking $x/3 = u$. This makes

$$A_n = \frac{27(2n+1)}{2} \int_0^1 u^3 P_n(u) du.$$

For the first few coefficients we get

$$A_0 = \frac{27}{2} \int_0^1 u^3 du = \frac{27}{8},$$

$$A_1 = \frac{81}{2} \int_0^1 u^4 du = \frac{81}{10},$$

$$A_2 = \frac{135}{4} \int_0^1 (3u^5 - u^3) du = \frac{135}{16}.$$

Similarly, one finds $A_3 = \frac{27}{5}$, $A_4 = \frac{243}{128}$. We have, then,

$$\frac{27}{8}P_0\left(\frac{x}{3}\right) + \frac{81}{10}P_1\left(\frac{x}{3}\right) + \frac{135}{16}P_2\left(\frac{x}{3}\right) + \frac{27}{5}P_3\left(\frac{x}{3}\right)$$

$$+ \frac{243}{128}P_4\left(\frac{x}{3}\right) + \cdots + A_nP_n\left(\frac{x}{3}\right) + \cdots$$

$$= \begin{cases} 0, & -3 \leqq x \leqq 0, \\ x^3, & 0 \leqq x \leqq 3, \end{cases}$$

where

$$A_n = \frac{27(2n+1)}{2}\int_0^1 u^3 P_n(u)du.$$

COMMENT. Let us see how well the partial sum made up of the first five terms of our series approximates to x^3 when x is, say, 2 and to zero when x is, say -3. When $x = 2$ the argument to be taken in each Legendre polynomial is 2/3. From Table III-2 we have by linear interpolation

$$P_2\left(\frac{2}{3}\right) \cong .1667, \quad P_3\left(\frac{2}{3}\right) \cong -.2592, \quad P_4\left(\frac{2}{3}\right) \cong -.4273.$$

We know, of course, that $P_0\left(\frac{2}{3}\right) = 1$ and $P_1\left(\frac{2}{3}\right) = \frac{2}{3}$. Multiplying these values by their respective coefficients in our expansion above, we get

$$2^3 \cong 7.975.$$

The error is 5/16 of 1 per cent.

When $x = -3$ the argument $x/3$ is -1; and we get

$$0 \cong \frac{27}{8} - \frac{81}{10} + \frac{135}{16} - \frac{27}{5} + \frac{243}{128} \cong .21.$$

The approximation here is not too good, since the series converges rather slowly at the end points.

IV-6. Express x^3 as a linear combination of the first four Legendre polynomials of argument $x/3$.   ★

The desired combination is readily and easily obtained by application of Prob. IV-4, taking $f(x) = x^3$ and $b = 3$:

$$x^3 = \sum_{n=0}^{\infty} A_n P_n\left(\frac{x}{3}\right), \qquad A_n = \frac{2n+1}{6} \int_{-3}^{3} x^3 P_n\left(\frac{x}{3}\right) dx.$$

As in Prob. IV-5, we let $x/3 = u$. Then in the present problem we have

$$A_n = \frac{27(2n+1)}{2} \int_{-1}^{1} u^3 P_n(u) du.$$

By Eq. (III-20.5) we get zero for A_0 and A_2, also for every A_n for which $n > 3$. It remains only to compute A_1 and A_3, which by Eq. (III-20.5) with $m = 3$, $n = 1$ and by Eq. (III-20.4) with $n = 3$ we find to be

$$A_1 = \left(\frac{81}{2}\right)\frac{3!\Gamma(3/2)}{2!(5)\Gamma(5/2)} = \frac{81}{5} \qquad \text{by Eq. (I-4.3)},$$

$$A_3 = \left(\frac{169}{2}\right)\frac{2^4(3!)^2}{7!} = \frac{54}{5}.$$

Thus, we have

$$x^3 = \frac{81}{5}P_1\left(\frac{x}{3}\right) + \frac{54}{5}P_3\left(\frac{x}{3}\right).$$

COMMENT. The right side of the last equation, according to our method of obtaining it, is to be regarded as an infinite series expansion of x^3 on the interval $-3 \leq x \leq 3$, all terms of the series vanishing except the two shown. But since it reduces to a polynomial of degree 3 which equals x^3 for more than 3 distinct values of x, the equation obtained holds for all values of x.

IV-7. Let a fixed point Q in space have rectangular coordinates (x_0, y_0, z_0) and let P: (x, y, z) be a variable point. Show that the reciprocal U of the distance

$$D = QP = \sqrt{(x - x_0)^2 + (y - y_0)^2 + (z - z_0)^2} = \sqrt{S}$$

satisfies Laplace's equation

$$\nabla^2 U = \frac{\partial^2 U}{\partial x^2} + \frac{\partial^2 U}{\partial y^2} + \frac{\partial^2 U}{\partial z^2} = 0 \qquad \text{(IV-7.1)}$$

at every finite point other than Q.   ⋆

We have

$$U = D^{-1} = S^{-1/2}$$

$$\frac{\partial U}{\partial x} = - (x - x_0)S^{-3/2},$$

$$\frac{\partial^2 U}{\partial x^2} = 3(x - x_0)^2 S^{-5/2} - S^{-3/2},$$

$$= S^{-5/2}[3(x - x_0)^2 - S^2]$$

$$= S^{-5/2}[2(x - x_0)^2 - (y - y_0)^2 - (z - z_0)^2].$$

Since U is unchanged when the differences are permuted, it follows that

$$\frac{\partial^2 U}{\partial y^2} = S^{-5/2}\left[2(y - y_0)^2 - (x - x_0)^2 - (z - z_0)^2 \right],$$

$$\frac{\partial^2 U}{\partial z^2} = S^{-5/2}\left[2(z - z_0)^2 - (x - x_0)^2 - (y - y_0)^2 \right].$$

Addition of these last three equations yields Eq. (IV-7.1), which is valid at all points in space where the derivatives involved exist, namely, all finite points other than Q.

REMARKS. 1. The function $U = 1/D$ belongs to the class of *harmonic functions*. A function H is said to be *harmonic* in a region R of space when (**a**) the partial derivatives $\partial^2 H/\partial x^2$, $\partial^2 H/\partial y^2$, $\partial^2 H/\partial z^2$ are all continuous at every point R, (**b**) the function H satisfies *Laplace's equation* $\nabla^2 H = 0$ everywhere in R. The function U is harmonic in the region consisting of all finite points except Q.

2. The function $U = 1/D$ represents, except for a constant factor, the *potential*, gravitational or electric, at P of a unit mass or unit electric charge at Q because the negated gradient of U, namely $-\nabla U$, represents at P (except for a constant factor) the gravitational force vector $\vec{F}$ or the electric force vector $\vec{E}$ due to the unit mass or charge at Q.

IV-8. Let R denote the distance from the origin O of the fixed point Q in Prob. IV-7. Let r be the spherical coordinate denoting distance from origin of the variable point P. As in Prob. IV-7 let $U = D^{-1}$ = reciprocal of the distance QP. Show that

$$U = \frac{1}{R} \sum_{n=0}^{\infty} \left(\frac{r}{R}\right)^n P_n(x), \qquad r < R, \qquad \text{(IV-8.1)}$$

$$U = \frac{1}{r} \sum_{n=0}^{\infty} \left(\frac{R}{r}\right)^n P_n(x), \qquad r > R, \qquad \text{(IV-8.2)}$$

where $x = \cos\theta$, θ being the angle ($0 \leq \theta \leq \pi$) of intersection of the vectors $\overrightarrow{OQ}$ and $\overrightarrow{OP}$, and $P_n(x)$ is the Legendre polynomial of degree n.   ★

We note, first of all, that the letter x is to be used in this problem *not* to denote the usual rectangular coordinate but merely as a convenient single-letter designation for $\cos\theta$. Likewise, the letter θ in this problem does *not* denote the usual polar or spherical angular coordinate but is used to designate the angle between the fixed position vector $\overrightarrow{OQ}$ and the variable position vector $\overrightarrow{OP}$.

The expansions to be demonstrated in this problem are apparently modifications of the expansion obtained in Prob. III-1. And since the distances r and R are involved together with the angle θ between the vectors r and R, it seems likely that the Law of Cosines can be used to obtain an expression in r, R, and θ to which we can apply Eq. (III-1.7). Consider, then, the triangle OQP where R denotes the side OQ, r denotes the side OP, D denotes the side QP and θ denotes the angle QOP. By the Law of Cosines we have

$$D = (r^2 + R^2 - 2rR \cos \theta)^{1/2}.$$

Remembering that x shall denote $\cos \theta$ and U shall denote D^{-1}, we may write

$$U = \frac{1}{R}\left[1 - 2x\left(\frac{r}{R}\right) + \left(\frac{r}{R}\right)^2 \right]^{-1/2} \qquad \text{(IV-8.3)}$$

and

$$U = \frac{1}{r}\left[1 - 2x\left(\frac{R}{r}\right) + \left(\frac{R}{r}\right)^2 \right]^{-1/2}. \qquad \text{(IV-8.4)}$$

Application of Eq. (III-1.7) is now apparent: for $r < R$ we identify the ratio r/R with h in Eq. (III-1.7); for $r > R$ we take $h = R/r$. Eqs. (IV-8.1) and (IV-8.2) hold by virtue of Prob. III-1.

Problems: Steady-State Heat-Flow Temperature Distribution

IV-9. Determine the steady-state temperature distribution T in a homogeneous spherical solid when one hemispherical half (excluding boundary) of the surface is maintained at 300° while the other half (boundary not included) is kept at 75°. Temperature on the separating great circle may be considered undefined.   ★

The equation for heat conduction (see, for example, Eckert and Drake, *Heat and Mass Transfer*, McGraw-Hill, 1959) in a solid is

$$\frac{\partial T}{\partial t} = \frac{k}{\rho c}(\nabla^2 T + Q'), \qquad \text{(IV-9.1)}$$

where T is temperature, t is time, k is the thermal conductivity, ρ is the density, c is specific heat, and Q' is rate of heat generation, $\nabla^2 T$ is the *Laplacian*

$$\nabla^2 T = \frac{\partial^2 T}{\partial x^2} + \frac{\partial^2 T}{\partial y^2} + \frac{\partial^2 T}{\partial z^2}.$$

In the present problem $Q' = 0$ because there is no heat generation and T is independent of time t, so that $\partial T / \partial t = 0$. Let us take the solid with center at the origin and use spherical coordinates (r, ϕ, θ) where r is the distance from origin, ϕ is colatitude from the positive z-axis (cone angle) and θ is the angle of sweep about the z-axis. Let the coordinate system be so oriented that the hemispherical surface $0 \leq \phi < \pi/2$ is the half which is kept at $300°$. Then by symmetry of surface temperature maintenance it follows that the temperature distribution T within the solid and on the surface is independent of θ. Eq. (IV-9.1) for the present problem becomes

$$\nabla^2 T = 0 \qquad\qquad \text{(IV-9.2)}$$

with $T = T(r, \phi)$. Laplace's Eq. (IV-9.2) when written in terms of spherical coordinates with T not dependent on θ becomes

$$\frac{\partial}{\partial r}\left(r^2 \frac{\partial T}{\partial r} \right) + \frac{1}{\sin \phi} \frac{\partial}{\partial \phi}\left(\sin \phi \frac{\partial T}{\partial \phi} \right) = 0. \qquad\qquad \text{(IV-9.3)}$$

So, our problem is to find T as a function of r and ϕ satisfying Eq. (IV-9.3) within the spherical solid and taking on the prescribed boundary values on the surface. As in Prob. II-50 we shall seek to obtain a solution for T as a composite of particular solutions of Eq. (IV-9.3), each particular solution being a product solution of the form FG where F is a function of r only and G is a function of ϕ only.

Assuming a solution of Eq. (IV-9.3) of the form

$$T = FG = f(r)g(\phi), \qquad\qquad \text{(IV-9.4)}$$

we have

$$\frac{\partial T}{\partial r} = GF', \qquad \frac{\partial^2 T}{\partial r^2} = GF'', \qquad\qquad \text{(IV-9.5)}$$

where F' and F'' denote dF/dr and d^2F/dr^2 respectively.

Similarly, we have

$$\frac{\partial T}{\partial \phi} = FG', \qquad \frac{\partial^2 T}{\partial \phi^2} = FG''. \tag{IV-9.6}$$

Then Eq. (IV-9.3) by virtue of Eqs. (IV-9.5) and (IV-9.6) becomes

$$2rGF' + r^2GF'' + FG' \cot\phi + FG'' = 0. \tag{IV-9.7}$$

Separation of variables in Eq. (IV-9.7) gives

$$\frac{2rF' + r^2F''}{F} = -\frac{G' \cot\phi + G''}{G}. \tag{IV-9.8}$$

By the same reasoning as was used in Prob. II-50 we conclude that the equated ratios in Eq. (IV-9.8) must both equal the same constant, since one is independent of ϕ while the other is independent of r. Is this constant value of the ratios negative or zero or positive? Let us see if we can determine which it must be. If we let C denote this constant, then from Eq. (IV-9.8) we have

$$G' \cot\phi + G'' = -CG. \tag{IV-9.9}$$

Consider the change in T as we move downwards in the upper half of the solid along a curve of which r remains constant. Then the change in T along such a curve is due wholly to the change in G. Now T is certainly positive throughout; and if we assume F and G both positive, then G decreases as we move downwards along the curve. This reasoning maintains the equality of Eq. (IV-9.4). Thus the derivative G' is negative. In the upper half cotan ϕ is positive. It is plausible to think that the second derivative G'' will be negative in the upper half, indicating that the negative first derivative is becoming more negative. The left side of Eq. (IV-9.9) will thus be negative in the upper half. We conclude that C should be positive. We let $C = k^2$, $k \neq 0$. Let us see, then, if we can find solutions for Eq.

(IV-9.2) of the form $T = FG = f(r)g(\phi)$ where F and G are both positive and such that

$$\frac{2rF' + r^2F''}{F} = -\frac{G'\cot\phi + G''}{G} = k^2, \qquad k \neq 0.$$
(IV-9.10)

Eq. (IV-9.10) requires that

$$r^2F'' + 2rF' - k^2F = 0, \tag{IV-9.11}$$

$$G'' + (\cot\phi)G' + k^2G = 0. \tag{IV-9.12}$$

If we put $k^2 = n(n + 1)$ we recognize Eq. (IV-9.12) to be Legendre's equation (III-0.4) with G playing the role of y. Thus a particular solution of Eq. (IV-9.12) is

$$G = C_nP_n(\cos\phi)$$

where C_n is an arbitrary constant.

When $k^2 = n(n + 1)$ the general solution of Eq. (IV-9.11) is

$$F = S_nr^n + \frac{B_n}{r^{n+1}}, \tag{IV-9.13}$$

where S_n and B_n are arbitrary constants. The second term on the right in Eq. (IV-9.13) becomes infinite at $r = 0$ and is thus unsuitable. Hence we let $B_n = 0$.

We have found that the product solutions of Eq. (IV-9.2) may be had in the form

$$T = FG = S_nC_nr^nP_n(\cos\phi). \tag{IV-9.14}$$

But no such particular solution will meet the boundary conditions

$$T(R, \phi) = \begin{cases} 300, & 0 \leqq \phi < \dfrac{\pi}{2} \\ 75, & \dfrac{\pi}{2} < \phi \leqq \pi. \end{cases} \tag{IV-9.15}$$

To do so would require that $P_n(\cos \phi)$ remain at one constant value for $0 \leq \phi < \pi/2$ and keep another constant value for $\pi/2 < \phi \leq \pi$. This is impossible: Legendre polynomials are continuous functions of the argument.

We can, however, meet the boundary conditions (IV-9.15) as follows. We let $T_E(\phi)$ denote the excess of the temperature T on the upper half of the surface over that of T on the lower half. On the bounding great circle between these halves we arbitrarily define T_E to be 225/2. We then have

$$T_E = \begin{cases} 225, & 0 \leq \phi < \pi/2, \\ 0, & \pi/2 < \phi \leq \pi, \\ 225/2, & \phi = \pi/2. \end{cases} \qquad \text{(IV-9.16)}$$

If we let the letter x denote $\cos \phi$ then $T_E(\phi)$ becomes $f(x)$ defined by

$$f(x) = \begin{cases} 225, & 0 < x \leq 1, \\ 0, & -1 \leq x < 0, \\ 225/2, & x = 0. \end{cases} \qquad \text{(IV-9.17)}$$

And now $f(x)$ may be expanded (by remark 4 in Prob. IV-1):

$$f(x) = \sum_{n=0}^{\infty} A_n P_n(x), \qquad A_n = \frac{2n+1}{2} \int_0^1 f(x) P_n(x) dx$$

$$= 225 \left[\frac{1}{2} + \frac{3}{4} P_1(x) - \frac{7}{16} P_3(x) + \frac{11}{32} P_5(x) - \cdots \right].$$

$$\text{(IV-9.18)}$$

Reverting to Eq. (IV-9.14), we choose constants $S_n C_n$, $n = 0, 1, 2, \cdots$ so that for each n we take

$$S_n C_n = \frac{A_n}{R^n}, \qquad \text{(IV-9.19)}$$

where A_n is the coefficient of $P_n(x)$ in Eq. (IV-9.18) and R is the radius of the solid. Composing an infinite series of such particular solutions of Eq. (IV-9.14), we take

$$T(r, \phi) = 75 + \sum_{n=0}^{\infty} A_n \left(\frac{r}{R}\right)^n P_n(\cos \phi) \tag{IV-9.20}$$

$$= 75 + 225 \left[\frac{1}{2} + \frac{3}{4}\left(\frac{r}{R}\right) P_1(\cos \phi) - \frac{7}{16}\left(\frac{r}{R}\right)^3 P_3(\cos \phi)\right.$$

$$\left. + \frac{11}{32}\left(\frac{r}{R}\right)^5 P_5(\cos \phi) - \cdots \right]. \tag{IV-9.21}$$

The right side of Eq. (IV-9.21) converges by Eq. (IV-9.18) to the prescribed surface temperature distribution when $r = R$. The right side of Eq. (IV-9.21) will also then be convergent for $r < R$ by virtue of the factor $(r/R)^n$. Each term on the right in Eq. (IV-9.21) individually satisfies Laplace's Eq. (IV-9.2). And the convergence is such that the series of Laplacians of the individual terms converges to zero for $r < R$. Thus Eq. (IV-9.21) provides the solution to our problem.

REMARK. As was pointed out earlier the temperature function T does the best that could be expected of it on the great circle separating the halves of the bounding surface in that it assumes there the average of the two different constant temperatures.

IV-10. If the temperature T on the surface of a spherical solid having radius R and center at the origin is maintained at $T_0(1 - \cos \phi)$ where T_0 is a constant and ϕ is the cone angle (co-latitude from positive z-axis) of spherical coordinates, determine the steady-state temperature distribution in the solid.   *

This problem is seen to be a variant of Prob. IV-9. So, we may appropriate, without repeating the arguments, the earlier results of that problem. In the present problem we may conclude, as in Prob.

IV-9, that the temperature distribution T will be independent of the coordinate θ and will satisfy Laplace's equation $\nabla^2 T = 0$ at all interior points. Then by Prob. IV-9 we have

$$T = T(r, \phi) = \sum_{n=0}^{\infty} A_n\left(\frac{r}{R}\right)^n P_n(x), \qquad x = \cos\phi,$$
(IV-10.1)

where x is *not* a rectangular coordinate but is (as indicated) just a designation for $\cos\phi$, where $P_n(x)$ is the Legendre polynomial of degree n and where the coefficients $A_0, A_1, A_2, \cdots$ are to be determined so that when $r = R$ the series on the right in Eq. (IV-10.1) converges to the given surface temperature maintenance.

On the surface, the temperature is prescribed to be $T_0(1 - x)$. As ϕ varies from π to 0, x will vary from -1 to 1. Accordingly, we have to determine by Eq. (III-18.2) the coefficients $A_0, A_1, A_2, \cdots$ of the expansion on the interval $-1 \leq x \leq 1$ of the function $T_0(1 - x)$ in a series of Legendre polynomials.

We have

$$A_n = \frac{2n + 1}{2} \int_{-1}^{1} T_0(1 - x) P_n(x)\, dx, \qquad n = 0, 1, 2, \cdots$$
(IV-10.2)

$$= \frac{(2n + 1)T_0}{2}\left[\int_{-1}^{1} P_n(x)dx - \int_{-1}^{1} xP_n(x)dx\right].$$
(IV-10.3)

The first integral on the right in Eq. (IV-10.3) vanishes, by Eq. (III-16.5), for all $n > 0$. The second integral, by Eq. (III-20.5), vanishes for all $n > 1$. Accordingly, all the coefficients indicated by Eq. (IV-10.2) vanish except A_0 and A_1. Since $P_0(x) = 1$ and $P_1(x) = x$, we get

$$A_0 = \frac{[2(0) + 1]T_0}{2} \int_{-1}^{1} (1 - x)dx = T_0,$$

$$A_1 = \frac{3T_0}{2} \int_{-1}^{1} (x - x^2)dx = -T_0,$$

$$A_n = 0, \qquad n = 2, 3, \cdots.$$

Putting these values for the coefficients in Eq. (IV-10.1), we have for the solution of our problem

$$T = T_0\left(1 - \frac{r}{R}\cos\phi\right).$$ (IV-10.4)

IV-11. Determine the steady-state temperature distribution T in a homogeneous hemispherical solid of radius R if the temperature on the surface is maintained at 20 on the hemispherical part of the surface (exclusive of its great circle boundary) and at 100 on the flat base of the solid (including the rim of the base). $\star$

We should be able to do this problem with appropriate modification of the procedure used in Prob. IV-9. Accordingly, we take the center of the base of our hemispherical solid at the origin with the hemispherical surface in that part of space where $0 \leqq \phi < \pi/2$. We also use the letter x to denote $\cos\phi$.

Since we must have

$$T(R, \phi) = g(x) = \begin{cases} 100, & x = \cos\dfrac{\pi}{2} = 0, \\ 20, & 0 < x \leqq 1 \end{cases}$$

and must have $T(r, \pi/2) = 100$, $0 \leqq r \leqq 1$, we see that we can attain solution of our problem by application of Prob. IV-1 as follows. We take

$$f(x) = \begin{cases} 80, & -1 \leqq x < 0, \\ 0, & x = 0, \\ -80, & 0 < x \leqq 1. \end{cases}$$

For $0 < x \leqq 1$ this makes $g(x) = 100 + f(x)$ and makes $g(0) = 0$. We do not consider negative values of x as far as $g(x)$ is concerned. However, we did define $f(x)$ for $-1 \leqq x < 0$ in order that we could make use of the expansion of $f(x)$ as given in Eq. (IV-1.4), taking

$C = -80$. Thus, applying Eq. (IV-1.4) with $C = -80$, we have on the hemispherical surface and on the rim of the base of the solid

$$T(R, \phi) = g(x)$$
$$= 100 - 80\left[\frac{3}{2}P_1(x) - \frac{7}{8}P_3(x) + \cdots + A_nP_n(x) + \cdots\right],$$

where every A_n is given by Eq. (IV-1.2).

And now all we have to do to obtain the solution of our problem is to follow through as in Eqs. (IV-9.19), (IV-9.20), (IV-9.21) in Prob. IV-9. Recalling that $x = \cos\phi$, the desired steady-state temperature distribution in the hemispherical solid we thus find to be

$$T(r, \phi) = 100 - 80\left[\frac{3}{2}\left(\frac{r}{R}\right)P_1(\cos\phi) - \frac{7}{8}\left(\frac{r}{R}\right)^3 P_3(\cos\phi) + \cdots\right.$$
$$\left. + A_n\left(\frac{r}{R}\right)^n P_n(\cos\phi) + \cdots\right],$$

where every A_n is given by Eq. (IV-1.2).

We observe that, by second remark in Prob. IV-1, the boundary condition $T(r, \pi/2) = 100$ is met at every point of the base of the solid by the obtained solution, since every term of the solution (except the additive constant 100) contains as factor a polynomial comprised only of odd powers of $\cos\phi$ and hence vanishes at $\phi = \pi/2$ regardless of the size of the ratio r/R. One also sees immediately that, for $0 \leqq \phi < \pi/2$, $T = 20$ when $r = R$.

IV-12.   The radius of the inner surface of a spherical shell solid is R_1. The radius of the outer surface is $R_2 > R_1$. The common center of the surfaces is at the origin. Temperature T on each surface is prescribed as a function of the cone angle ϕ (co-latitude from positive z-axis):

$$T(R_1, \phi) = g_1(\phi), \quad T(R_2, \phi) = g_2(\phi), \quad g_2(\phi) \neq g_1(\phi),$$

where $g_1(\phi)$ and $g_2(\phi)$ are such functions of $x = \cos \phi$ as in Prob. III-18. Determine the steady-state temperature distribution T at all points of space between the two bounding surfaces.　　★

This problem is like Probs. IV-9 and IV-10 in that T satisfies Laplace's equation $\nabla^2 T = 0$ at all interior points of the region occupied by the solid and in that T does not vary with the sweep coordinate θ. But the present problem differs in that the region of space occupied is the region bounded by two spherical surfaces instead of having a single bounding surface. Moreover the boundary values to be taken by T on the two bounding surfaces are different functions of ϕ. This will require some modification of the procedure used in Probs. IV-9 and IV-10, where T was required to take on only one prescribed boundary functional value on the sole bounding surface. If we take $R = R_2$ in Eq. (IV-10.1) and then determine the coefficients A_n, $n = 0, 1, 2, \cdots$ so that the series converges to $g_2(\phi)$ when $r = R_2$, the series will converge when $r = R_1$ because $R_1 < R_2$. But the function to which the series converges on the surface $r = R_1$, is then defined by the very series itself and cannot coincide with an arbitrarily preassigned function $g_1(\phi)$ except in the single case of perchance having taken

$$g_1(\phi) = \sum_{n=0}^{\infty} A_n (R_1/R_2)^n P_n(\cos \phi), \quad \text{where } A_n, \quad n = 0, 1, 2, \cdots$$

are the coefficients of the Legendre polynomial series for $g_2(\phi)$.

It appears, then, that our problem is to determine *two* sets of coefficients: one to make T agree with $g_2(\phi)$ when $r = R_2$, one to make $T = g_1(\phi)$ when $r = R_1$. Scanning the procedure of Prob. IV-9, we find that there we discarded one set of coefficients, namely the B_n, $n = 0, 1, 2, \cdots$ in the terms of the form B_n/r^{n+1} for the reason that terms of such character become infinite at $r = 0$. But in the present problem we have $0 < R_1 \leqq r \leqq R_2$, so that we do not have

$r = 0$ at any point of the region occupied. Let us then start with the series solution

$$T = T(r, \phi) = \sum_{n=0}^{\infty} \left(E_n r^n + \frac{F_n}{r^{n+1}} \right) P_n(x), \qquad x = \cos \phi,$$
$$(IV-12.1)$$

from Prob. IV-9, writing E_n for $S_n C_n$ and F_n for $B_n C_n$.

When $r = R_2$ we require this series to converge to $f_2(x) = g_2(\phi)$. The coefficients in Eq. (IV-12.1) must then, by Eq. (III-18.2), be such that

$$E_n R_2{}^n + \frac{F_n}{R_2{}^{n+1}} = \frac{2n+1}{2} \int_{-1}^{1} f_2(x) P_n(x) dx, \qquad n = 0, 1, 2, \cdots$$

$$= \frac{2n+1}{2} \int_{\pi}^{0} g_2(\phi) P_n(\cos \phi)(-\sin \phi) d\phi$$

$$= \frac{2n+1}{2} \int_{0}^{\pi} g_2(\phi) P_n(\cos \phi)(\sin \phi) d\phi. \qquad (IV-12.2)$$

We further require that the series in Eq. (IV-12.1) shall converge to $g_1(\phi)$ when $r = R_1$. This, by similar application of Eq. (III-18.2), means that the coefficients in Eq. (IV-12.1) must also be such that

$$E_n R_1 + \frac{F_n}{R_1{}^{n+1}} = \frac{2n+1}{2} \int_{0}^{\pi} g_1(\phi) P_n(\cos \phi) \sin \phi \, d\phi.$$
$$(IV-12.3)$$

For each pair of numbers E_n and F_n, $n = 0, 1, 2, \cdots$ we have by Eqs. (IV-12.2) and (IV-12.3) a pair of simultaneous linear equations to be satisfied. E_n and F_n are thus determined by this pair of equations. Accordingly, we may say that the desired steady-state temperature distribution is given by Eq. (IV-12.1) where the pairs of constants E_n and F_n, $n = 0, 1, 2, \cdots$ are determined by Eqs. (IV-12.2) and (IV-12.3).

REMARKS. 1. The question of convergence of the series in Eq. (IV-12.1) when r is *between* R_1 and R_2 can be met as follows.

Write the series on the right in Eq. (IV-12.1) as the sum of two series:

$$\sum_{n=0}^{\infty} E_n r^n P_n(x) + \sum_{n=0}^{\infty} \frac{F_n}{r^{n+1}} P_n(x). \qquad \text{(IV-12.4)}$$

The determination by Eq. (IV-12.2) of the coefficients to make the series converge when $r = R_2$ makes the first series in Eq. (IV-12.4) convergent for all r such that $r < R_2$. Similarly, determination of the coefficients to make the series converge when $r = R_1$ makes the second series in Eq. (IV-12.4) convergent for all r such that $r > R_1$. But we determined the coefficients so as to have the series convergent *both* for $r = R_2$ and $r = R_1$. It follows that both series in Eq. (IV-12.4) are convergent for $R_1 \leqq r \leqq R_2$, which means that the same is true of the series in Eq. (IV-12.1).

2. At the risk of being guilty of pointing out something that may be considered obvious, we observe that no pair of equations in the coefficients E_n and F_n, $n = 0, 1, 2, \cdots$ can fail to have solution. The determinant of the coefficients in every such pair is $R_2{}^n/R_1{}^{n+1} - R_1{}^n/R_2{}^{n+1}$, which can vanish only when $R_1 = R_2$.

3. When $g_1(\phi)$ and $g_2(\phi)$ are both constant, say $g_1(\phi) = C_1$ and $g_2(\phi) = C_2 > C_1$, we should expect from considerations of symmetry to find T to be a function of r alone increasing from C_1 to C_2 as r varies from R_1 to R_2. Indeed, this turns out to be so. By application of Eq. (III-18.2) together with Eq. (III-16.5) we find

$$E_0 + \frac{F_0}{R_1} = C_1, \qquad E_0 + \frac{F_0}{R_2} = C_2,$$

$$E_n = F_n = 0, \qquad n = 1, 2, 3, \cdots.$$

And solution for both surface temperatures constant reduces to

$$T = T(r) = \frac{1}{R_2 - R_1}\left[R_2 C_2 - R_1 C_1 - \frac{R_1 R_2 (C_2 - C_1)}{r} \right].$$
$$\text{(IV-12.5)}$$

IV-13. A homogeneous spherical shell solid has inner radius $R_1 = 1$ and outer radius $R_2 = 2$. Center of the surfaces is at the origin. Temperature on the inner surface is kept at 80 and on the outer surface is maintained at $700(1 - \cos \phi)$, where ϕ is the cone angle (colatitude from positive z-axis). Assuming that the temperature T within the shell has reached steady-state temperature distribution, compute the value T at a point midway between the bounding surfaces and having $\phi = \arccos .75$. ★

This problem is a direct application of Prob. IV-12. It appears that we may also be able to appropriate to some extent from Prob. IV-10. Applying Prob. IV-12 with $g_1(\phi) = f_1(x) = 80$ and $g_2(\phi) = f_2(x) = 700(1 - \cos \phi)$, $R_1 = 1$, $R_2 = 2$, we have

$$T = \sum_{n=0}^{\infty} \left(E_n r^n + \frac{F_n}{r^{n+1}} \right) P_n(x), \qquad x = \cos \phi, \quad \text{(IV-13.1)}$$

where

$$E_n + F_n = \frac{2n + 1}{2} \int_{-1}^{1} 80 P_n(x) dx, \qquad \text{(IV-13.2)}$$

$$2^n E_n + \frac{F_n}{2^{n+1}} = \frac{2n + 1}{2} \int_{-1}^{1} 700(1 - x) P_n(x) dx. \quad \text{(IV-13.3)}$$

The right side of Eq. (IV-13.2) vanishes by Eq. (III-16.5) for every $n > 0$ and equals 80 when $n = 0$, since $P_0(x) = 1$. The right side of Eq. (IV-13.3) vanishes, as in Prob. IV-10, for every $n > 1$. And for $n = 0$ and $n = 1$ the right side of Eq. (IV-13.3) equals 700 and -700 respectively. We have, then, the following pairs of simultaneous equations to solve.

$$E_0 + F_0 = 80, \qquad E_0 + \frac{F_0}{2} = 700; \qquad \text{(IV-13.4)}$$

$$E_1 + F_1 = 0, \qquad 2E_1 + \frac{F_1}{4} = -700; \qquad \text{(IV-13.5)}$$

$$E_n + F_n = 0, \qquad 2^n E_n + \frac{F_n}{2^{n+1}} = 0, \qquad n > 1.$$

$$\text{(IV-13.6)}$$

Solution of Eqs. (IV-13.4), (IV-13.5) and (IV-13.6) yields

$$E_0 = 1320, \qquad F_0 = -1240;$$

$$E_1 = -400, \qquad F_1 = 400;$$

$$E_n = F_n = 0, \qquad n > 1.$$

Putting these results in Eq. (IV-13.1), we get

$$T = 1320 - \frac{1240}{r} - 400\left(r - \frac{1}{r^2}\right)\cos\phi. \qquad \text{(IV-13.7)}$$

Equation (IV-13.7) gives the temperature distribution for steady-state temperature distribution at every point of the shell including the two bounding surfaces. A direct check of Eq. (IV-13.7), independent of the preceding work of solution, shows that **(a)** T satisfies Laplace's equation $\nabla^2 T = 0$ in spherical coordinates, namely,

$$\frac{\partial^2 T}{\partial r^2} + \frac{1}{r^2}\frac{\partial^2 T}{\partial\phi^2} + \frac{1}{r^2\sin^2\phi}\frac{\partial^2 T}{\partial\theta^2} + \frac{2}{r}\frac{\partial T}{\partial r} + \frac{\cotan\phi}{r^2}\frac{\partial T}{\partial\phi} = 0;$$

and **(b)** $T = 80$ when $r = 1$, $T = 700(1 - \cos\phi)$ when $r = 2$.

Finally, we have to evaluate T at a point where $\phi = \text{arc cos }.75$ and $r = \frac{3}{2}$. We get from Eq. (IV-13.7)

$$T\left(\frac{3}{2}, \cos^{-1}\frac{3}{4}\right) = 1320 - \frac{2480}{3} - \frac{950}{3} \cong 176.67.$$

Problem: Gravitational Potential of a Circular Lamina

IV-14. Determine the gravitational potential V of a homogeneous circular lamina of radius R. ⋆

The potential V of a continuous distribution of matter (such as a material surface or material curve or solid) is naturally defined via an integral as generalization from the definition of potential M/D of a mass-particle M as in the second remark in Prob. IV-7. In the case of a material surface S, for example, we regard S as the sum of many small portions ΔS_i, $i = 1, 2, \cdots, n$. We multiply each element of area ΔS_i by the density σ at a point thereof and we define the potential V of S at a point P in space to be the limit (when it exists uniquely) of the sum of terms $\sigma \Delta S_i / D_i$, $i = 1, 2, \cdots, n$, where D_i denotes the distance between P and a point of ΔS_i:

$$V = V(P) = \iint\limits_S \frac{\sigma dS}{D}. \qquad \text{(IV-14.1)}$$

Since the reciprocated distance $1/D$, regarded as a function of the coordinates of P, satisfies Laplace's equation (as shown in Prob. IV-7) at all points of space except at the fixed point from which D is measured, we find by differentiation under the integral sign (partially with respect to the coordinates of P) that the potential V as given by Eq. (IV-14.1) satisfies Laplace's equation at all points P of space not on S. For we have at any P not on S

$$\nabla^2 V = \nabla^2 \left[\iint\limits_S \frac{\sigma dS}{D} \right]$$

$$= \iint\limits_S \sigma \nabla^2 \left(\frac{1}{D} \right) dS$$

$$= \iint\limits_S (\sigma)(\text{zero}) dS$$

$$= 0.$$

In the present problem we take the circular laminar (denoted now by A) in the xy-plane with center at the origin. The density of σ is constant: $\sigma = k$. Since the distribution of mass is symmetrical about the z-axis, it is clear that (in terms of spherical coordinates) the potential V is a function only of two coordinates r and ϕ, where r

denotes distance of P from origin and ϕ is the cone angle (colatitude) coordinate of P:

$$V = V(r, \phi).$$

From here on the problem is mathematically the same (with appropriate modifications) as Prob. IV-9, since V satisfies Laplace's equation at points P not on S. Accordingly, we shall try to get V as a series of the form

$$V = \sum_{n=0}^{\infty} \left[C_n r^n + \frac{B_n}{r^{n+1}} \right] P_n(\cos \phi). \qquad \text{(IV-14.2)}$$

If we can determine coefficients A_n and B_n, $n = 1, 2, \cdots$ so that V meets the conditions of our problem, then the series expansion (IV-14.2) will provide the desired solution. The question is: "What condition or conditions of the present problem will give us a hold on the coefficients to be used in Eq. (IV-14.2)?"

Perhaps the procedure of Prob. IV-9 will give us a clue. There we found that the solution turned out to be like Eq. (IV-14.2) above except that every B_n vanished while every term $C_n P_n(\cos \phi)$ was of the form $A_n (r/R)^n P_n(\cos \phi)$ where R was the radius of the solid. And we were able to determine A_n from a boundary condition which permitted a series expansion when $r = R$. But the particular situation $r = R$ does not seem capable of yielding any hold on the desired coefficients in the present problem. Let us, then, try specializing the *other* factor in each term of the expansion, namely, $P_n(\cos \phi)$. A moment's reflection indicates that when P is any point of the positive z-axis, then $\cos \phi = 1$, making every $P_n(\cos \phi) = 1$. And for a point P in the upper z-axis we should be able to compute the potential in Eq. (IV-14.1) as a function of the distance h of P from the origin 0.

For a point P in the positive z-axis at distance h from 0 we have by Eq. (IV-14.1), using rectangular coordinates,

$$V(0, 0, h) = k \iint_A \frac{dA}{\sqrt{h^2 + (x^2 + y^2)}}.$$

This integral is readily evaluated via an iterated integral in polar coordinates. The result, with M denoting mass of lamina, is

$$V(0, 0, h) = \frac{2M}{R^2}(\sqrt{h^2 + R^2} - h). \qquad \text{(IV-14.3)}$$

And now if we expand the right side of Eq. (IV-14.3) in powers of h we will have the right side of the desired Eq. (IV-14.2) for the case where $r = h$ and every $P_n(\cos \phi) = 1$. Thus, we can identify the coefficients of the powers of h in the expansion of the right side of Eq. (IV-14.3) with the coefficients C_n and B_n desired in Eq. (IV-14.2).

We may expand by the binomial formula the right side of Eq. (IV-14.3) in positive powers of h when $h < R$ and in negative powers of h when $h > R$ as follows. For $h < R$ we have

$$V(0, 0, h) = \frac{2M}{R^2}\left\{R\left[1 + \left(\frac{h}{R}\right)^2\right]^{1/2} - h\right\}$$

$$= \frac{2M}{R^2}\left[R - h + \frac{1}{2}\frac{h^2}{R} - \frac{1}{2^2}\frac{1}{2!}\frac{h^4}{R^3} + \frac{1 \cdot 1 \cdot 3}{2^3 3!}\frac{h^6}{R^5} - \cdots\right],$$

$$h < R. \quad \text{(IV-14.4)}$$

Thus, in Eq. (IV-14.2) when $r < R$ we have every $B_n = 0$ because negative powers of h (identified as r) do not occur in Eq. (IV-14.4), which is what Eq. (IV-14.2) becomes (with $r = h$) when $\cos \phi = 1$ so that every $P_n(\cos \phi) = 1$. Identifying the C_n in Eq. (IV-14.2) with the corresponding coefficients of powers of h in Eq. (IV-14.4), we have

$$V = V(r, \phi) = \frac{2M}{R}\left[P_0(\cos \phi) - \left(\frac{r}{R}\right)P_1(\cos \phi) + \frac{1}{2}\left(\frac{r}{R}\right)^2 P_2(\cos \phi)\right.$$

$$- \frac{1 \cdot 1}{2^2 2!}\left(\frac{r}{R}\right)^4 P_4(\cos \phi) + \frac{1 \cdot 1 \cdot 3}{2^3 3!}\left(\frac{r}{R}\right)^6 P_6(\cos \phi)$$

$$\left. - \cdots\right], \qquad r < R, \quad z > 0. \qquad \text{(IV-14.5)}$$

Similarly, when $h > R$ we have

$$V(0, 0, h) = \frac{2M}{R^2} \left\{ h \left[1 + \left(\frac{R}{h} \right)^2 \right]^{1/2} - h \right\}$$

$$= \frac{2M}{R^2} \left[\frac{1}{2h} R^2 - \frac{1 \cdot 1}{2^2 2!} \frac{R^4}{h^3} + \frac{1 \cdot 1 \cdot 3}{2^3 3!} \frac{R^6}{h^5} - \cdots \right], \quad h > R,$$

$$(\text{IV-14.6})$$

so that now every $C_n = 0$ in Eq. (IV-14.2), since positive powers of h (identified as r) do not occur in Eq. (IV-14.6). Then we get

$$V = V(r, \phi)$$

$$= \frac{2M}{R} \left[\frac{1}{2} \left(\frac{R}{r} \right) P_0(\cos \phi) - \frac{1 \cdot 1}{2^2 2!} \left(\frac{R}{r} \right)^3 P_2(\cos \phi) \right.$$

$$\left. + \frac{1 \cdot 1 \cdot 3}{2^3 3!} \left(\frac{R}{r} \right)^5 P_4(\cos \phi) - \cdots \right], \quad r > R, \quad z > 0.$$

$$(\text{IV-14.7})$$

Equations (IV-14.5) and (IV-14.7) provide the solution of our problem for all points having $z > 0$ except those for which $r = R$. For a point on the lower half of the z-axis, where h is *negative* (with $h \neq R$), $\sqrt{h^2} = -h$, so that all terms in Eqs. (IV-14.4) and (IV-14.6) remain the same with one exception: the second term in the brackets in Eq. (IV-14.4) becomes h in place of $-h$. Accordingly, we may say that Eq. (IV-14.7) gives the potential V at all points P having $r > R$ with $z \neq 0$, while Eq. (IV-14.5) with the term $- (r/R)P_1(\cos \phi)$ replaced by $(r/R)P_1(\cos \phi)$ gives V at all points having $r < R$ when z is negative.

When $h = R$, neither of the expansions employed in Eqs. (IV-14.4) and (IV-14.6) is possible. Nor are they available when $h = 0$. If it is required to know V at a specified point where $z = 0$ or where $r = R$, the value of V for such a point may be obtained (by approximate integration, if need be) directly from the integral

$$V = k \iint_A \frac{dA}{D}.$$

Problem: Potential of an Electric Charge Distribution

IV-15.   Distribution of electric charge on the surface T of an insulated conductor which lies wholly interior to the spherical surface $S: r = R$ is symmetric about the z-axis and is such that on S the potential V of the charge distribution equals $C(1 + \cos^4 \phi)$ where C is a constant and ϕ is the cone angle (co-latitude) spherical coordinate. Obtain formula for V valid at all points exterior to or on S.   ⋆

As already indicated, the potential V is a function only of r and ϕ: $V = V(r, \phi)$. As in Eq. (IV-14.1) in Prob. IV-14, the potential is defined as the integral over T of σ/D, where σ now denotes density (assumed to be continuous on T) of charge per unit area on T. Then as in Prob. IV-14 one finds that V satisfies Laplace's equation at all points exterior to T. Accordingly, we look for a formula for V of the form given in Eq. (IV-14.2) in Prob. IV-14. However, we must take every $C_n = 0$. In the region consisting of the exterior of S, the formula for V as given by Eq. (IV-14.2) in Prob. IV-14 cannot contain any terms in positive powers of r. Any such term or sum of such terms would require $V(r, \phi)$ to become infinite in magnitude as $r \to \infty$. For, if V be expressed as an integral over T, the maximum of the integrand σ/D approaches the limit zero as $r \to \infty$.

So, we seek a series solution of the form

$$V(r, \phi) = \sum_{n=0}^{\infty} \frac{B_n}{r^{n+1}} P_n(\cos \phi) \qquad (IV-15.1)$$

uniformly convergent for $r \geqq R$ and equal to $C(1 + \cos^4 \phi)$ on S where $r = R$. Letting $u = \cos \phi$, we thus require

$$C(1 + u^4) = \sum_{n=0}^{\infty} \frac{B_n}{R^{n+1}} P_n(u).$$

By Prob. III-18 this requirement is met by having

$$\frac{B_n}{R^{n+1}} = \frac{2n+1}{2} \int_{-1}^{1} C(1 + u^4)P_n(u)du,$$

$$B_n = \frac{(2n+1)CR^{n+1}}{2}\left[\int_{-1}^{1} P_n(u)du + \int_{-1}^{1} u^4 P_n(u)du\right].$$

As in Prob. IV-10, the first integral vanishes for all $n > 0$ by Eq. (III-16.5) while the second integral vanishes by Eq. (III-20.5) for all $n > 4$ as well as for $n = 1$ and $n = 3$. Computing B_0 and B_2 and B_4, taking the Legendre polynomials involved from Table III-1, we find that

$$B_0 = \frac{6CR}{5}, \quad B_2 = + \frac{4CR^3}{7}, \quad B_4 = \frac{8CR^5}{35}.$$

Thus, the solution sought for in Eq. (IV-15.1) reduces to a closed form:

$$V = V(r, \phi)$$
$$= C\left[\frac{6}{5}\left(\frac{R}{r}\right)P_0(\cos\phi) + \frac{4}{7}\left(\frac{R}{r}\right)^3 P_2(\cos\phi) + \frac{8}{35}\left(\frac{R}{r}\right)^5 P_4(\cos\phi)\right].$$

Problems: Specific Series Expansion

IV-16. Show that

$$\frac{1 - h^2}{(1 - 2xh + h^2)^{3/2}} = \sum_{n=0}^{\infty} (2n+1)P_n(x)h^n, \quad |x| \leq 1, \quad |h| < 1, \quad\text{(IV-16.1)}$$

where $P_n(x)$ denotes the Legendre polynomial of degree n. ⋆

Apparently the expansion to be established will follow from appropriate manipulation of the characteristic expansion (III-1.7).

By the properties of the expansion established in Prob. III-1 we may differentiate Eq. (III-1.7) termwise, obtaining

$$\frac{x - h}{(1 - 2xh + h^2)^{3/2}} = \sum_{n=0}^{\infty} nP_n(x)h^{n-1}, \qquad |x| \leqq 1, \quad |h| < 1.$$
$$(\text{IV-16.2})$$

And this Eq. (IV-16.2) will remain valid when both sides are multiplied by $2h$:

$$\frac{2xh - 2h^2}{(1 - 2xh + h^2)^{3/2}} = \sum_{n=0}^{\infty} 2nP_n(x)h^n, \qquad |x| \leqq 1, \quad |h| < 1.$$
$$(\text{IV-16.3})$$

Addition of Eq. (IV-16.3) to Eq. (III-1.7) yields the expansion to be demonstrated.

IV-17. Show that

$$\sum_{n=0}^{\infty} \frac{x^{n+1}P_n(x)}{n + 1} = \frac{1}{2} \log_e \left[\frac{1 + x}{1 - x} \right], \qquad |x| < 1, \quad (\text{IV-17.1})$$

where $P_n(x)$ is the Legendre polynomial of degree n.   ★

We start with the expansion

$$\sum_{n=0}^{\infty} P_n(x)h^n = (1 - 2xh + h^2)^{-(1/2)} \qquad (\text{IV-17.2})$$

established in Prob. III-1. If we choose an x such that $|x| < 1$ and hold it fast, then we may integrate termwise the series on the left in Eq. (IV-17.2) from $h = 0$ to an arbitrary h such that $|h| < 1$ and the series of such integrals will converge to the integral over the same interval of the expression on the right in Eq. (IV-17.2). Doing so, we get

$$\sum_{n=0}^{\infty} \frac{P_n(x)h^{n+1}}{n+1} = \int_0^h \frac{dh}{\sqrt{h^2 - 2xh + 1}}$$

$$= \int_0^h \frac{dh}{\sqrt{(h-x)^2 + (1-x^2)}}$$

$$= \log_e [(h-x) + \sqrt{(h-x)^2 + (1-x^2)}]$$
$$- \log_e (1-x)$$

$$= \log_e \frac{(h-x) + \sqrt{h^2 - 2xh + 1}}{1-x}. \qquad \text{(IV-17.3)}$$

Equation (IV-17.3) holds for any x such that $|x| < 1$ and for any h with $|h| < 1$. In particular Eq. (IV-17.3) is valid when h and x are any two numbers which are *equal* to each other and less in absolute value than unity.

Thus when $h = x$ with $|x| < 1$, we have

$$\sum_{n=0}^{\infty} \frac{x^{n+1} P_n(x)}{n+1} = \log_e \frac{\sqrt{1-x^2}}{1-x}$$

$$= \log_e \frac{\sqrt{1+x}\sqrt{1-x}}{1-x}$$

$$= \frac{1}{2} \log_e \left[\frac{1+x}{1-x} \right]. \qquad \text{(IV-17.4)}$$

The series on the left in Eq. (IV-17.4) converges absolutely and uniformly for $|x| < 1$ since on this interval $|P_n(x)| \leqq 1$ by Prob. III-12. Consequently, this series may be rearranged so as to be exhibited as a power series

$$a_0 + a_1 x + a_2 x^2 + a_3 x^3 + \cdots$$

convergent for $|x| < 1$. By Eq. (IV-17.4) this power series converges for $|x| < 1$ to the function on the right in Eq. (IV-17.3) and therefore is identical with the Maclaurin series of this function for $|x| < 1$. But it is known (as shown in calculus texts) that the Maclaurin series expansion of this function converges thereto for $|x| < 1$. It follows that Eq. (IV-17.4) is valid for $|x| < 1$, as was to be shown.

Problem: Infinite Product Expression for $\pi/2$

IV-18. Show that $\pi/2$ can be expressed as an infinite product of definite integrals involving the Legendre polynomials as follows:

$$\prod_{n=1}^{\infty} \int_{-1}^{1} 2nxP_n(x)P_{n-1}(x)dx = \frac{\pi}{2}, \qquad \text{(IV-18.1)}$$

where $P_n(x)$ denotes the Legendre polynomial of degree n.   ★

It is clear that the value of each integral in the infinite product will be a rational number, since each integrand is a polynomial in x with rational coefficients, as follows at once from the definition of Legendre polynomials given in the introduction to Chapter III. Thus Eq. (IV-18.1) expresses the irrational number $\pi/2$ in terms of rationals. We should not be surprised if the infinite product in Eq. (IV-18.1), once the integrals are evaluated as rational numbers, turns out to be similar (perhaps the same as) Wallis's product for $\pi/2$ in Prob. II-28.

Apart from the factor $2n$, which is a constant as far as the integration is concerned, each integrand $xP_n(x)P_{n-1}(x)$ suggests that perhaps a transformation of itself whereby the factor x is eliminated will allow us to evaluate the resulting integral (or integrals). The transformation we seek is at hand in the recurrence formula Eq. (III-2.5) whereby, taking $m - 1 = n$, we have

$$xP_n(x) = \frac{n+1}{2n+1}P_{n+1}(x) + \frac{n}{2n+1}P_{n-1}(x).$$

Thus the nth factor in the product on the left in Eq. (IV-18.1) becomes the sum of two integrals:

$$2n\left[\frac{n+1}{2n+1}\int_{-1}^{1}P_{n+1}(x)P_{n-1}(x)dx + \frac{n}{2n+1}\int_{-1}^{1}\left[P_{n-1}(x)\right]^{2}dx\right].$$

As we had anticipated, we can readily evaluate these two integrals: the first one vanishes by the orthogonality property established in Prob. III-16 while the value of the second one is $2/[2(n-1)+1]$ $= 2/(2n-1)$ by Prob. III-17. Accordingly, the infinite product in Eq. (IV-18.1) is

$$\prod_{n=1}^{\infty}\frac{4n^{2}}{4n^{2}-1}.$$

This is none other than Wallis's product, which we proved in Prob. 11-28 to be convergent to $\pi/2$. Eq. (IV-18.1) is thus established.

Problems: Application of Gauss's Mechanical Quadrature Formula with Pertinent Table

IV-19. Using Table IV-1 and taking $n = 6$, compute approximately $\log_{e} 7$ by Gauss's mechanical quadrature formula

$$\int_{p}^{q}f(x)dx \cong \frac{q-p}{2}\sum_{i=1}^{n}a_{i}f\left(\frac{q-p}{2}x_{i}+\frac{q+p}{2}\right),$$

where $x_{i} = $ the ith zero of the Legendre polynomial $P_{n}(x)$ and

$$a_{i} = \frac{1}{P'_{n}(x)}\int_{-1}^{1}\frac{P_{n}(x)}{x-x_{i}}dx. \quad \star$$

Before carrying out the computations involved in this problem let us observe (see, for instance, Hobson, *The Theory of Spherical and Ellipsoidal Harmonics*, Cambridge, 1931; see also Lowan, Davids, and Levenson, *Table of the Zeros of the Legendre Polynomials of Order 1–16 and the Weight Coefficients for Gauss's Mechanical Quadrature Formula*, Bulletin of the American Mathematical Society, Vol. 48, 1942) that Gauss's method of mechanical quadrature for obtaining an approximate value for a definite integral is advantageous in that it requires half the number of ordinate computations as required by most methods of approximate integration to get the same degree of closeness to the actual value of the integral.

We write

$$\log_e 7 = \int_1^x \frac{1}{x} dx.$$

Thus we have $p = 1$, $q = 7$, $f(x) = \dfrac{1}{x}$, so that

$$\log_e 7 \cong \frac{7-1}{2} \sum_{i=1}^{6} \frac{a_i}{3x_i + 4}$$

$$\cong 3 \left[\frac{a_1}{3x_1 + 4} + \frac{a_2}{3x_2 + 4} + \cdots + \frac{a_6}{3x_6 + 4} \right]$$

$$\cong 3 \left[\frac{.46791}{4.71586} + \frac{.36076}{5.98363} + \frac{.17132}{6.79741} \right.$$

$$\left. + \frac{.46791}{3.28414} + \frac{.36076}{2.01637} + \frac{.17132}{1.20259} \right]$$

$$\cong 1.9457.$$

REMARK.   Tables of natural logarithms give $\log_e 7$ as 1.9459 to four decimal places. Our approximation is in error by a little more than a hundredth of 1 per cent.

IV-20. Compute an approximate value for π by application of Gauss's mechanical quadrature formula (see Prob. IV-19) to the integral on the right side of

$$\frac{\pi}{4} = \int_0^1 \frac{dx}{1 + x^2}. \qquad \star$$

Using Table IV-1 as in Prob. IV-19, we take $n = 5$. We have

$$p = 0, \quad q = 1, \quad f(x) = 1/(1 + x^2), \quad f\left(\frac{q - p}{2} x_i + \frac{q + p}{2}\right)$$

$$= 1 \bigg/ \left[1 + \left(\frac{1}{2}x_i + \frac{1}{2}\right)^2\right], \text{ so that}$$

$$\frac{\pi}{4} \cong \frac{1}{2} \sum_{i=1}^{5} \frac{a_i}{1 + \left(\frac{x_i}{2} + \frac{1}{2}\right)^2}$$

$$\pi \cong 8 \sum_{i=1}^{5} \frac{a_i}{4 + (x_i + 1)^2}$$

$$\cong 8 \left[\frac{.56888}{5} + \frac{.47863}{6.3670} + \frac{.23693}{7.6333}\right.$$

$$\left. + \frac{.47863}{4.2133} + \frac{.23693}{4.0088}\right]$$

$$\cong 3.1416.$$

TABLE IV-1*

Zeros of the Legendre polynomials $P_n(x)$, $n = 2, 3, 4, 5, 6$ and the corresponding weight coefficients a_i for Gauss's mechanical quadrature formula

$$x_i = i^{\text{th}} \text{ zero of } P_n(x);$$

$$a_i = \frac{1}{P_n'(x_i)} \int_{-1}^{1} \frac{P_n(x)}{x - x_i} dx.$$

The x_i and a_i are numbered so that

(1) when n is even: $x_{i+n/2} = -x_i$, $a_{i+n/2} = a_i$;

(2) when n is odd: $x_{i+(n-1)/2} = -x_i$, $a_{i+(n-1)/2} = a_i,$ i $> 1.$

$n = 2$			$n = 5$	
$x_1 \cong .57735$	$a_1 = 1$		$x_1 = 0$	$a_1 \cong .56888$
$x_2 = -x_1$	$a_2 = a_1$		$x_2 \cong .53847$	$a_2 \cong .47863$
			$x_3 \cong .90618$	$a_3 \cong .23693$
$n = 3$			$x_4 = -x_2$	$a_4 = a_2$
$x_1 = 0$	$a_1 \cong .88888$		$x_5 = -x_3$	$a_5 = a_3$
$x_2 \cong .77460$	$a_2 \cong .55555$			
$x_3 = -x_2$	$a_3 = a_2$		$n = 6$	
			$x_1 \cong .23862$	$a_1 \cong .46791$
$n = 4$			$x_2 \cong .66121$	$a_2 \cong .36076$
$x_1 \cong .33998$	$a_1 \cong .65215$		$x_3 \cong .93247$	$a_3 \cong .17132$
$x_2 \cong .86114$	$a_2 \cong .34785$		$x_4 = -x_1$	$a_4 = a_1$
$x_3 = -x_1$	$a_3 = a_1$		$x_5 = -x_2$	$a_5 = a_2$
$x_4 = -x_2$	$a_4 = a_2$		$x_6 = -x_3$	$a_6 = a_3$

* Taken (with slight change in numbering of the x_i and a_i) from Lowan, Davids, and Levenson, "Table of the Zeros of the Legendre Polynomials of Order 1–16, and the Weight Coefficients for Gauss's Mechanical Quadrature," *Bulletin*, Amer. Math. Soc., Vol. 48, 1942; also found reprinted in *Tables of Functions and Zeros of Functions* National Bureau of Standards, Applied Math. Series 37.

5 BESSEL FUNCTIONS

INTRODUCTION

Bessel's differential equation is

$$x^2 y'' + xy' + (x^2 - p^2)y = 0, \qquad \text{(V-0.1)}$$

where y' and y'' denote respectively the derivatives dy/dx and d^2y/dx^2 and where p is a constant. Bessel's equation is a particular case of the equation

$$(1 + Ax^H)y'' + \frac{1}{x}(B + Cx^H)y' + \frac{1}{x^2}(D + Ex^H)y = 0,$$

being the case thereof in which

$$A = 0, \quad B = 1, \quad C = 0, \quad D = -p^2, \quad E = 1, \quad H = 2.$$

Although Bessel's equation (V-0.1) is a differential equation of second order as far as derivatives are concerned, it is often referred to as Bessel's equation of order p, the designation "order" referring to the value of the parameter p, not to the highest order derivative in the equation. Equation (V-0.1) actually denotes a family of equations, there being an individual member of the family for each value, real or complex, of the parameter. In this book we shall not be concerned with any Bessel equation of complex order, that is, a

Bessel equation which p is a complex number $a + bi$, a and b being real and i denoting $\sqrt{-1}$. We observe that the Bessel equation of order $-p$ is the same as the Bessel equation of order p.

A solution $y = F_1(x)$ of Bessel's equation (V-0.1) which is continuous for all values of x is called a Bessel function of the first kind. A solution $y = F_2(x)$ which is continuous for all values of x except $x = 0$ and which becomes infinite in absolute value as $x \to 0$ is known as a Bessel function of the second kind. Since a Bessel function of the first kind and one of the second kind cannot satisfy identically (for all x except $x = 0$) a linear equation

$$AF_1(x) + BF_2(x) = 0$$

where A and B are constants, the general solution of Bessel's equation is

$$y = C_1 F_1(x) + C_2 F_2(x),$$

where F_1 is a Bessel function of the first kind and F_2 is a Bessel function of the second kind and where C_1 and C_2 are arbitrary constants.

Solutions of Bessel's equation (V-0.1) are usually expressed in terms of one or two of four standardized *Bessel functions*

$$J_p(x), \quad Y_p(x), \quad I_p(x), \quad K_p(x).$$

The standardized function $J_p(x)$ is defined for all real orders p as follows:

$$J_p(x) = \sum_{k=0}^{\infty} \frac{(-1)^k x^{p+2k}}{2^{p+2k} k! \Gamma(p + k + 1)}, \quad p \neq -1, -2, -3, \cdots$$

$$\text{(V-0.2)}$$

$$J_{-n}(x) = (-1)^n J_n(x), \qquad n = 1, 2, 3, \cdots, \qquad \text{(V-0.3)}$$

where $\Gamma(p + k + 1)$ is the Gamma function (see Chap. I). Bessel functions $J_p(x)$ of order $p = 0, 1, 2, 3, \cdots$ are of considerable importance in applications, especially $J_0(x)$ and $J_1(x)$. From (V-0.1) we have, by virtue of Prob. I-10,

$$J_0(x) = 1 - \frac{x^2}{2^2} + \frac{x^4}{2^4(2!)^2} - \frac{x^6}{2^6(3!)^2} + \cdots, \qquad \text{(V-0.4)}$$

$$J_1(x) = \frac{x}{2} - \frac{x^3}{2^3 2!} + \frac{x^5}{2^5 2! 3!} - \frac{x^7}{2^7 3! 4!} + \cdots. \qquad \text{(V-0.5)}$$

The standardized function $Y_p(x)$ is defined as follows:

$$Y_p(x) = \frac{1}{\sin p\pi}\Big[(\cos p\pi)J_p(x) - J_{-p}(x)\Big], \qquad p \text{ nonintegral,}$$
$$\text{(V-0.6)}$$

$$Y_n(x) = \frac{1}{\pi}\Bigg\{2\Big[\gamma + \log_e \frac{x}{2}\Big]J_n(x) - \sum_{k=0}^{n-1}\frac{(n-k-1)!\,x^{-n+2k}}{2^{-n+2k}k!}$$

$$- \sum_{k=0}^{\infty}\frac{(-1)^k[\phi(n+k) + \phi(k)]x^{n+2k}}{2^{n+2k}k!(n+k)!}\Bigg\},$$
$$n = 1, 2, 3, \cdots, \quad \text{(V-0.7)}$$

where γ denotes Euler's constant (see Prob. I-32) and where $\phi(0) = 0$ and $\phi(k) = 1 + \frac{1}{2} + \frac{1}{3} + \cdots + \frac{1}{k}$, $k = 1, 2, 3, \cdots$.

For $Y_0(x)$ the finite summation in Eq. (V-0.7) is omitted. We remark here that $\phi(k)$ as defined here is related to the value, for positive integers, of the logarithmic derivative of the Gamma function, namely $\psi(x) = \Gamma'(x)/\Gamma(x)$ considered in Probs. I-32 and I-33. The relationship is this: $\phi(k) = \psi(k+1) + \gamma$.

For the general solution of Bessel's equation (V-0.1) we will take, as is customary,

$$y = Z_p(x) = \begin{cases} C_1 J_p(x) + C_2 J_{-p}(x), & p \text{ nonintegral} \\ C_1 J_n(x) + C_2 Y_n(x), & p = n = 0, 1, 2, 3, \cdots \end{cases}$$
$$\text{(V-0.8)}$$

where C_1 and C_2 are arbitrary constants.

In applied mathematics there are problems which involve differential equations having solutions in terms of Bessel functions of pure imaginary argument. Real solutions are then expressed in terms of modified functions. The standardized *modified Bessel functions* are denoted by $I_p(x)$ and $K_p(x)$. They are defined as follows, x being real and i denoting $\sqrt{-1}$:

$$I_p(x) = i^{-p} J_p(ix), \tag{V-0.9}$$

$$K_p(x) = \frac{\pi}{2 \sin p\pi} \left[I_{-p}(x) - I_p(x) \right], \qquad p \text{ nonintegral}, \tag{V-0.10}$$

$$K_n(x) = \frac{\pi}{2} i^{n+1} \left[J_n(ix) + i Y_n(ix) \right], \qquad n = 0, 1, 2, 3, \cdots. \tag{V-0.11}$$

When the solution of a differential equation can be expressed in terms of Bessel functions of pure imaginary argument ix, we will take as general solution

$$y = Z_p(ix) = \begin{cases} C_1 I_p(x) + C_2 I_{-p}(x), & p \text{ nonintegral} \\ C_1 I_n(x) + C_2 K_n(x), & p = n = 0, 1, 2, 3, \cdots. \end{cases} \tag{V-0.12}$$

There are also problems which involve differential equations having solutions in terms of Bessel functions of argument $i^{3/2}x$, where x is real and $i = \sqrt{-1}$. The standardized Bessel functions of such argument are $J_p(i^{3/2}x)$ and $i^{-p}K_p(i^{3/2}x)$. Since these functions have complex values for real x, they are often indicated as follows:

$$J_p(i^{3/2}x) = \operatorname{ber}_p x + i \operatorname{bei}_p x = i^p I_p(i^{1/2}x) \tag{V-0.13}$$

$$i^{-p}K_p(i^{1/2}x) = \operatorname{ker}_p x + i \operatorname{kei}_p x. \tag{V-0.14}$$

It is, however, customary to omit the subscript from these latter designations when the order p is zero and to write simply

$$J_0(i^{3/2}x) = \operatorname{ber} x + i \operatorname{bei} x, \tag{V-0.15}$$

$$K_0(i^{1/2}x) = \operatorname{ker} x + i \operatorname{kei} x. \tag{V-0.16}$$

The complex function ber $x + i$ bei x is often expressed in terms of its modulus and its amplitude:

$$\text{ber } x + i \text{ bei } x = M_0(x)e^{i\theta_0(x)} \tag{V-0.17}$$

where

$$M_0(x) = \left[(\text{ber } x)^2 + (\text{bei } x)^2 \right]^{1/2}, \qquad \theta_0 = \arc\tan \frac{\text{bei } x}{\text{ber } x}.$$

Similarly, one may write

$$\text{ber}_p x + i \text{ bei}_p x = M_p(x)e^{i\theta_p(x)}, \tag{V-0.18}$$

where

$$M_p(x) = \left[(\text{ber}_p x)^2 + (\text{bei}_p x)^2 \right]^{1/2}, \qquad \theta_p = \arc\tan \frac{\text{bei}_p x}{\text{ber}_p x}.$$

Another class of complex functions found to be convenient in certain problems is the class of functions known as the *Hankel functions* $H_p^{(1)}(x)$ and $H_p^{(2)}(x)$ defined as follows:

$$H_p^{(1)}(x) = J_p(x) + i Y_p(x), \tag{V-0.19}$$

$$H_p^{(2)}(x) = J_p(x) - i Y_p(x). \tag{V-0.20}$$

Comparison of (V-0.19) with (V-0.11) shows that the Bessel function $K_n(x)$ may be indicated alternatively as

$$K_n(x) = \frac{\pi}{2} i^{n+1} H_n^{(1)}(x), \qquad n = 0, 1, 2, 3, \cdots. \tag{V-0.21}$$

A class of functions known as *Struve functions* $\mathbf{H}_p(x)$ is the the class defined by

$$\mathbf{H}_p(x) = \sum_{k=0}^{\infty} \frac{(-1)^k x^{p+2k+1}}{\Gamma(k + \frac{3}{2})\Gamma(p + k + \frac{3}{2})}, \qquad p > -\frac{1}{2}. \tag{V-0.22}$$

$\mathbf{H}_p(x)$ is a particular solution of

$$x^2y'' + xy' + (x^2 - p^2)y = \frac{x^{p+1}}{2^{p-1}\Gamma(\tfrac{1}{2})\Gamma(p + \tfrac{1}{2})}, \quad \text{(V-0.23)}$$

whose general solution is

$$y = \begin{cases} C_1J_p(x) + C_2J_{-p}(x) + \mathbf{H}_p(x), & p \text{ nonintegral} \\[2mm] C_1J_n(x) + C_2Y_n(x) + \mathbf{H}_n(x), & p = n = 0, 1, 2, 3, \cdots. \end{cases} \quad \text{(V-0.24)}$$

Bessel functions are also called *cylindrical harmonics* because they furnish one ingredient for product solutions of problems in which it is required to find a function which is harmonic, that is, satisfies Laplace's equation and has continuous second-order partial derivatives, within a right circular cylinder and which takes on assigned boundary values on the surface of the cylinder.

The problems and exercises in this chapter are concerned with some of the properties and the mutual relations of the Bessel functions and with their relations to other functions, in particular their relations to trigonometric functions, exponential functions and Legendre polynomials.

At the end of the chapter is a selection of tables.

Problems: Differentiation Formulas

V-1. Show that

$$\frac{d}{dx}\left[x^pJ_p(x)\right] = x^pJ_{p-1}(x). \quad \star \quad \text{(V-1.1)}$$

When p is such that x^p is not real for x negative, we shall exclude negative values of x from consideration. When p is such that negative powers of x are contained in any of the series expansions involved, then $x = 0$ will naturally be excluded.

Case 1. p is not a negative integer. Then by Eq. (V-0.2) we have

$$x^pJ_{p-1}(x) = \sum_{k=0}^{\infty} \frac{(-1)^k(x^p)x^{p-1+2k}}{2^{p-1+2k}k!\,\Gamma(p + k)}. \quad \text{(V-1.2)}$$

Also by Eq. (V-0.2) we get

$$\frac{d}{dx}\left[x^p J_p(x)\right] = \sum_{k=0}^{\infty} \frac{(-1)^k(2p+2k)x^{2p+2k-1}}{2^{p+2k}k!\,\Gamma(p+k+1)}. \quad \text{(V-1.3)}$$

Termwise differentiation is valid in obtaining Eq. (V-1.3) since the resulting series converges uniformly on any chosen finite interval of positive values of x (on any chosen finite interval whatsoever in case negative values of x and $x = 0$ are admissible). By application of Eq. (I-4.1) to each term on the right in Eq. (V-1.3) we get

$$\frac{d}{dx}\left[x^p J_p(x)\right] = \sum_{k=0}^{\infty} \frac{(-1)^k(2p+2k)x^{2p+2k-1}}{2^{p+2k}k!\,(p+k)\Gamma(p+k)}. \quad \text{(V-1.4)}$$

The right side of Eq. (V-1.2) is seen to be the same as the right side of Eq. (V-1.4). Thus Eq. (V-1.1) is established for Case 1.

Case 2. $p = -n,\ n = 1, 2, 3, \cdots$. We have

$$\frac{d}{dx}\left[x^p J_p(x)\right] = \frac{d}{dx}\left[x^{-n} J_{-n}(x)\right]$$

$$= (-1)^n \frac{d}{dx}\left[x^{-n} J_n(x)\right] \qquad \text{by Eq. (V-0.3)}$$

$$= (-1)^n [-x^{-n} J_{n+1}(x)] \qquad \text{by Prob. V-3}$$

$$= (-1)^{n+1}[x^{-n}(-1)^{n+1} J_{-n-1}(x)] \qquad \text{by Eq. (V-0.3)}$$

$$= x^p J_{p-1}(x).$$

REMARK. It is legitimate here to apply Prob. V-3 with $p > 0$, since the demonstration of that case of Prob. V-3 does not depend on Prob. V-1.

V-2. Show that Eq. (V-1.1) holds also for Y_p with Y_{p-1}:

$$\frac{d}{dx}(x^p Y_p) = x^p Y_{p-1}. \qquad \star \qquad \text{(V-2.1)}$$

For p nonintegral we have by Eq. (V-0.6)

$$\frac{d}{dx}(x^p Y_p) = \frac{1}{\sin p\pi}\left[(\cos p\pi)\frac{d}{dx}(x^p J_p) - \frac{d}{dx}(x^p J_{-p})\right]$$

$$= \frac{1}{\sin p\pi}\left[(\cos p\pi)x^p J_{p-1} + x^p J_{1-p}\right] \qquad \begin{array}{l}\text{by Eqs. (V-1.1)}\\ \text{and (V-3.1)}\end{array}$$

$$= \frac{x^p\{[-\cos(p-1)\pi]J_{p-1} + J_{-(p-1)}\}}{-\sin(p-1)\pi}$$

$$= x^p Y_{p-1}.$$

Equation (V-2.1) may be established when $p = n = 1, 2, 3, \cdots$ directly from Eq. (V-0.7) in the manner used in Prob. V-1 to establish the identity for J_p, or by taking into account the fact that $Y_p(x)$ is a continuous function (for fixed $x \neq 0$) of the parameter p.

REMARK. In similar manner the identities established in the next several problems for the J-functions can be shown to hold also for the Y-functions.

V-3. Show that

$$\frac{d}{dx}\left[x^{-p}J_p(x)\right] = -x^{-p}J_{p+1}(x). \qquad \star \qquad \text{(V-3.1)}$$

Equation (V-3.1) is established in the same manner as Eq. (V-1.1) except for one item, which arises from the fact that the series for $x^{-p}J_p(x)$ begins with a constant term whose derivative is zero. Consequently, the series for the left side of Eq. (V-3.1) may be written as a summation beginning with $k = 1$ instead of $k = 0$. But the series for the right side of Eq. (V-3.1) is a summation beginning at $k = 0$. However, this latter summation becomes identical with the summation for the left side of Eq. (V-3.1) simply by using a new index of summation $k_1 = k + 1$, thus establishing Eq. (V-3.1).

V-4. Show that

$$\frac{d}{dx}\left[J_p(x)\right] = J_{p-1}(x) - \frac{p}{x}J_p(x). \qquad \star \qquad \text{(V-4.1)}$$

Carrying out the differentiation indicated on the left in Eq. (V-1.1), we find that Eq. (V-1.1) becomes

$$x^p \frac{d}{dx} J_p(x) + J_p(x) p x^{p-1} = x^p J_{p-1}(x). \qquad \text{(V-4.2)}$$

Solving Eq. (V-4.2) for the derivative of $J_p(x)$, we get Eq. (V-4.1).

V-5. Show that

$$\frac{dJ_p(x)}{dx} = \frac{p}{x} J_p(x) - J_{p+1}(x). \qquad \star \qquad \text{(V-5.1)}$$

Carrying out the differentiation indicated on the left in Eq. (V-3.1) and then solving the resulting equation for $dJ_p(x)/dx$, as was done in Prob. V-4, we get Eq. (V-5.1).

V-6. Show that

$$\frac{dJ_p(x)}{dx} = \frac{1}{2} \left[J_{p-1}(x) - J_{p+1}(x) \right]. \qquad \star \qquad \text{(V-6.1)}$$

Equation (V-6.1) is obtained by simply adding the identities (V-4.1) and (V-5.1).

Problem: Recursion Formulas

V-7. Show that

$$J_p(x) = \frac{x}{2p} \left[J_{p+1}(x) + J_{p-1}(x) \right], \qquad p \neq 0. \qquad \star \qquad \text{(V-7.1)}$$

Equation (V-7.1) follows at once from the identities (V-4.1) and (V-5.1) simply by subtracting the latter from the former and then transposing the term in $J_p(x)$.

REMARK. Identity (V-7.1) is the analogue in Bessel functions $J_p(x)$ of the identity (III-2.5) in Legendre polynomials.

Problems: Differentiation Formulas

V-8. Verify the first formula in Table V-1 for $q = J$, namely

$$\frac{d}{dx}\left[x^p J_p(ax) \right] = ax^p J_{p-1}(ax). \qquad \star \qquad \text{(V-8.1)}$$

The identity in question is evidently a slight generalization of the identity established in Prob. V-1. All that needs to be done is to let $u = ax$ and apply Eq. (V-1.1) written in terms of the letter u:

$$\begin{aligned}
\frac{d}{dx}\left[x^p J_p(ax) \right] &= \frac{d}{dx}\left[\left(\frac{u}{a}\right)^p J_p(u) \right] \\
&= \frac{d}{du}\left[\frac{u^p}{a^p} J_p(u) \right]\frac{du}{dx} \\
&= a^{-p}\frac{d}{du}\left[u^p J_p(u) \right] a \\
&= a^{1-p}\left[u^p J_{p-1}(u) \right] \qquad \text{by Eq. (V-1.1)} \\
&= a^{1-p}\left[(ax)^p J_{p-1}(ax) \right] \\
&= ax^p J_{p-1}(ax).
\end{aligned}$$

REMARK. Similar verifications may be made of each of the other formulas involving the argument ax in Table V-1. We may also write

$$\int ax^p J_{p-1}(ax)dx = x^p J_p(ax) + C.$$

V-9. Develop and establish formula for the nth derivative $d^n J_p(x)/dx^n$ as generalization of the formula for the first derivative $dJ_p(x)/dx$ established in Prob. V-6. $\qquad \star$

Let us first work out formula for the second derivative $d^2 J_p/dx^2$ and also for the third derivative $d^3 J_p/dx^3$. That will probably give us sufficient basis from which to surmise the general formula for the

n^{th} derivative and upon which we can construct demonstration to establish the general formula.

Starting with Eq. (V-6.1) and differentiating both sides thereof, we get

$$\frac{d^2J_p}{dx^2} = \frac{1}{2}\left[\frac{dJ_{p-1}}{dx} - \frac{dJ_{p+1}}{dx}\right]. \qquad (\text{V-9.1})$$

Now we apply Eq. (V-6.1) to each of the two derivatives on the right in Eq. (V-9.1):

$$\frac{d^2J_p}{dx^2} = \frac{1}{2}\left\{\frac{1}{2}(J_{p-2} - J_p) - \frac{1}{2}(J_p - J_{p+2})\right\}$$

$$= \frac{1}{2^2}(J_{p-2} - 2J_p + J_{p+2}). \qquad (\text{V-9.2})$$

It looks as if we can surmise the general formula for the nth derivative already from the formula for second derivative in Eq. (V-9.2). But let us strengthen (we trust) our surmise by going on to the third derivative. Differentiating Eq. (V-9.2) and applying Eq. (V-6.1) to the resulting derivatives on the right and collecting terms, we find that

$$\frac{d^3J_p}{dx^3} = \frac{1}{2^3}(J_{p-3} - 3J_{p-1} + 3J_{p+1} - J_{p+3}). \qquad (\text{V-9.3})$$

We surmise that

$$\frac{d^nJ_p}{dx^n} = \frac{1}{2^n}\Bigg[J_{p-n} - C(n, 1)J_{p-n+2} + C(n, 2)J_{p-n+4} - \cdots$$
$$+ (-1)^kC(n, k)J_{p-n+2k} + \cdots (-1)^nJ_{p+n}\Bigg],$$
$$(\text{V-9.4})$$

where $C(n, k)$ denotes the coefficient of the $(k + 1)^{\text{th}}$ term of the expansion of the nth power of a binomial. $C(n, k)$ is the number of combinations of n things taken k at a time:

$$C(n, k) = \frac{n!}{k!(n - k)!}. \qquad (\text{V-9.5})$$

The demonstration of the truth of Eq. (V-9.4) is essentially the same as that of the binomial expansion theorem to be found in any good algebra text. Let us sketch it, however, adding the extra arguments pertinent to the present situation.

We have Eq. (V-9.4) now established for $n = 1, 2, 3$. Assume it true for a positive integer n and differentiate it. It is at once apparent that, where every derivative dJ_j/dx on the right is replaced by $(\frac{1}{2})(J_{j-1} - J_{j+1})$, then the $(n + 1)^{\text{th}}$ derivative $d^{n+1}J_p/dx^{n+1}$ will have the coefficient $(\frac{1}{2})^{n+1}$. Moreover, one sees (as in the proof of the binomial theorem in algebra texts) that the $(k + 1)^{\text{th}}$ term in $d^{n+1}J_p/dx^{n+1}$ comes from the kth and the $(k + 1)^{\text{th}}$ terms of d^nJ_p/dx^n. From them, apart from the factor $\frac{1}{2}$, we obtain, by Eq. (V-6.1), as their contribution to $d^{n+1}J_p/dx^{n+1}$,

$$(-1)^k C(n, k - 1)\left[J_{p-n+2k-3} - J_{p-n+2k-1} \right]$$
$$+ (-1)^k C(n, k)\left[J_{p-n+2k-1} - J_{p-n+2k+1} \right].$$

The portion of this contribution which makes up the $(k + 1)^{\text{th}}$ term in $d^{n+1}J_p/dx^{n+1}$ is

$$\left[(-1)^{k-1}(-1)C(n, k - 1) + (-1)^k C(n, k) \right] J_{p-n+2k-1},$$

which combines by application of Eq. (V-9.5) to

$$(-1)^k C(n + 1, k) J_{p-(n+1)+2k}.$$

This last expression is the same as that occurring for the $(k + 1)^{\text{th}}$ term in Eq. (V-9.4) with n replaced by $n + 1$. Thus, whenever Eq. (V-9.5) holds for a positive integer n, it holds for $n + 1$. This, combined with the fact that it holds for $n = 1, 2, 3$ validates Eq. (V-9.4) for derivatives of all orders.

V-10. Show that

$$\frac{d}{dx}(x \operatorname{ber}' x) = - x \operatorname{bei} x, \tag{V-10.1}$$

$$\frac{d}{dx}(x \operatorname{bei}' x) = x \operatorname{ber} x, \tag{V-10.2}$$

where, as given in the introduction to this chapter, ber x and bei x are defined by

$$J_0(i^{3/2}x) = \text{ber } x + i \text{ bei } x, \qquad i = \sqrt{-1}, \qquad \text{(V-10.3)}$$

and the primes on ber and on bei mean derivative with respect to x. Also given is a particular solution of

$$xy'' + y' - ixy = 0, \qquad \text{(V-10.4)}$$

namely

$$y = \text{ber } x + i \text{ bei } x. \qquad * \qquad \text{(V-10.5)}$$

If we observe that Eq. (V-10.4) may be written as

$$\frac{d}{dx}(xy') = ixy,$$

then, by Eq. (V-10.5), we have

$$\frac{d}{dx}\left[x(\text{ber}'x + i \text{ bei}'x) \right] = ix(\text{ber } x + i \text{ bei } x), \qquad \text{(V-10.6)}$$

that is,

$$\frac{d}{dx}(x \text{ ber}'x) + i \frac{d}{dx}(x \text{ bei}'x) = -x \text{ bei } x + ix \text{ ber } x.$$
$$\text{(V-10.7)}$$

And now the equations to be demonstrated, namely, Eqs. (V-10.1) and (V-10.2), follow at once from Eq. (V-10.7) by separation of reals and imaginaries.

REMARK. Equation (V-10.4) is solved in Prob. VI-6.

Problem: Evaluation of Integrals Involving Bessel Functions

V-11. Evaluate

$$\int_0^a x\left[(\text{ber}'x)^2 + (\text{bei}'x)^2 \right] dx. \qquad *$$

We write this integral as the sum of two integrals

$$\int_0^a x(\text{ber}'x)^2 dx + \int_0^a x(\text{bei}'x)^2 dx.$$

Each of these two integrals is readily evaluated via integration by parts, the procedure for the second being quite the same as for the first. So, we shall present the details only for the first integral, which we will denote by R_1. We write

$$R_1 = \int_0^a x\,\text{ber}'x \cdot \text{ber}'x\,dx$$

and integrate by parts, taking $u = x\,\text{ber}'x$ and $dv = \text{ber}'x\,dx$. We get

$$R_1 = \left[x\,\text{ber}'x\,\text{ber}\,x \right]_{x=0}^{x=a} - \int_0^a \text{ber}\,x \cdot \frac{d}{dx}(x\,\text{ber}'x)dx$$

$$= a\,\text{ber}'a\,\text{ber}\,a - \int_0^a (\text{ber}\,x)(-x\,\text{bei}\,x)dx$$

by Eq. (V-10.1). Thus,

$$R_1 = a\,\text{ber}\,a\,\text{ber}'a + \int_0^a x\,\text{ber}\,x\,\text{bei}\,x\,dx. \qquad \text{(V-11.1)}$$

In exactly the very same manner we find via Eq. (V-10.2) in Prob. V-10 that the second integral

$$R_2 = a\,\text{bei}\,a\,\text{bei}'a - \int_0^a x\,\text{ber}\,x\,\text{bei}\,x\,dx. \qquad \text{(V-11.2)}$$

Addition of Eqs. (V-11.1) and (V-11.2) yields the value requested:

$$\int_0^a x\left[(\text{ber}'x)^2 + (\text{bei}'x)^2 \right]dx = a(\text{ber}\,a\,\text{ber}'a + \text{bei}\,a\,\text{bei}'a).$$

$$\text{(V-11.3)}$$

Problem: Differentiation Formulas

V-12. Show that, for each real number x, the modulus of $(d/dx)[J_0(i^{3/2}x)]$ is equal to the modulus of $J_1(i^{3/2}x)$.   ★

By formula (C) in Table V-1 we have

$$\frac{d}{dx} J_0(i^{3/2}x) = - i^{3/2} J_1(i^{3/2}x). \qquad \text{(V-12.1)}$$

The factor $- i^{3/2}$ on the right in Eq. (V-12.1) has modulus unity. Thus, the modulus of the function on the right in Eq. (V-12.1) is the same as that of $J_1(i^{3/2}x)$. Consequently,

$$\left| \frac{d}{dx} J_0(i^{3/2}x) \right| = \left| J_1(i^{3/2}x) \right|, \qquad \text{(V-12.2)}$$

as was to be demonstrated.

The equality expressed in Eq. (V-12.2) can be written in another form, by Eqs. (V-0.13) and (V-0.15), as follows:

$$|\text{ber}'x + i\,\text{bei}'x| = |\text{ber}_1 x + i\,\text{bei}_1 x| = M_1(x). \quad \text{(V-12.3)}$$

REMARK. Although $J_0'(i^{3/2}x)$ and $J_1(i^{3/2}x)$ have the same modulus for each real x, they do not have the same amplitude by virtue of the rotating factor $- i^{3/2}$ on the right in Eq. (V-12.1).

Problems: Specific Evaluations

V-13. Evaluate $J_1(.1)$ correct to four decimal places via Eq. (V-0.5), namely

$$J_1(x) = \frac{x}{2} - \frac{x^3}{2^2 2!} + \frac{x^5}{2^5 2! 3!} - \frac{x^7}{2^7 3! 4!} + \cdots . \qquad \star$$

$$\text{(V-13.1)}$$

The terms in the series in Eq. (V-13.1) alternate in sign. For any chosen x they are such that eventually (from some term on) each term is less in magnitude than its predecessor. And the limit of the nth term is zero as $n \to \infty$. So, the error committed by taking a partial sum of the series as approximate value for $J_1(x)$ is less in amount than the size of the first term not included in the partial sum. When we take $x = .1$, the third term of the series is less in size than 10^{-7}. Consequently, the partial sum made up of only the first two terms will give $J_1(.1)$ correct to six decimal places. And we have

$$J_1(.1) \cong \frac{.1}{2} - \frac{.001}{16}$$

$$\cong .0499375.$$

Since the third term is positive, the value of $J_1(.1)$ is slightly more than the approximate value just indicated. Thus, we may conclude that, to six decimal places,

$$J_1(.1) = .049938.$$

V-14. Compute $Y_3(.8)$ to four decimal places via Eq. (V-0.7). ⋆

The term preceding the two summations in the formula for $Y_n(x)$ as given by Eq. (V-0.7) can be computed to as many significant figures as are available for γ and natural logarithms and $J_n(x)$. If we take $J_3(.8)$ from our four-place Table V-3, our accuracy in the computation of this part of $Y_3(.8)$ is limited to four significant figures. But, as we shall see, only two will be needed.

On the other hand, the coefficients in the finite summation in the formula for $Y_n(x)$ are all rational numbers, so that the contribution of this summation to $Y_3(.8)$ can be computed with any desired degree of accuracy whatsoever.

The infinite summation in the formula for $Y_n(x)$ is an alternating series. When written for $Y_3(.8)$ the terms decrease in size immediately from the first term. Consequently, the error committed in using a partial sum of this series will be less in magnitude than the size of the first term not retained. It looks as if it will suffice to retain only the first two terms of this series. Let us check this hasty judgement by an appraisal of the size of the third term. The third term (term where $k = 2$) in this series when $n = 3$ and $x = .8$ is

$$\frac{(-1)^2[(1 + \tfrac{1}{2} + \tfrac{1}{3} + \tfrac{1}{4} + \tfrac{1}{5}) + (1 + \tfrac{1}{2})](.8)^7}{2^7 \cdot 2!5!},$$

which is readily found by a rough appraisal to be much less in size than .00005.

We are now ready for our computation. We have

$$Y_3(.8) \cong \frac{1}{\pi} \left\{ 2\left[\gamma + \log_e (.4) \right] J_3(.8) - \left[\frac{2^3(2!)}{(.8)^3} + \frac{2}{.8} + \frac{.8}{2(2!)} \right] \right.$$

$$- \left[\frac{(1 + \frac{1}{2} + \frac{1}{3})(.8)^3}{2^3(3!)} \right.$$

$$\left. \left. - \frac{[(1 + \frac{1}{2} + \frac{1}{3} + \frac{1}{4}) + 1](.8)^5}{2^5(4!)} \right] \right\}$$

$$\cong \frac{1}{\pi} \left\{ - .006917 - 33.95 - .018240 \right\}.$$

The middle number within the braces is exact. The other two are correct to six decimal places. Thus, we may write

$$Y_3(.8) \cong \frac{- 33.975157}{3.14159} \cong - 10.8146,$$

which will be found to agree to four decimal places with the entry for $Y_3(.8)$ in published tables. (See, for example, Watson, *Theory of Bessel Functions*, Cambridge, 1944.)

Problem: Differentiation Formulas

V-15. Show that the *Wronskian*

$$W_p = W_p(x) = \begin{vmatrix} J_p & Y_p \\ J_p' & Y_p' \end{vmatrix} = J_p Y_p' - J_p' Y_p$$

equals C/x, where C is a constant and where (as indicated) the argument for each of the involved functions is x.   ★

We can obtain an expression for W_p from the fact that $y = J_p$ and $y = Y_p$ both satisfy the Bessel equation

$$x^2 y'' + xy' + (x^2 - p^2)y = 0.$$

We have

$$x^2 J_p'' + x J_p' + (x^2 - p^2) J_p = 0, \qquad \text{(V-15.1)}$$

$$x^2 Y_p'' + x Y_p' + (x^2 - p^2) Y_p = 0. \qquad \text{(V-15.2)}$$

And we get W_p by multiplying Eq. (V-15.1) by Y_p, multiplying Eq. (V-15.2) by J_p, then subtracting and solving for W_p:

$$W_p = -x(J_p Y_p'' - Y_p J_p''). \qquad \text{(V-15.3)}$$

Differentiation of W_p, namely $(d/dx)(J_p Y_p' - Y_p J_p')$, shows that the expression in brackets on the right in Eq. (V-15.3) is dW_p/dx. So, Eq. (V-15.3) becomes

$$W_p = -x \frac{dW_p}{dx}. \qquad \text{(V-15.4)}$$

Solution of Eq. (V-15.4) yields $W_p = C/x$, where C is a constant.

REMARKS. Evaluation of C can be effected by determining the limit of the product xW_p as $x \to 0$. It is found thereby that $C = 2/\pi$, so that we have

$$J_p Y_p' - Y_p J_p' = \frac{2}{\pi x}. \qquad \text{(V-15.5)}$$

Any two solutions $y = F(x)$ and $y = G(x)$ of Bessel's equation (V-0.1) which satisfy the Wronskian relation

$$FG' - F'G = C/x$$

are said to constitute a fundamental system.

Problem: Recursion Formulas

V-16. Show that

$$J_{p+1}(x) Y_p(x) - J_p(x) Y_{p+1}(x) = \frac{2}{\pi x}. \qquad \star \qquad \text{(V-16.1)}$$

The identity to be established follows from Eq. (V-15.5) by application of Eq. (V-5.1) written for Y_p instead of J_p, such replacement being valid by the remark in Prob. V-8. Starting with Eq. (V-15.5), namely

$$J_p \frac{dY_p}{dx} - Y_p \frac{dJ_p}{dx} = \frac{2}{\pi x},$$

and replacing each of the derivatives therein by its equivalent from Eq. (V-5.1) with Y_p for J_p, we have

$$J_p(x)\left[\frac{p}{x}Y_p(x) - Y_{p+1}(x)\right] - \left[\frac{p}{x}J_p(x) - J_{p+1}(x)\right]Y_p(x) = \frac{2}{\pi x},$$

which simplifies to Eq. (V-16.1).

REMARK. The equation corresponding to Eq. (V-16.1) which is satisfied by the modified Bessel functions is

$$I_{p+1}(x)K_p(x) + I_p(x)K_{p+1}(x) = \frac{1}{x}. \qquad \text{(V-16.2)}$$

Problems: Functions of Orders $n + \frac{1}{2}$

V-17. Show that $J_{-1/2}(x) = \sqrt{\dfrac{2}{\pi x}} \cos x.$ ★   (V-17.1)

Taking $p = -1/2$ in Eq. (V-0.2) we have

$$J_{-1/2}(x) = \sum_{k=0}^{\infty} \frac{(-1)^k x^{2k-(1/2)}}{2^{2k-1/2}k!\Gamma(k + \frac{1}{2})}$$

$$= \sqrt{\frac{2}{x}} \sum_{k=0}^{\infty} \frac{(-1)^k x^{2k}}{2^{2k}k!\Gamma(k + \frac{1}{2})}. \qquad \text{(V-17.2)}$$

Now we apply Eq. (I-12.2) to the denominator factor $\Gamma(k + \frac{1}{2})$. We also observe as in Prob. I-16, using k for n, that

$$2^k k! = (2k)(2k - 2)(2k - 4) \cdots (4)(2).$$

We get

$$J_{-1/2}(x) = \sqrt{\frac{2}{\pi x}} \sum_{k=0}^{\infty} \frac{(-1)^k x^{2k}}{(2k)!}. \qquad \text{(V-17.3)}$$

The series indicated by the summation in Eq. (V-17.3) is seen to be the well-known Maclaurin series for cos x. Thus Eq. (V-17.1) is established:

$$J_{-1/2}(x) = \sqrt{\frac{2}{\pi x}} \cos x. \tag{V-17.4}$$

V-18. Show that $J_{1/2}(x) = \sqrt{2/\pi x} \sin x$.   ★

Taking $p = \frac{1}{2}$ in Eq. (V-0.2) and then multiplying both sides by $\sqrt{x/2}$, we get

$$\sqrt{\frac{x}{2}} J_{1/2}(x) = \sum_{k=0}^{\infty} \frac{(-1)^k x^{2k+1}}{2^{2k+1} k! \Gamma(k + \frac{3}{2})}. \tag{V-18.1}$$

Replacing $\Gamma(k + \frac{3}{2})$ by $[(2k + 1)(2k - 1)(2k - 3) \cdots (3)(1)\sqrt{\pi}]$ $/2^{k+1}$ from Prob. I-13, we have

$$J_{1/2}(x) = \sqrt{\frac{2}{\pi x}} \sum_{k=0}^{\infty} \frac{(-1)^k x^{2k+1}}{2^k k! (2k + 1)(2k - 1)(2k - 3) \cdots (3)(1)}. \tag{V-18.2}$$

Since $2^k k! = (2k)(2k - 2)(2k - 4) \cdots (4)(2)$, we find, by Probs. I-16 and I-10, that Eq. (V-18.2) becomes

$$J_{1/2}(x) = \sqrt{\frac{2}{\pi x}} \sum_{k=0}^{\infty} \frac{(-1)^k x^{2k+1}}{(2k + 1)!}. \tag{V-18.3}$$

The summation in Eq. (V-18.3) is seen to be the well-known Maclaurin series for sin x, namely

$$x - \frac{x^3}{3!} + \frac{x^5}{5!} - \cdots.$$

Thus,

$$J_{1/2}(x) = \sqrt{\frac{2}{\pi x}} \sin x. \tag{V-18.4}$$

REMARK. In similar manner one finds from Eq. (V-0.2) together with Eq. (V-0.9) that

$$I_{1/2}(x) = \sqrt{\frac{2}{\pi x}} \sinh x, \qquad \text{(V-18.5)}$$

$$I_{-1/2}(x) = \sqrt{\frac{2}{\pi x}} \cosh x. \qquad \text{(V-18.6)}$$

Problem: Differentiation Formulas

V:19. Show that the formula established in Prob. V-3, namely $(d/dx)[x^{-p}J_p(x)] = -x^{-p}J_{p+1}(x)$, can be generalized to

$$\left(\frac{d}{xdx}\right)^n \left\{x^{-p}J_p(x)\right\} = (-1)^n x^{-p-n}J_{p+n}(x), \qquad n = 1, 2, 3, \cdots. \quad \star$$
$$\text{(V-19.1)}$$

We multiply both sides of Eq. (V-3.1) by $1/x$:

$$\frac{1}{x}\frac{d}{dx}\left[x^{-p}J_p(x)\right] = (-1)x^{-p-1}J_{p+1}(x). \qquad \text{(V-19.2)}$$

Then we indicate the combined operations of taking derivative and multiplying by $1/x$ as follows:

$$\left(\frac{d}{xdx}\right)\left\{x^{-p}J_p(x)\right\} = (-1)x^{-p-1}J_{p+1}(x). \qquad \text{(V-19.3)}$$

We apply this operator to both sides of Eq. (V-19.3) and make use of Eq. (V-3.1) to transform the resulting right side:

$$\begin{aligned}
\left(\frac{d}{xdx}\right)\left[\frac{d}{xdx}\left\{x^{-p}J_p(x)\right\}\right] &= \left(\frac{d}{xdx}\right)\left\{(-1)x^{-p-1}J_{p+1}(x)\right\} \\
&= \left(-\frac{1}{x}\right)\frac{d}{dx}\left[x^{-p-1}J_{p+1}(x)\right] \\
&= \left(-\frac{1}{x}\right)\left[-x^{-p-1}J_{p+2}(x)\right] \\
&= (-1)^2 x^{-p-2}J_{p+2}(x). \qquad \text{(V-19.4)}
\end{aligned}$$

Indicating the iterated combined operation on the left in Eq. (V-19.4) by $(d/xdx)^2$, we may write Eq. (V-19.4) as

$$\left(\frac{d}{xdx}\right)^2\left\{x^{-p}J_p(x)\right\} = (-1)^2x^{-p-2}J_{p+2}(x). \qquad \text{(V-19.5)}$$

It is now apparent that by mathematical induction we can arrive at Eq. (V-19.1).

Problem: Functions of Order $n + \frac{1}{2}$

V-20. Apply the formula (V-19.1) to the expression for $J_{1/2}(x)$ obtained in Prob. V-18 to obtain $J_{5/2}(x)$ as a closed expression in terms of x and $\sin x$ and $\cos x$.   ★

Taking $n = 2$ in Eq. (V-19.1) with the sides interchanged and then replacing $J_{1/2}(x)$ by the expression obtained for it in Prob. V-18, we find that

$$(-1)^2x^{-5/2}J_{5/2}(x) = \left(\frac{d}{xdx}\right)^2\left\{x^{-1/2}J_{1/2}(x)\right\},$$

$$J_{5/2}(x) = x^{5/2}\left[\frac{d}{xdx}\left(\frac{d}{xdx}\left\{\sqrt{\frac{2}{\pi}}\frac{\sin x}{x}\right\}\right)\right]$$

$$= x^{5/2}\sqrt{\frac{2}{\pi}}\frac{d}{xdx}\left(\frac{1}{x}\left\{\frac{\cos x}{x} - \frac{\sin x}{x^2}\right\}\right)$$

$$= x^{3/2}\sqrt{\frac{2}{\pi}}\frac{d}{dx}\left(\frac{\cos x}{x^2} - \frac{\sin x}{x^3}\right)$$

$$= x^{3/2}\sqrt{\frac{2}{\pi}}\left(\frac{3\sin x}{x^4} - \frac{3\cos x}{x^3} - \frac{\sin x}{x^2}\right)$$

$$= \sqrt{\frac{2}{\pi x}}\left(\frac{3\sin x}{x^2} - \frac{3\cos x}{x} - \sin x\right).$$

REMARK. Similarly one can obtain a closed expression for $J_{n+1/2}(x)$, where n is any positive integer, consisting of $\sqrt{2/\pi x}$ times a finite sum of fractions whose denominators are nonnegative integral powers of x while the numerators are integral multiples of sin x or cos x.

Problem: Alternation of Zeros

V-21. Figure V-1 suggests that **(a)** $J_0(x)$ has infinitely many distinct zeros (values of x where $J_0(x) = 0$), **(b)** the same is true of $J_1(x)$, **(c)** the zeros of $J_0(x)$ alternate with those of $J_1(x)$. Given **(a)** and **(b)**, prove **(c)**.   ★

Let x' and x'' denote any two *consecutive* zeros of $J_0(x)$. By Eq. (V-3.1) with $p = 0$ we have

$$\frac{dJ_0(x)}{dx} = -J_1(x). \qquad (\text{V-21.1})$$

Since $J_0(x') = J_1(x'') = 0$ and since J_0 is differentiable at all x, it follows by Rolle's theorem (between two zeros of a differentiable function lies at least one zero of its derivative) that the derivative dJ_0/dx must vanish at least once between x' and x'', say at x^*. Then by Eq. (V-21.1) we have $J_1(x^*) = 0$. By similar argument using Eq. (V-1.1) one finds that between every pair of consecutive zeros of $J_1(x)$ must lie a zero of $J_0(x)$. Consequently, the zeros of $J_1(x)$ alternate with those of $J_0(x)$.

REMARKS. 1. The existence of infinitely many distinct zeros is true not only for $J_0(x)$ and $J_1(x)$ but for every $J_p(x)$ and every $Y_p(x)$ whatever be the order of p. (For proof see, for instance, Watson, *Theory of Bessel Functions*, Cambridge, 1944.)

2. If we move away from the origin to avoid such discontinuities as may come from x^{-h}, the property of alternation of zeros just established for $J_0(x)$ and $J_1(x)$ holds by essentially the same proof for every pair of Bessel functions $J_p(x)$ and $J_q(x)$ whose orders differ by unity.

3. It is really not surprising to learn that $J_p(x)$ has infinitely many distinct zeros, since the series formula for $J_p(x)$ in Eq. (V-0.2)

is so similar to the Maclaurin series for sin x or for cos x. Moreover, the presence of the two additional denominator factors 2^{p+2k} and $\Gamma(p + k + 1)$ is apparently what accounts for the decrease in absolute value of $J_p(x)$ as $x \to \infty$.

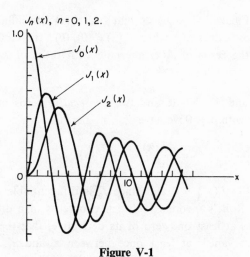

Figure V-1

Bessel Functions of the First Kind

Problems: Generating Functions

V-22. Demonstrate: A generating function for the family of Bessel functions

$$J_n(x), \qquad n = 0, \pm 1, \pm 2, \pm 3, \cdots$$

is the function

$$G(h, x) = e^{\frac{x}{2}\left(h - \frac{1}{h}\right)}. \qquad \star$$

This problem is the analogue of Prob. III-1. Accordingly, our problem is to express $G(h, x)$ in an infinite series in integral powers of h and to show that the coefficient of h^n is $J_n(x)$ and of h^{-n} is $J_{-n}(x)$.

To expand $G(h, x)$ in positive and negative powers of h we naturally try making use of Maclaurin's series for e^u, namely

$$e^u = \sum_{k=0}^{\infty} \frac{u^k}{k!}, \quad -\infty < u < +\infty. \quad \text{(V-22.1)}$$

Then, by Eq. (V-22.1), we may write

$$G(h, x) = e^{\frac{x}{2}h} e^{\left(-\frac{x}{2}\right)\left(\frac{1}{h}\right)}$$

$$= \left(\sum_{k=0}^{\infty} \left[\left(\frac{x}{2}\right)^k \frac{1}{k!} h^k\right]\right)\left(\sum_{k=0}^{\infty} \left[\left(-\frac{x}{2}\right)^k \frac{1}{k!} h^k\right]\right). \quad \text{(V-22.2)}$$

Since both series in Eq. (V-22.2) converge by Eq. (V-22.1) for all x together with all $h \neq 0$ and, for each choice of x, converge uniformly on any closed finite interval of values of h not containing $h = 0$, we may multiply termwise the two series and collect together terms in like powers of h; and the resulting series so formed will converge to $G(h, x)$ for all x together with all $h \neq 0$. For $n \geq 0$ the term in h^n in the multiplied series is made up of the collection of all those pairs of products of terms—one in each of the two series being multiplied together—such that the power of h in the first series on the right in Eq. (V-22.2) exceeds the absolute of the power of h in the second series by n. Thus, for the coefficient A_n of h^n when $n \geq 0$ we get

$$A_n = \sum_{k=0}^{\infty} \left\{\left[\left(\frac{x}{2}\right)^{k+n} \frac{1}{(k+n)!}\right]\left[\left(-\frac{x}{2}\right)^k \frac{1}{k!}\right]\right\} \quad \text{(V-22.3)}$$

$$= \sum_{k=0}^{\infty} \frac{(-1)^k x^{n+2k}}{2^{n+2k} k! (n+k)!}$$

$$= \sum_{k=0}^{\infty} \frac{(-1)^k x^{n+2k}}{2^{n+2k} k! \Gamma(n+k+1)} \quad \text{by Eq. (I-10.1)}$$

$$= J_n(x) \quad \text{by Eq. (V-0.2).}$$

For the coefficient A_{-n} of h^{-n} we find that the only change from the right side of Eq. (V-22.3) is that k changes places with $k + n$ with the result that we get $(-1)^{n+k}$ in place of $(-1)^k$. This means that

$$A_{-n} = (-1)^n J_n(x) = J_{-n}(x) \text{ by Eq. (V-0.3).}$$

We have, therefore, shown that

$$e^{\frac{x}{2}\left(h - \frac{1}{h}\right)} = \sum_{n=-\infty}^{+\infty} J_n(x)h^n \tag{V-22.4}$$

for all x together with all $h \neq 0$.

V-23. It is known that

$$e^{xh}J_0(\sqrt{1 - x^2}\, h) = \sum_{n=0}^{\infty} [P_n(x)/n!]h^n, \tag{V-23.1}$$

where J_0 denotes the Bessel Function of order zero and $P_n(x)$ is the Legendre polynomial of degree n. (See, for example, Rainville, *Special Functions*, Macmillan, 1960.) The function on the left in Eq. (V-23.1) is thus a generating function for the set of Legendre polynomials as defined in the introduction to Chapter III. Verify that the coefficients of the first several powers of h in the Maclaurin series for $e^{xh}J_0(\sqrt{1 - x^2}\, h)$ considered as a function of h with x held constant are as indicated in Eq. (V-23.1).   ★

To work out the first several terms of the expansion of $e^{xh}J_0(\sqrt{1 - x^2}\, h)$ via the successive derivatives of this product function appears to be a forbidding task. But there is another way which is not so onerous. It is to multiply the Maclaurin series for e^{xh} by the series for the J_0-function as given in Eq. (V-0.4). We find, by substituting the argument $\sqrt{1 - x^2}\, h$ for x in Eq. (V-0.4), that

$$J_0(\sqrt{1 - x^2}\, h) = 1 - \frac{(1 - x^2)h^2}{2^2} + \frac{(1 - x^2)^2 h^4}{2^4(2!)^2}$$

$$- \frac{(1 - x^2)^3 h^6}{2^6(3!)^2} + \cdots. \tag{V-23.2}$$

This expansion is valid for all h and for all x having $|x| \leqq 1$. (We confine ourselves to real x and real h.) The Maclaurin series for e^{xh} is well known:

$$e^{xh} = 1 + xh + \frac{x^2 h^2}{2!} + \frac{x^3 h^3}{3!} + \cdots \qquad \text{(V-23.3)}$$

and is valid for all x and h. Since the series expansions (V-23.2) and (V-23.3) have the common interval of convergence $-\infty < h < +\infty$, no matter what value be taken for x in them with $|x| \leqq 1$, the formal term-by-term algebraic product of the series in Eqs. (V-23.2) and (V-23.3) will converge for all h, together with any choice of x, to the product of the functions on the left in these equations. Thus we get

$$e^{xh} J_0(\sqrt{1-x^2}\,h) = 1 + xh + \left[\frac{x^2}{2!} - \frac{1-x^2}{2^2}\right] h^2$$

$$+ \left[\frac{x^3}{3!} - \frac{x(1-x^2)}{2^2}\right] h^3$$

$$+ \left[\frac{x^4}{4!} - \frac{x^2(1-x^2)}{2^2 2!} + \frac{(1-x^2)^2}{2^4(2!)^2}\right] h^4$$

$$+ \left[\frac{x^5}{5!} - \frac{x^3(1-x^2)}{2^2 3!} + \frac{x(1-x^2)^2}{2^4(2!)^2}\right] h^5 + \cdots.$$

Multiplying and dividing the coefficient of h^n by $n!$ and simplifying, we have

$$e^{xh} J_0(\sqrt{1-x^2}\,h) = [1] + [x]h + \frac{[\frac{3}{2}x^2 - \frac{1}{2}]}{2!} h^2 + \frac{[\frac{5}{2}x^3 - \frac{3}{2}x]}{3!} h^3$$

$$+ \frac{[\frac{3\cdot5}{8}x^4 - \frac{1\cdot5}{4}x^2 + \frac{3}{8}]}{4!} h^4$$

$$+ \frac{[\frac{6\cdot3}{8}x^5 - \frac{3\cdot5}{4}x^3 + \frac{1\cdot5}{8}]}{5!} h^5 + \cdots.$$

We see that the coefficients in brackets are the Legendre polynomials $P_0, P_1, P_2, P_3, P_4, P_5$ (as determined in Probs. III-4, III-5, III-6).

Problems: Orthogonality Property and Related Property

V-24. Show that the family

$$\sqrt{x}J_0(\alpha_n x), \qquad n = 1, 2, 3, \cdots$$

is orthogonal on the interval B: $0 \leq x \leq 1$, where α_1, α_2, α_3, $\cdots$ are the positive zeros of the Bessel function $J_0(x)$.   ★

We have to show that

$$\int_0^1 xJ_0(\alpha_m x)J_0(\alpha_n x)dx = 0, \qquad m \neq n. \qquad \text{(V-24.1)}$$

Comparison with the analogous problem in Legendre polynomials, Prob. III-16, suggests the possibility of similar procedure here. We seek, then, to express the integral above in a form which will be amenable to integration. And to do so we start, as in Prob. III-16, from differential equations satisfied by $J_0(\alpha_m x)$ and $J_0(\alpha_n x)$.

By formulas (A) and (C) in Table V-1 we have

$$\frac{d}{dx}\left[x\frac{d}{dx}J_0(\alpha_m x)\right] = -\alpha_m^2 xJ_0(\alpha_m x), \qquad \text{(V-24.2)}$$

$$\frac{d}{dx}\left[x\frac{d}{dx}J_0(\alpha_n x)\right] = -\alpha_n^2 xJ_0(\alpha_n x). \qquad \text{(V-24.3)}$$

Multiplying Eq. (V-24.2) by $J_0(\alpha_n x)$, Eq. (V-24.3) by $J_0(\alpha_m x)$ and then subtracting, we have

$$J_0(\alpha_n x)\frac{d}{dx}\left[x\frac{d}{dx}J_0(\alpha_m x)\right] - J_0(\alpha_m x)\frac{d}{dx}\left[x\frac{d}{dx}J_0(\alpha_n x)\right]$$
$$= (\alpha_n^2 - \alpha_m^2)xJ_0(\alpha_m x)J_0(\alpha_n x). \qquad \text{(V-24.4)}$$

At this point the procedure becomes simpler than in Prob. III-16. For scrutiny of the left side of Eq. (V-24.4) shows that we do not need (as in Prob. III-16) to resort to integration by parts: the left side of Eq. (V-24.4) as it stands is, in fact, the derivative of a difference. Thus, Eq. (V-24.4) reduces at once to

$$\frac{d}{dx}\left[xJ_0(\alpha_n x)\frac{d}{dx}J_0(\alpha_m x) - xJ_0(\alpha_m x)\frac{d}{dx}J_0(\alpha_n x) \right]$$
$$= (\alpha_n^2 - \alpha_m^2)xJ_0(\alpha_m x)J_0(\alpha_n x). \qquad \text{(V-24.5)}$$

Equation (V-24.5) can be still further reduced by applying Eq. (V-3.1) to the derivatives within the brackets as follows. Let $u = kx$ where k denotes an arbitrary constant. Then by Eq. (V-3.1) written in terms of u with $p = 0$ we have

$$\frac{dJ_0(u)}{dx} = \frac{dJ_0(u)}{du}\frac{du}{dx} = -kJ_1(kx).$$

Thus, Eq. (V-24.5) becomes

$$\frac{d}{dx}\left[-\alpha_m xJ_0(\alpha_n x)J_1(\alpha_m x) + \alpha_n xJ_0(\alpha_m x)J_1(\alpha_n x) \right]$$
$$= (\alpha_n^2 - \alpha_m^2)xJ_0(\alpha_m x)J_0(\alpha_n x). \qquad \text{(V-24.6)}$$

Interchanging sides in Eq. (V-24.6) and then integrating both sides over the interval B: $0 \leq x \leq 1$ and recalling that $J_0(\alpha_m) = J_0(\alpha_n) = 0$ and that $J_1(0) = 0$, we get

$$(\alpha_n^2 - \alpha_m^2)\int_0^1 xJ_0(\alpha_m x)J_0(\alpha_n x)dx = 0. \qquad \text{(V-24.7)}$$

Since $\alpha_n \neq \alpha_m$ when $n \neq m$, it follows that the integral in Eq. (V-24.7) must vanish when $n \neq m$, thus establishing Eq. (V-24.1).

REMARK. The orthogonality property expressed in Eq. (V-24.1) is often described by saying that the family $\{J_0(\alpha_n x)\}$ is orthogonal on the interval $0 \leq x \leq 1$ with respect to the weight function $w(x) = x$.

V-25. Show that the family $\sqrt{x}J_0(\alpha_n x)$, $n = 1, 2, 3, \cdots$ is such that the integral over the interval $0 \leq x \leq 1$ of the square of the nth member is $(1/2)[J_1(\alpha_n)]^2$:

$$\int_0^1 x \left[J_0(\alpha_n x) \right]^2 dx = \frac{1}{2} \left[J_1(\alpha_n) \right]^2. \qquad \star \qquad \text{(V-25.1)}$$

Apparently here we cannot parallel the work in the corresponding problem, namely Prob. III-17, in the Legendre polynomials, since we have nothing to correspond to Rodrigues's formula. However, a review of Prob. V-24 shows that all the steps carried out there through Eq. (V-24.6) will hold when α_m and α_n are replaced respectively by any two unequal constants g and h. Making such replacement and integrating both sides of Eq. (V-24.6) over the interval $0 \leq x \leq 1$, we get

$$\int_0^1 x J_0(gx) J_0(hx) dx = \frac{h J_0(g) J_1(h) - g J_0(h) J_1(g)}{h^2 - g^2}. \quad \text{(V-25.2)}$$

How can we make use of Eq. (V-25.2) to evaluate the integral of $x[J_0(\alpha_n x)]^2$? We would like to evaluate the integral in Eq. (V-25.2) when $g = h = \alpha_n$. But when $g = h$, both the numerator and denominator on the right in Eq. (V-25.2) equal zero. However, that is the very thing which provides the clue, as follows. The right side of Eq. (V-25.2) is a function of two arguments g and h. If we hold one of them fixed, say g, at α_n, then the right side of Eq. (V-25.2) is a function of h alone which becomes the indeterminate form $0/0$ when $h = g = \alpha_n$. Otherwise the right side of Eq. (V-25.2), regarded as a function of h alone with $g = \alpha_n$ is a continuous and differentiable function of h. So, we may apply L'Hospital's Rule:

$$\int_0^1 x \left[J_0(\alpha_n x) \right]^2 dx = \lim_{h \to \alpha_n} \left\{ \frac{\dfrac{d}{dh} \left[h J_0(\alpha_n) J_1(h) - \alpha_n J_0(h) J_1(\alpha_n) \right]}{\dfrac{d}{dh} \left(h^2 - \alpha_n^2 \right)} \right\}$$

$$= \lim_{h \to \alpha_n} \left\{ \frac{J_0(\alpha_n)[h J_1'(h) + J_1(h)] - \alpha_n J_0'(h) J_1(\alpha_n)}{2h} \right\}.$$

Replacing $J_0(\alpha_n)$ by zero and applying Eq. (V-3.1) with $h = x$ and $p = 0$ to replace the derivative $J_0(h)$ by $- J_1(h)$, we get

$$\int_0^1 x \left[J_0(\alpha_n x) \right]^2 dx = \lim_{h \to \alpha_n} \left\{ \frac{\alpha_n [J_1(\alpha_n)]^2}{2h} \right\}$$

$$= \frac{1}{2} \left[J_1(\alpha_n) \right]^2.$$

Problems: Expansion of a Given Function in Bessel Functions

V-26. If on the interval B: $0 \leq x \leq 1$ the function $f(x)$ is bounded, is continuous except for a finite number of simple discontinuities and is such that the curve $y = f(x)$ is rectifiable (has finite length), then $f(x)$ can be expanded in a series of the form

$$f(x) = \sum_{n=1}^{\infty} A_n J_0(\alpha_n x) \qquad \text{(V-26.1)}$$

valid at every point of continuity of $f(x)$ on the open interval $0 < x < 1$, where α_n, $n = 1, 2, 3, \cdots$ are the positive zeros of $J_0(x)$. Assuming the series expansion in Eq. (V-26.1) is termwise integrable over B when each term thereof is multiplied by an arbitrary $J_0(\alpha_k x)$, show that the coefficients A_n, $n = 1, 2, 3, \cdots$ are given by

$$A_n = \frac{2}{[J_1(\alpha_n)]^2} \int_0^1 x f(x) J_0(\alpha_n x) dx. \qquad \star \qquad \text{(V-26.2)}$$

This problem is analogous to Prob. III-18. Multiplying both sides of Eq. (V-26.1) by $x J_0(\alpha_k x)$ and integrating over B, we get

$$\int_0^1 x f(x) J_0(\alpha_k x) dx = \sum_{n=1}^{\infty} \left[A_n \int_0^1 x J_0(\alpha_n x) J_0(\alpha_k x) dx \right]. \quad \text{(V-26.3)}$$

Every integral on the right in Eq. (V-26.3) vanishes by the orthogonality property shown in Prob. V-24 except the term where $n = k$. The value of this lone nonvanishing term by Prob. V-25 is $(A_k/2)[J_1(\alpha_k)]^2$. Thus Eq. (V-26.3) yields the formula (V-26.1).

REMARKS. The convergence properties of the expansion (V-26.1) having coefficients given by (V-26.2) are essentially the same as those of the expansion in Prob. III-18. In particular, the series in Eq. (V-26.1) converges to the arithmetic mean of the two functional limits at each point x_0 of discontinuity on the open interval $0 < x < 1$. At $x = 1$, however, the sum of the series is zero (because $J_0(\alpha_n) = 0$ for $n = 1, 2, 3, \cdots$) regardless of the value of $f(x)$ at $x = 1$. If $f(x)$ is bounded and continuous and has limited total fluctuation on the open interval $0 < x < 1$, then at $x = 0$, the series in Eq. (V-26.1) converges to $f(0^+)$, that is, to $\lim_{x \to 0^+} [f(x)]$. If $f(x)$ has continuous second derivative on B and if $f(1) = 0$, then the series in Eq. (V-26.1) is uniformly convergent to $f(x)$ on B.

For proof of these remarks see, for example, Watson, *Theory of Bessel Functions*, Cambridge, 1944. Compare also Kaplan, *Advanced Calculus*, Addison-Wesley, 1952.

V-27. Show that for a function $f(x)$ to be expanded on the interval $0 \leqq x \leqq L$ the expansion formula of Prob. V-26 becomes

$$f(x) = \sum_{n=1}^{\infty} A_n J_0\left(\frac{\alpha_n}{L} x\right), \qquad (\text{V-27.1})$$

where

$$A_n = \frac{2}{[L J_1(\alpha_n)]^2} \int_0^L x f(x) J_0\left(\frac{\alpha_n}{L} x\right) dx. \qquad \star \quad (\text{V-27.2})$$

We first write the expansion formula of Prob. V-26 in terms of a letter t and with $f(x)$ replaced by $g(t)$. Then the linear transformation $x = Lt$ makes $t = x/L$ and $dt = dx/L$, so that the expansion formula of Prob. V-26 becomes Eq. (V-27.1) with coefficients given by Eq. (V-27.2). Compare with Prob. IV-4.

REMARK. Expansions analogous to those of Probs. V-26 and V-27 may be obtained in the form

$$f(x) = B_{p1} J_p(\alpha_{p1} x) + B_{p2} J_p(\alpha_{p2} x) + B_{p3} J_p(\alpha_{p3} x) + \cdots,$$

where α_{pj} is the jth positive zero of $J_p(x)$. The details of the treatment are essentially the same as those involved in Prob. V-26 and V-27 and problems preparatory thereto. It turns out that for a suitably restricted $f(x)$ we may have

$$f(x) = \sum_{j=1}^{\infty} B_{pj} J_p(\alpha_{pj} x), \qquad 0 \leq x \leq L, \qquad \text{(V-27.3)}$$

where

$$B_{pj} = \frac{2}{[LJ_{p+1}(\alpha_{pj})]^2} \int_0^L xf(x) J_p\left(\frac{\alpha_{pj}}{L} x\right) dx. \qquad \text{(V-27.4)}$$

Problems: Evaluation of Integrals Involving Bessel Functions

V-28. Evaluate the integral $\int_0^b x J_0(ax) dx, \qquad a \neq 0.$ ★

This integral is readily evaluated by applying to its integrand the formula (A) in Table V-1, taking $q_p(ax) = J_p(ax)$ with $p = 1$. Thus, we have

$$\int_0^b x J_0(ax) dx = \int_0^b \frac{1}{a} \frac{d}{dx}\left[x J_1(ax) \right] dx \qquad \text{(V-28.1)}$$

$$= \frac{1}{a}\left[x J_1(ax) \right]_{x=0}^{x=b} \qquad \text{(V-28.2)}$$

$$= \frac{b}{a} J_1(ab). \qquad \text{(V-28.3)}$$

V-29. Evaluate the integral $M = \int_0^b x^2 J_0(ax) dx$ where a is a constant unequal to zero. ★

Letting $ax = t$, we have

$$M = \frac{1}{a^3} \int_0^{ab} t^2 J_0(t) dt = \frac{1}{a^3} \int_0^{ab} t \cdot t J_0(t) dt.$$

Integrating by parts, we take $u = t$ and $dv = t J_0(t) dt$, so that by Eq. (V-1.1) we have $v = t J_1(t)$. Then, as in Prob. V-28,

$$M = \frac{1}{a^3}\left[a^2 b^2 J_1(ab) - \int_0^{ab} t J_1(t) dt \right].$$

We integrate by parts once more, taking $u = t$ and $dv = J_1(t)dt$. Then, by Eq. (V-3.1), we have $v = -J_0(t)$. Thus

$$\int_0^b x^2 J_0(ax)dx = \frac{1}{a^3}\left[a^2b^2 J_1(ab) + abJ_0(ab) - \int_0^{ab} J_0(t)dt \right].$$
(V-29.1)

The integral of $J_0(t)$ can be approximately evaluated with any desired degree of accuracy by termwise integration of the series for $J_0(t)$ as given by Eq. (V-0.4). Termwise integration is valid since the series converges uniformly over any finite interval. It is readily found that

$$\int_0^{ab} J_0(t)dt = ab - \frac{a^3b^3}{3\cdot 2^2} + \frac{a^5b^5}{5\cdot 2^4(2!)^2} - \frac{a^7b^7}{7\cdot 2^6(3!)^2} + \cdots.$$
(V-29.2)

Thus, the value of the integral $x^2 J_0(ax)$ over $0 \leq x \leq b$ is given by the right side of Eq. (V-29.1) where the value of the integral $J_0(t)$ over $0 \leq t \leq ab$ is given by Eq. (V-29.2).

V-30. Evaluate the integral $H = \int_0^b x^3 J_0(ax)dx$ where a is a constant $\neq 0$. ★

We let $ax = t$. Then we have

$$H = \frac{1}{a^4}\int_0^{ab} t^3 J_0(t)dt.$$

We integrate by parts taking $u = t^2$ and $dv = tJ_0(t)dt$. Then

$$\frac{dv}{dt} = tJ_0(t) = \frac{d}{dt}\left[tJ_1(t) \right] \qquad \text{by Eq. (V-1.1)}$$

Thus $v = tJ_1(t) + C$ where C is a constant. It will not matter what value we take for C, since we are integrating between limits. We choose $C = 0$, so that $v = tJ_1(t)$. Accordingly,

$$H = \frac{1}{a^4}\left\{ \left[t^3 J_1(t) \right]_0^{ab} - 2\int_0^{ab} t^2 J_1(t)dt \right\}$$

$$= \frac{1}{a^4}\left[a^3b^3 J_1(ab) - 2\int_0^{ab} t^2 J_1(t)dt \right].$$

We apply Eq. (V-1.1) once more, this time replacing $t^2J_1(t)$ by $(d/dt)[t^2J_2(t)]$. Thus we have

$$\int_0^b x^3J_0(ax)dx = \frac{1}{a^4}\left[a^3b^3J_1(ab) - 2a^2b^2J_2(ab) \right]. \quad \text{(V-30.1)}$$

REMARK. In similar manner one can evaluate any integral of the type $\int_0^b x^nJ_0(ax)dx$ where n is any odd positive integer. One finds, for example that

$$\int_0^b x^5J_0(ax)dx = \frac{1}{a^6}\left[a^5b^5J_1(ab) - 4a^4b^4J_2(ab) + 8a^3b^3J_3(ab) \right].$$

V-31. Evaluate the integral $Q = \int_0^b x^4J_0(ax)dx$ where a is a constant $\neq 0$. ★

The initial steps of procedure are the same as in Prob. V-28 and V-29. The final step, however, will be seen to be somewhat different. Letting $ax = t$, we have

$$Q = \frac{1}{a^5}\int_0^{ab} t^3 \cdot tJ_0(t)dt.$$

Integrating by parts with $u = t^3$ and $dv = tJ_0(t)dt$, whereby we have $v = tJ_1(t)$ by Eq. (V-1.1), we get

$$Q = \frac{1}{a^5}\left[a^4b^4J_1(ab) - 3\int_0^{ab} t \cdot t^2J_1(t)dt \right].$$

We integrate by parts once more, this time taking $u = t$ and $dv = t^2J_1(t)dt$, so that by Eq. (V-1.1) we have $v = t^2J_2(t)$. Accordingly,

$$Q = \frac{1}{a^5}\left[a^4b^4J_1(ab) - 3a^3b^3J_2(ab) + 3\int_0^{ab} t^2J_2(t)dt \right]. \quad \text{(V-31.1)}$$

The integral remaining to be evaluated can be transformed so as to be amenable to integration as follows. We express $J_2(t)$ in terms of $J_1(t)$ and $J_0(t)$ by Eq. (V-7.1), namely,

$$J_p(t) = \frac{t}{2p}\left[J_{p+1}(t) + J_{p-1}(t)\right],$$

taking $p = 1$ and then solving for $J_2(t)$:

$$J_2(t) = \frac{2}{t}J_1(t) - J_0(t).$$

Thus Eq. (V-31.1) becomes

$$Q = \frac{1}{a^5}\left[a^4b^4J_1(ab) - 3a^3b^3J_2(ab) + 6\int_0^{ab}tJ_1(t)dt - 3\int_0^{ab}t^2J_0(t)dt\right].$$
$$\text{(V-31.2)}$$

Both the integrals remaining to be evaluated were evaluated in Prob. V-29. There it was found that

$$\int_0^{ab}tJ_1(t)dt = \int_0^{ab}J_0(t)dt - abJ_0(ab); \qquad \text{(V-31.3)}$$

$$\int_0^{ab}t^2J_0(t)dt = a^2b^2J_1(ab) - \int_0^{ab}tJ_1(t)dt$$

$$= a^2b^2J_1(ab) + abJ_0(ab) - \int_0^{ab}J_0(t)dt. \quad \text{(V-31.4)}$$

Substituting in Eq. (V-31.2) the values of the two remaining integrals as given by Eqs. (V-31.3) and (V-31.4), we get

$$\int_0^b x^4J_0(ax)dx = \frac{1}{a^5}\left[(a^4b^4 - 3a^2b^2)J_1(ab) - 3a^3b^3J_2(ab) - 9abJ_0(ab)\right.$$

$$\left. + 9\int_0^{ab}J_0(t)dt\right],$$

where the integral of $J_0(t)$ over $0 \leq t \leq ab$ is given by Eq. (V-29.2).

REMARK. Comparison of Probs. V-29 and V-31 with Probs. V-28 and V-30 indicates that integrals of the form

$$\int_0^b x^n J_0(ax)dx$$

can be evaluated in terms of closed forms when n is an odd positive integer, but that such is not the case when n is an even positive integer. In the latter case, then, it would seem advisable to express the integrand $x^n J_0(ax)$ immediately as an infinite series via Eq. (V-0.4) and integrate termwise.

V-32. Show that for $p \geqq 0$,

$$\int_0^x x\left[J_p(ax)\right]^2 dx = \frac{x^2}{2}\left\{\left[J_p(ax)\right]^2 + J_{p-1}(ax)J_{p+1}(ax)\right\},$$
(V-32.1)

where a is a constant.   ★

If we let $y = J_p(ax)$, then it will be seen that a point of departure is furnished by the differential equation satisfied by y, namely (see remark below)

$$x^2 y'' + xy' + (a^2 x^2 - p^2)y = 0. \qquad \text{(V-32.2)}$$

This equation can be put in a form which will permit evaluation of the given integral as follows. If we multiply both sides of Eq. (V-32.2) by $2y'$ and then add $2a^2 xy^2$ to both sides and then interchange sides, we find that Eq. (V-32.2) becomes

$$2a^2 xy^2 = \frac{d}{dx}\left[x^2(y')^2 + (a^2 x^2 - p^2)y^2\right]. \qquad \text{(V-32.3)}$$

The left side of Eq. (V-32.3), except for the factor $2a^2$, is the integrand of the given integral. And the right side of Eq. (V-32.3) is a derivative. So, integrating over the interval zero to x, we get

$$2a^2 \int_0^x xy^2 dx = \left[x^2(y')^2 + (a^2 x^2 - p^2)y^2\right]_{x=0}^{x=x}. \qquad \text{(V-32.4)}$$

At $x = 0$, the first term in brackets in Eq. (V-32.4) vanishes for every $p \geqq 0$. For p positive the second term also vanishes at $x = 0$, because $y = J_p(ax)$ vanishes there by Eq. (V-0.2). When $p = 0$, then $a^2x^2 - p^2$ vanishes at $x = 0$. So, for $p \geqq 0$ we have

$$\int_0^x x\left[J_p(ax)\right]^2 dx = \frac{1}{2a^2}\left\{x^2\left[J_p'(ax)\right]^2 + (a^2x^2 - p^2)\left[J_p(ax)\right]^2\right\}.$$
(V-32.5)

There remains only to transform the right side of Eq. (V-32.5) into the right side of Eq. (V-32.1). It is evident that the transformation will involve two of the formulas listed in Table V-1 as follows. By (E) in Table V-1 we have

$$\left[J_p'(ax)\right]^2 = a^2\left[J_{p-1}(ax)\right]^2 - \frac{2ap}{x}J_p(ax)J_{p-1}(ax) + \frac{p^2}{x^2}J_p(ax).$$
(V-32.6)

And (L) in Table V-1 when multiplied on both sides by $(2ap/x)J_{p-1}(ax)$ gives us

$$\frac{2ap}{x}J_p(ax)J_{p-1}(ax) = a^2\left[J_{p-1}(ax)\right]^2 + a^2J_{p-1}(ax)J_{p+1}(ax).$$
(V-32.7)

By virtue of Eqs. (V-32.6) and (V-32.7) we find that Eq. (V-32.5) becomes Eq. (V-32.1), as was to be shown.

REMARK.   Eq. (V-32.2) above is solved in Prob. VI-1.

V-33. Show that

$$\int_0^\infty J_0(x)dx = 1,$$
(V-33.1)

assuming that the integrals $\int_0^\infty J_n(x)dx$, $n = 0, 1, 2, \cdots$ are convergent and given that

$$\int_0^\pi \frac{\cos n\phi \, d\phi}{a - ib \cos \phi} = \frac{\pi i^n}{\sqrt{a^2 + b^2}}\left[\frac{\sqrt{a^2 + b^2} - a}{b}\right]^n,$$

$$b \neq 0, \quad i = \sqrt{-1}; \quad \text{(V-33.2)}$$

$$J_n(x) = \frac{(-i)^n}{\pi} \int_0^\pi e^{ix \cos \phi} \cos n\phi \, d\phi, \qquad i = \sqrt{-1},$$
$$n = 0, 1, 2, 3, \cdots. \qquad \star \qquad \text{(V-33.3)}$$

It is not immediately apparent how we can use Eqs. (V-33.2) and (V-33.3) so as to arrive at Eq. (V-33.1). However, since the two integrals are to be involved to evaluate the given integral, it turns out that the procedure is to set up an iterated integral involving the exponential function e^{-ax} as follows.

If a is any constant $\geqq 0$, then the integral

$$\int_0^\infty e^{-ax} J_n(bx) dx \qquad \text{(V-33.3a)}$$

will be convergent. And we will have

$$\int_0^\infty e^{-ax} J_n(bx) dx = \frac{(-i)^n}{\pi} \int_0^\infty \left[\int_0^\pi e^{ibx \cos \phi} \cos n\phi \, d\phi \right] e^{-ax} dx.$$
$$\text{(V-33.4)}$$

If, now, we consider x and ϕ as rectangular plane coordinates (with ϕ playing the role usually denoted by y), then by arguments like those used in Prob. I-11 to evaluate $\Gamma(\frac{1}{2})$ we may consider the iterated integral in Eq. (V-33.4) equivalent to a double integral taken over the strip region R in the $x\phi$-plane comprised of all points (x, ϕ) for which $x \geqq 0$, $0 \leqq \phi \leqq \pi$. And then the double integral taken over R may be set equal to an interated integral in which the order of integration is reversed from that in Eq. (V-33.4). Thus, Eq. (V-33.4) is equivalent to

$$\int_0^\infty e^{-ax} J_n(bx) dx = \frac{(-i)^n}{\pi} \int_0^\pi \left[\int_0^\infty e^{-(a - ib \cos \phi)x} dx \right] \cos n\phi \, d\phi.$$
$$\text{(V-33.5)}$$

The inner integral on the right in Eq. (V-33.5) for $a > 0$ has the value

$$\frac{e^{-(a - ib \cos \phi)x}}{-(a - ib \cos \phi)} \Bigg|_{x = 0}^{x = \infty} = \frac{1}{a - ib \cos \phi}.$$

Thus, for $a > 0$, we have

$$\int_0^\infty e^{-ax} J_n(bx) dx = \frac{(-1)^n}{\pi} \int_0^\pi \frac{\cos n\phi \, d\phi}{a - ib \cos \phi},$$

which by application of Eq. (V-33.2) becomes

$$\int_0^\infty e^{-ax} J_n(bx) dx = \frac{1}{\sqrt{a^2 + b^2}} \left[\frac{\sqrt{a^2 + b^2} - a}{b} \right]^n, \quad b \neq 0, \quad a \geqq 0.$$
$$\text{(V-33.6)}$$

Equation (V-33.6) has been attained under the qualification that $a > 0$. But the integral on the left is convergent for $a \geqq 0$, as observed in the second paragraph. Moreover, the integral in Eq. (V-33.6), when regarded as function of the parameter a, with b held constant, is a continuous function of a for $a \geqq 0$. Consequently, its value for $a = 0$ is given by the right side of Eq. (V-33.6) when $a = 0$, since the right side of Eq. (V-33.6), regarded as function of a, is continuous for all a. Taking $a = 0$ and $n = 0$ in Eq. (V-33.6), we have

$$\int_0^\infty J_0(bx) dx = \frac{1}{b}, \qquad \text{(V-33.7)}$$

and taking $b = 1$,

$$\int_0^\infty J_0(x) dx = 1. \qquad \text{(V-33.8)}$$

REMARK. 1. An integral more general than the integral in Eq. (V-33.6) is

$$\int_0^\infty e^{-ax} x^p J_p(bx) dx = \frac{2^p \Gamma(p + \frac{1}{2})}{\Gamma(\frac{1}{2})} \cdot \frac{b^p}{(b^2 + a^2)^{p + (1/2)}}.$$

It will be observed that when $p = 0$ and $a = 0$ we have the integral in Eq. (V-33.7).

2. The result established in Eq. (V-33.6) will also hold true in the

complex domain when the real constant a is replaced by the pure imaginary ai. Making this replacement we find that Eq. (V-33.6) for the case $n = 0$ becomes

$$\int_0^\infty e^{-aix} J_0(bx) dx = \begin{cases} \dfrac{1}{\sqrt{b^2 - a^2}}, & b^2 > a^2, \\[2ex] \dfrac{-i}{\sqrt{a^2 - b^2}}, & a^2 > b^2. \end{cases} \tag{V-33.9}$$

If, now, we make the replacement

$$e^{-aix} = \cos ax - i \sin ax$$

and then write the integral on the left in Eq. (V-33.9) as the sum of two integrals (one multiplied by $-i$) and then equate real and imaginary parts in Eq. (V-33.9), we obtain

$$\int_0^\infty J_0(bx) \cos ax \, dx = \frac{1}{\sqrt{b^2 - a^2}}, \qquad b^2 > a^2, \tag{V-33.10}$$

$$\int_0^\infty J_0(bx) \sin ax \, dx = 0, \qquad b^2 > a^2, \tag{V-33.11}$$

$$\int_0^\infty J_0(bx) \cos ax \, dx = 0, \qquad a^2 > b^2, \tag{V-33.12}$$

$$\int_0^\infty J_0(bx) \sin ax \, dx = \frac{1}{\sqrt{a^2 - b^2}}, \qquad a^2 > b^2, \quad a > 0. \tag{V-33.13}$$

Problems: Approximations for Small and Large Arguments

V-34. For $p \neq -1, -2, -3, \cdots$ show that for values of x near zero $J_p(x)$ is given approximately by $x^p / 2^p \Gamma(p + 1)$; more precisely, show that

$$\lim_{x \to 0} \left[\frac{J_p(x)}{x^p} \right] = \frac{1}{2^p \Gamma(p + 1)}, \qquad p \neq -1, -2, -3, \cdots. \qquad \star \tag{V-34.1}$$

By Eq. (V-0.2) we have

$$\lim_{x \to 0} \left[\frac{J_p(x)}{x^p} \right] = \lim_{x \to 0} \left[\frac{1}{x^p} \sum_{k=0}^{\infty} \frac{(-1)^k x^{p+2k}}{2^{p+2k} k! \Gamma(p+k+1)} \right] \quad \text{(V-34.2)}$$

$$= \frac{1}{2^p} \lim_{x \to 0} \sum_{k=0}^{\infty} \frac{(-1)^k x^{2k}}{2^{2k} k! \Gamma(p+k+1)}. \quad \text{(V-34.3)}$$

At this point we apply Eq. (I-4.1), namely, $\Gamma(x+1) = x\Gamma(x)$ to each denominator factor $\Gamma(p+k+1)$ as follows:

$$\begin{aligned}
\Gamma(p+k+1) &= (p+k)\Gamma(p+k) \\
&= (p+k)(p+k-1)\Gamma(p+k-1) \\
&= (p+k)(p+k-1)(p+k-2)\Gamma(p+k-2) \\
&= (p+k)(p+k-1)(p+k-2) \cdots \\
&\qquad\qquad\qquad\qquad (p+1)\Gamma(p+1).
\end{aligned}$$
$$\text{(V-34.4)}$$

Thus, every denominator in the summation in Eq. (V-34.2) has the factor $\Gamma(p+1)$. Thus, Eq. (V-34.3) becomes

$$\lim_{x \to 0} \left[\frac{J_p(x)}{x^p} \right]$$

$$= \frac{1}{2^p \Gamma(p+1)} \lim_{x \to 0} \sum_{k=0}^{\infty} \frac{(-1)^k x^{2k}}{2^{2k} k! (p+k)(p+k-1) \cdots (p+1)}.$$
$$\text{(V-34.5)}$$

The series indicated by the summation in Eq. (V-34.5) is not only convergent for all x; it is, moreover, uniformly convergent on any chosen finite interval of values of x. This means that the function

$f(x)$ denoted by the series is continuous at every x, in particular at $x = 0$. Therefore,

$$\lim_{x \to 0} f(x) = f(0)$$

= the first term (the constant term) of the series

= 1.

Eq. (V-34.1) is thus established.

REMARKS. 1. If p is a positive integer n, then $\Gamma(p + 1) = \Gamma(n + 1) = n!$ by Prob. I-10, and we have

$$\lim_{x \to 0} \left[\frac{J_n(x)}{x^n} \right] = \frac{1}{2^n n!}, \qquad n = 1, 2, 3, \cdots.$$

2. By virtue of the definition given in Eq. (V-0.9), namely $I_p(x) = i^{-p} J_p(ix)$, one finds that Eq. (V-34.1) holds for $I_p(x)$ as well as $J_p(x)$.

3. By procedure quite the same as was used in establishing Eq. (V-34.1) one finds by use of the definition of $Y_0(x)$, namely Eq. (V-0.7) with $n = 0$, that

$$\lim_{x \to 0^+} \left[\frac{Y_0(x)}{\log_e x} \right] = \frac{2}{\pi} \lim_{x \to 0^+} \left[J_0(x) \left\{ \frac{\log_e x}{\log_e x} - \frac{\log_e 2}{\log_e x} \right\} \right]$$

$$= \frac{2}{\pi}.$$

We may indicate this property of $Y_0(x)$ by saying that $Y_0(x)$ is "like" $(2/\pi)\log_e x$ as $x \to 0^+$. Or it may be indicated in this manner:

$$Y_0(x) \sim \frac{2}{\pi} \log_e x \text{ as } x \to 0^+.$$

One may indicate in similar manner the properties expressed in Eq. (V-34.1) and Remark 1:

$$J_p(x) \sim \frac{x^p}{2^p \Gamma(p + 1)} \qquad \text{as } x \to 0, \quad p \neq -1, -2, -3, \cdots,$$

in particular,

$$J_n(x) \sim \frac{x^n}{2^n n!} \qquad \text{as } x \to 0, \quad n = 1, 2, 3, \cdots.$$

4. One can similarly find what $Y_n(x)$ is "like" as $x \to 0^+$ for $n = 1, 2, 3, \cdots$. For convenience let us denote the tripartite formula (V-0.7) as

$$Y_n(x) = \frac{1}{\pi}(A - B - C),$$

where A denotes the term containing $\log_e (x/2)$, C is the infinite series, and B is the finite summation. Then we find that

$$\lim_{x \to 0^+}\left[x^n Y_n(x)\right] = \frac{1}{\pi}\left\{ \lim_{x \to 0^+} (Ax^n) - \lim_{x \to 0^+} (Bx^n) - \lim_{x \to 0^+} (Cx^n)\right\}\cdot$$

$$\text{(V-34.6)}$$

The first and third limits on the right are readily seen to be zero. The second limit is seen to be equivalent to the limit of the product of the first term in B with x^n. Thus Eq. (V-34.6) becomes

$$\lim_{x \to 0^+}\left[x^n Y_n(x)\right] = -\frac{1}{\pi} \lim_{x \to 0^+}\left[x^n \frac{(n-1)! x^{-n}}{2^{-n}}\right]$$

$$= -\frac{2^n(n-1)!}{\pi}\cdot$$

Thus,

$$Y_n(x) \sim -\frac{2^n(n-1)!}{\pi x^n} \text{ as } x \to 0^+, \qquad n = 1, 2, 3, \cdots.$$

5. If p is positive and nonintegral, the determination of what $Y_p(x)$ is "like" as $x \to 0^+$ is made by way of Eq. (V-0.6):

$$\lim_{x \to 0^+}\left[x^p Y_p(x)\right] = \frac{1}{\sin p\pi}\left\{ \lim_{x \to 0^+}\left[(\cos p\pi)x^p J_p(x)\right]\right.$$

$$\left. - \lim_{x \to 0^+}\left[x^p J_{-p}(x)\right]\right\}$$

$$= \frac{1}{\sin p\pi}\left\{ \text{zero} - \frac{1}{2^{-p}\Gamma(-p+1)}\right\}\cdot$$

Accordingly, we may say that

$$Y_p(x) \text{ is ``like''} \frac{2^p}{(\sin p\pi)x^p\Gamma(1-p)} \text{ as } x \to 0^+.$$

6. It can be shown that, for large values of x, the functions J_p and Y_p are "like" elementary functions as follows:

$$J_p(x) \sim \sqrt{\frac{2}{\pi x}} \cos\left(x - \frac{\pi}{4} - \frac{p\pi}{2}\right), \qquad x \to +\infty;$$

$$Y_p(x) \sim \sqrt{\frac{2}{\pi x}} \sin\left(x - \frac{\pi}{4} - \frac{p\pi}{2}\right), \qquad x \to +\infty.$$

V-35. Evaluate the following limit:

$$\lim_{x \to 0}\left[x^{1/3}J_{-1/3}(x)\right]. \qquad \star$$

We evaluate this limit via Prob. V-34 with $p = -1/3$:

$$\lim_{x \to 0}\left[x^{1/3}J_{-1/3}(x)\right] = \lim_{x \to 0}\left[\frac{J_{-1/3}(x)}{x^{-1/3}}\right]$$

$$= \frac{1}{2^{-1/3}\Gamma(2/3)} \qquad \text{by Prob. V-34}$$

$$= \frac{\sqrt[3]{2}}{\frac{3}{2}\Gamma(5/3)} \qquad \text{by Eq. (I-4.2)}$$

$$\cong .930 \qquad \text{by Table I-1.}$$

Problems: Integral Expression of $J_p(x)$ and of $J_n(x)$

V-36. Show that, for $p > -\frac{1}{2}$,

$$J_p(x) = \frac{2(x/2)^p}{\sqrt{\pi}\Gamma(p + \frac{1}{2})} \int_0^1 (1 - t^2)^{p-(1/2)} \cos xt\, dt. \qquad \star$$

$$\text{(V-36.1)}$$

For convenience in demonstration let us write

$$H = \int_0^1 (1 - t^2)^{p-(1/2)} \cos xt\, dt. \qquad \text{(V-36.2)}$$

The integral in Eq. (V-36.2) is readily seen to be convergent for $p > -1/2$.

A method for demonstrating Eq. (V-36.1) which naturally suggests itself is evaluation of H by means of termwise integration (if valid) of an infinite series and then comparison of the resulting series (multiplied by the coefficient before the integral in Eq. (V-36.1)) with the series for $J_p(x)$ as given by Eq. (V-0.2). We expand the factor $\cos xt$ into its Maclaurin series, namely

$$\cos xt = \sum_{k=0}^{\infty} \frac{(-1)^k (xt)^{2k}}{(2k)!}, \tag{V-36.3}$$

where $0! = 1$ according to the usual convention. The series in Eq. (V-36.3) converges to $\cos xt$ for any fixed choice of x uniformly on any chosen finite interval of values of t. Consequently, when $\cos xt$ is replaced in Eq. (V-36.2) by the series in Eq. (V-36.3) the series which results by multiplication of each term by $(1 - t^2)^{p-(1/2)}$ may be integrated termwise over the interval $0 \leq t \leq 1$, even when the individual integrals are improper (albeit convergent, however). Thus, we find that Eq. (V-36.2) becomes

$$H = \sum_{k=0}^{\infty} \left\{ \frac{(-1)^k x^{2k}}{(2k)!} \int_0^1 (1 - t^2)^{p-(1/2)} t^{2k} dt \right\}. \tag{V-36.4}$$

Each of the integrals in Eq. (V-36.4) can be evaluated by the formula developed in Prob. II-17. To each integral in Eq. (V-36.4) we apply Eq. (II-17.1), identifying t with x and taking $a = 1$, $b = 2k$, $c = 2$, $d = p - \frac{1}{2}$. Accordingly, we find that

$$H = \sum_{k=0}^{\infty} \left\{ \frac{(-1)^k x^{2k}}{(2k)!} \frac{\Gamma\left(\dfrac{2k+1}{2}\right)\Gamma\left(p + \dfrac{1}{2}\right)}{(2k+2p)\Gamma(k+p)} \right\}$$

$$= \frac{1}{2} \sum_{k=0}^{\infty} \frac{(-1)^k x^{2k} \Gamma(k + \frac{1}{2})\Gamma(p + \frac{1}{2})}{(2k)!(k+p)\Gamma(k+p)}. \tag{V-36.5}$$

By Eq. (I-4.1) we have

$$(k + p)\Gamma(k + p) = \Gamma(k + p + 1). \qquad \text{(V-36.6)}$$

And by the Legendre duplication formula established in Prob. I-18 we have, by Eq. (I-18.4),

$$
\begin{aligned}
\Gamma(k + \tfrac{1}{2}) &= \frac{\sqrt{\pi}\,\Gamma(2k)}{2^{2k-1}\Gamma(k)} \\
&= \frac{\sqrt{\pi}(2k)\Gamma(2k)}{2^{2k}k\Gamma(k)} \\
&= \frac{\sqrt{\pi}(2k)!}{2^{2k}k!} \qquad \text{by Eqs. (I-4.1) and (I-10.1).} \qquad \text{(V-36.7)}
\end{aligned}
$$

Using Eqs. (V-36.6) and (V-36.7) in Eq. (V-36.5), we find that

$$H = \frac{\sqrt{\pi}}{2}\Gamma\left(p + \frac{1}{2}\right)\sum_{k=0}^{\infty}\frac{(-1)^k x^{2k}}{2^{2k}k!\,\Gamma(p + k + 1)}. \qquad \text{(V-36.8)}$$

Returning now to the right side of Eq. (V-36.1), we multiply H, as given by Eq. (V-36.8), by the coefficient before the integral in Eq. (V-36.1), obtaining the following for the right side of Eq. (V-36.1):

$$\sum_{k=0}^{\infty}\frac{(-1)^k x^{p+2k}}{2^{p+2k}k!\,\Gamma(p + k + 1)},$$

which is seen to be none other than $J_p(x)$ as given by Eq. (V-0.2). Thus, Eq. (V-36.1) is established for $p > -1/2$.

REMARK. If, in particular, we take $p = 0$ in Eq. (V-36.1), we have

$$J_0(x) = \frac{2}{\pi}\int_0^1 \frac{\cos xt}{\sqrt{1 - t^2}}\,dt \qquad \text{(V-36.9)}$$

.

by virtue of the fact that $\Gamma(\frac{1}{2}) = \sqrt{\pi}$, as established in Prob. I-11.

V-37. If n is zero or a positive integer, show that the function

$$y = f(x) = \int_0^\pi \cos(x \sin \phi - n\phi)d\phi \qquad (V-37.1)$$

satisfies Bessel's equation of order n, namely,

$$x^2 y'' + xy' + (x^2 - n^2)y = 0. \qquad \star \qquad (V-37.2)$$

By Leibniz's rule for differentiation under the integral sign with respect to the parameter x we have from Eq. (V-37.1)

$$y' = -\int_0^\pi \sin(x \sin \phi - n\phi) \sin \phi \, d\phi, \qquad (V-37.3)$$

$$y'' = -\int_0^\pi \cos(x \sin \phi - n\phi) \sin^2 \phi \, d\phi. \qquad (V-37.4)$$

Application of integration by parts to the integral for y' yields an integrated part which vanishes both at $\phi = \pi$ and at $\phi = 0$, leaving

$$y' = -\int_0^\pi \cos \phi \cdot \cos(x \sin \phi - n\phi) \cdot (x \cos \phi - n)d\phi.$$
$$(V-37.5)$$

Substitution for y, y', and y'' as given respectively by Eqs. (V-37.1), (V-37.5), and (V-37.4) in the left side of Eq. (V-37.2), with the integrand factor $\sin^2 \phi$ in (V-37.4) replaced by $1 - \cos^2 \phi$, yields an expression for the left side of Eq. (V-37.2) which is seen to reduce to

$$n \int_0^\pi \cos(x \sin \phi - n\phi) \cdot (x \cos \phi - n)d\phi.$$

Inspection of this integral reveals that its integrand is of the form cos u du, so that the left side of Eq. (V-37.2), when y is the function defined by Eq. (V-37.1), equals

$$n\left[\sin\left(x\sin\phi - n\phi\right)\right]_{\phi=0}^{\phi=\pi} = 0.$$

Thus the function y defined by Eq. (V-37.1) satisfies Bessel's equation of order n when n is a positive integer or zero.

REMARKS.   1. The function y defined by Eq. (V-37.1) is a continuous function of x for all x. This means in particular that y is finite at $x = 0$ and hence is a solution of the first kind of Bessel's equation. Consequently, y is equal to $J_n(x)$ multiplied by a constant factor C. It can be shown that $C = \pi$, which makes

$$J_n(x) = \frac{1}{\pi}\int_0^\pi \cos\left(x\sin\phi - n\phi\right)d\phi, \qquad n = 0, 1, 2, \cdots.$$
$$(V\text{-}37.6)$$

Since the cosine is an even function, we may also write

$$J_n(x) = \frac{1}{\pi}\int_0^\pi \cos\left(n\phi - x\sin\phi\right)d\phi, \qquad n = 0, 1, 2, \cdots.$$
$$(V\text{-}37.7)$$

In particular, when $n = 0$, we may write

$$J_0(x) = \frac{1}{\pi}\int_0^\pi \cos\left(x\sin\phi\right)d\phi. \qquad (V\text{-}37.8)$$

Compare the closed-form for $J_0(x)$ as given by Eq. (V-37.8) with the closed-form integral formula for $J_0(x)$ in Eq. (V-36.9).

2. It was by means of a definite integral that Bessel first defined the functions that have come to be associated with his name. In his paper (1826), *Untersuchung des Theils der planetarischen Störungen welcher aus der Bewegung der Sonne entsteht*, he took

$$J_n(x) = \frac{1}{2\pi}\int_0^{2\pi} \cos\left(n\phi - x\sin\phi\right)d\phi, \qquad n = 0, 1, 2, \cdots.$$

And from this definition he derived many of the further properties of the functions so defined.

3. Functions like the integrand in (V-37.8), namely cos $(x \sin \phi)$, have interesting expansions in terms of Bessel functions. Four such expansions due to Jacobi are as follows (for proof see, for instance, Watson, *Theory of Bessel Functions*, 1944, p. 22):

$$\cos (x \sin \phi) = J_0(x) + 2 \sum_{n=1}^{\infty} J_{2n}(x) \cos 2n\phi,$$

$$\cos (x \cos \phi) = J_0(x) + 2 \sum_{n=1}^{\infty} (- 1^n J_{2n}(x) \cos 2n\phi,$$

$$\sin (x \sin \phi) = 2 \sum_{n=0}^{\infty} J_{2n+1}(x) \sin (2n + 1)\phi,$$

$$\sin (x \cos \phi) = 2 \sum_{n=0}^{\infty} (- 1)^n J_{2n+1}(x) \cos (2n + 1)\phi.$$

Problems: Relations to Legendre Polynomials

V-38. Demonstrate:

$$\int_{-1}^{1} \cos (xt)P_n(t)dt = \left(\cos \frac{n\pi}{2} \right) \sqrt{\frac{2\pi}{x}} J_{n+(1/2)}(x), \quad \text{(V-38.1)}$$

where $P_n(t)$ denotes the Legendre polynomial of degree n.   ★

When n is odd, the right side of Eq. (V-38.1) vanishes by virtue of the cosine factor. And the left side vanishes because the integrand is an odd function of t, being the product of the even cosine function and a polynomial of odd degree. Thus, Eq. (V-38.1) is true for all odd n, albeit trivially so.

When n is even and $\neq 0$, we take $n = 2j$. Then, by Eq. (V-0.2), we have

$$J_{n+(1/2)}(x) = J_{2j+(1/2)}(x) = \sqrt{\frac{x}{2}} \sum_{k=0}^{\infty} \frac{(-1)^k x^{2j+2k}}{2^{2j+2k} k! \Gamma(2j+k+\frac{3}{2})}.$$
(V-38.2)

For the cosine factor on the right in Eq. (V-38.1) we have

$$\cos \frac{n\pi}{2} = \cos \frac{(2j)\pi}{2} = \cos (j\pi) = (-1)^j. \qquad \text{(V-38.3)}$$

Thus, by virtue of Eq. (V-38.2) with Eq. (V-38.3), the right side of Eq. (V-38.1) becomes

$$\sqrt{\pi} \sum_{k=0}^{\infty} \frac{(-1)^{j+k} x^{2j+2k}}{2^{2j+2k} k! \Gamma(2j+k+\frac{3}{2})}. \qquad \text{(V-38.4)}$$

Let us now see if we can express the left side of Eq. (V-38.1) as an infinite series identical with Eq. (V-38.4). The Maclaurin series for $\cos xt$ in powers of t converges, for any chosen value of x, for all t and converges uniformly on any chosen finite interval of values of t, in particular on the interval $-1 \leqq t \leqq 1$. Accordingly, when each term of this series is multiplied by $P_n(x)$, the resulting series is termwise integrable over $-1 \leqq t \leqq 1$. Thus, for the left side of Eq. (V-38.1) we may write

$$\int_{-1}^{1} \left\{ 1 - \frac{x^2 t^2}{2!} + \frac{x^4 t^4}{4!} - \cdots \frac{(-1)^j x^{2j} t^{2j}}{(2j)!} + \frac{(-1)^{j+1} x^{2j+2} t^{2j+2}}{(2j+2)!} + \cdots \right.$$
$$\left. + \frac{(-1)^{j+k} x^{2j+2k} t^{2j+2k}}{(2j+2k)!} + \cdots \right\} P_{2j}(t) dt. \qquad \text{(V-38.5)}$$

Multiplying each term within the braces by $P_{2j}(t)$ and observing that x is a constant as far as the integration is concerned and integrating termwise, we see that each integral can be evaluated by the results obtained in Prob. III-20.

By Eq. (III-20.5) we get the value zero for each of the integrals in which the factor t^{2r} is such that $2r < 2j$. When the factor t^{2r} is such that $2r \geq 2j$, we evaluate each of the integrals by the third formula on the right in Eq. (III-20.5). Thus, we find that the left side of Eq. (V-38.1) as given by Eq. (V-38.5) becomes

$$\int_{-1}^{1} \cos (xt) P_{2j}(t) dt$$

$$= \sum_{k=0}^{\infty} \frac{(-1)^{j+k} x^{2j+2k} (2j+2k)! \Gamma\left(\frac{2k+1}{2}\right)}{(2j+2k)! 2^{2j-1} (2k)! (4j+2k+1) \Gamma\left(\frac{4j+2k+1}{2}\right)}$$

$$= \sum_{k=0}^{\infty} \frac{(-1)^{j+k} x^{2j+2k} \Gamma\left(\frac{2k+1}{2}\right)}{2^{2j} (2k)! (2j+k+\frac{1}{2}) \Gamma(2j+k+\frac{1}{2})}.$$

We now apply Eq. (I-12.1) together with Eq. (I-11.1) to the numerator factor $\Gamma([2k+1]/2)$, at the same time writing out $(2k)!$ in expanded form. Thus, for $n = 2j$, the value of the integral equals the term of the summation for which $k = 0$ plus

$$\sum_{k=1}^{\infty} \frac{(-1)^{j+k} x^{2j+2k} \dfrac{2k-1}{2} \dfrac{2k-3}{2} \dfrac{2k-5}{2} \cdots \dfrac{3}{2} \dfrac{1}{2} \sqrt{\pi}}{2^{2j} (2k)(2k-1)(2k-2)(2k-3) \cdots (4)(3)(2)(1) \Gamma\left(2j+k+\dfrac{3}{2}\right)}.$$

Cancelling common factors from numerator and denominator and taking out the factor 2 from each of the denominator factors $2k$, $2k-2$, $2k-4$, $\cdots$, 4, 2, we find that this last summation reduces to

$$\sqrt{\pi} \sum_{k=0}^{\infty} \frac{(-1)^{j+k} x^{2j+2k}}{2^{2j+2k} k! \Gamma(2j+k+\frac{3}{2})}. \qquad \text{(V-38.6)}$$

Comparison of Eq. (V-38.6) with Eq. (V-38.4) shows that Eq. (V-38.1) holds for $n = 2j, j = 1, 2, 3, \cdots$.

Finally, we have yet to demonstrate that Eq. (V-38.1) is true for $n = 0$. In this case the equation to be demonstrated is

$$\int_{-1}^{1} \cos{(xt)}\, dt = \sqrt{\frac{2\pi}{x}} J_{1/2}(x), \qquad \text{(V-38.7)}$$

because $P_0(x) = 1$. (See Prob. III-4.) Upon integrating the left side of Eq. (V-38.7) we find that Eq. (V-38.7) becomes

$$\frac{1}{x} \sin{(xt)} \Big|_{t=-1}^{t=1} = \sqrt{\frac{2\pi}{x}} J_{1/2}(x),$$

that is,

$$\frac{2 \sin{x}}{x} = \sqrt{\frac{2\pi}{x}} J_{1/2}(x). \qquad \text{(V-38.8)}$$

Eq. (V-38.8) yields

$$J_{1/2}(x) = \sqrt{\frac{2}{\pi x}} \sin{x},$$

which agrees with the expression for $J_{1/2}(x)$ found in Prob. V-18. Thus, Eq. (V-38.1) holds also for $n = 0$.

V-39. Demonstrate

$$\int_{-1}^{1} e^{xt} P_n(t)\, dt = \sqrt{\frac{2\pi}{x}} I_{n+(1/2)}(x), \qquad \text{(V-39.1)}$$

where $P_n(t)$ is the Legendre polynomial of degree n. ★

We will express each side of Eq. (V-39.1) as a power series and show that the series are the same.

By Eqs. (V-0.9) and (V-0.2) the right side of Eq. (V-39.1) is

$$\sqrt{\frac{2\pi}{x}} (i^{-n-1/2}) \sum_{k=0}^{\infty} \frac{(-1)^k (ix)^{n+(1/2)+2k}}{2^{n+(1/2)+2k} k! \Gamma(n + \frac{1}{2} + k + 1)},$$

that is,

$$\sqrt{\pi} \sum_{k=0}^{\infty} \frac{x^{n+2k}}{2^{n+2k} k! \Gamma(n+k+\frac{3}{2})}. \tag{V-39.2}$$

On the left side in Eq. (V-39.1) we replace e^{xt} by its Maclaurin series in powers of t. Then we multiply each term of this series by $P_n(t)$ and integrate termwise. Termwise integration is valid because the series to be integrated converges uniformly on any chosen finite interval of values of t. Thus, for the left side of Eq. (V-39.1) with $n > 0$ we have

$$\int_{-1}^{1} \left\{ \left[1 + xt + \frac{x^2 t^2}{2!} + \cdots + \frac{x^{n-1} t^{n-1}}{(n-1)!} + \sum_{j=0}^{\infty} \frac{x^{n+j} t^{n+j}}{(n+j)!} \right] P_n(t) \right\} dt.$$

$$\tag{V-39.3}$$

By Eq. (III-20.5) we get the value zero, when we integrate termwise, for each of the integrals in which the exponent on t is less than n. We also get zero for each of the integrals involving t^{n+j} in which j is odd. There remains to evaluate, by the third formula on the right in Eq. (III-20.5), each integral involving t^{n+j} in which j is one of the numbers $0, 2, 4, 6, \cdots$. Thus, the left side of Eq. (V-39.1), as given by Eq. (V-39.3) with $j = 2k$ and $m = n + 2k$, equals

$$\sum_{k=0}^{\infty} \frac{x^{n+2k}(n+2k)! \Gamma\left(\frac{2k+1}{2}\right)}{(n+2k)! 2^{n-1}(2k)!(2n+2k+1)\Gamma\left(n+k+\frac{1}{2}\right)},$$

that is,

$$\sum_{k=0}^{\infty} \frac{x^{n+2k} \Gamma\left(\frac{2k+1}{2}\right)}{2^n (2k)! \left(n+k+\frac{1}{2}\right)\Gamma\left(n+k+\frac{1}{2}\right)}. \tag{V-39.4}$$

By Eq. (I-4.1) the product of the third and fourth factors in the denominator of Eq. (V-39.4) may be replaced by $\Gamma(n + k + \frac{3}{2})$. When this is done and when $(2k)!$ is replaced by $(2k)(2k - 1)$ $(2k - 2)(2k - 3) \cdots (3)(2)(1)$ and when, by Eqs. (I-11.1) and (I-12.1), we put

$$\Gamma\left(\frac{2k + 1}{2}\right) = \frac{2k - 1}{2}\frac{2k - 3}{2}\frac{2k - 5}{2}\cdots\frac{3}{2}\frac{1}{2}\sqrt{\pi},$$

we find that the left side of Eq. (V-39.1) as given by Eq. (V-39.4) is equal to

$$\sqrt{\pi}\sum_{k=0}^{\infty}\frac{x^{n+2k}}{2^{n+2k}k!\Gamma(n + k + \frac{3}{2})}, \tag{V-39.5}$$

which is the same series as we found in Eq. (V-39.2) for the other side of Eq. (V-39.1). Thus Eq. (V-39.1) is established for $n > 0$.

When $n = 0$, the equation to be established is

$$\int_{-1}^{1}e^{xt}dt = \sqrt{\frac{2\pi}{x}}I_{1/2}(x). \tag{V-39.6}$$

By Eq. (V-18.5) the right side of Eq. (V-39.6) equals $(2/x)\sinh x$. The left side of Eq. (V-39.6) equals

$$\int_{-1}^{1}\left[1 + xt + \frac{x^2t^2}{2!} + \frac{x^3t^3}{3!} + \cdots\right]dt$$

which by termwise integration (valid as indicated above) becomes

$$2 + \frac{2}{3}\frac{x^2}{2!} + \frac{2}{5}\frac{x^4}{4!} + \frac{2}{7}\frac{x^6}{6!} + \cdots. \tag{V-39.7}$$

The series indicated by (7) may be written

$$\frac{2}{x}\left[x + \frac{x^3}{3!} + \frac{x^5}{5!} + \frac{x^7}{7!} + \cdots\right]. \tag{V-39.8}$$

The series enclosed by the brackets in (V-39.8) is seen to be the Maclaurin series for sinh x. Thus, the left side of Eq. (V-39.6) as given by (V-39.8) equals the right side of Eq. (V-39.6) as given by Eq. (V-18.5), namely $(2/x)$ sinh x. Accordingly, Eq. (V-39.1) holds also for $n = 0$.

V-40. Demonstrate:

$$\lim_{n \to \infty} P_n\left(\cos \frac{x}{n}\right) = J_0(x), \qquad \text{(V-40.1)}$$

where $P_n(u)$ denotes the Legendre polynomial of degree n in u. $\qquad$ ★

Looking for a point of departure for our demonstration, we find from Probs. III-3 and V-37 that $P_n(x)$ and $J_n(x)$ can each be expressed as $1/\pi$ times an integral of a function of ϕ taken over the interval $0 \leqq \phi \leqq \pi$. We have

$$P_n(x) = \frac{1}{\pi} \int_0^\pi (x + \sqrt{x^2 - 1} \cos \phi)^n d\phi, \qquad \text{(V-40.2)}$$

$$J_0(x) = \frac{1}{\pi} \int_0^\pi \cos (x \sin \phi) \, d\phi. \qquad \text{(V-40.3)}$$

We will take the integral for $P_n(x)$ in Eq. (V-40.2) and replace the parameter x by $\cos (x/n)$ and then, assuming that the integral approaches a limit value as $n \to \infty$, see if we can show that the limit is the integral for $J_0(x)$ in Eq. (V-40.3). Thus,

$$\lim_{n \to \infty} P_n\left(\cos \frac{x}{n}\right) = \frac{1}{\pi} \lim_{n \to \infty} \int_0^\pi \left(\cos \frac{x}{n} + \sqrt{\left(\cos \frac{x}{n}\right)^2 - 1} \cos \phi\right)^n d\phi$$

$$= \frac{1}{\pi} \lim_{n \to \infty} \int_0^\pi \left(\cos \frac{x}{n} + i \sin \frac{x}{n} \cos \phi\right)^n d\phi. \qquad \text{(V-40.4)}$$

If it be assumed that, for a fixed value of x, the integrand in

Eq. (V-40.4) approaches a limit as $n \to \infty$, then the limit will be approached uniformly for $0 \leq \phi \leq \pi$, and we will have

$$\lim_{n \to \infty} P_n\left(\cos\frac{x}{n}\right) = \frac{1}{\pi}\int_0^\pi \left\{\lim_{n \to \infty}\left[\left(\cos\frac{x}{n} + i\sin\frac{x}{n}\cos\phi\right)^n\right]\right\} d\phi.$$

(V-40.5)

The integrand in Eq. (V-40.5) is seen to be the indeterminate form 1^∞, which we now seek to evaluate. To do so, we find it convenient to put

$$z = (\cos ux + i\sin ux\cos\phi)^{1/u}, \qquad u = 1/n. \quad \text{(V-40.6)}$$

If, now, we can show that z approaches a limit as $u \to 0$ continuously, it will follow that the same limit is approached by

$$\left(\cos\frac{x}{n} + i\sin\frac{x}{n}\cos\phi\right)^n$$

as $n \to \infty$, where n takes on only integral values while becoming infinite.

From Eq. (V-40.6) we have, by using the logarithmic form,

$$\lim_{u \to 0}\left[\log_e z\right] = \lim_{u \to 0}\left[\frac{\log_e(\cos ux + i\sin ux\cos\phi)}{u}\right].$$

(V-40.7)

The ratio in brackets on the right in Eq. (V-40.7) is the indeterminate form $0/0$. We can, therefore, apply L'Hospital's Rule to the right side of Eq. (V-40.7), obtaining

$$\lim_{u \to 0}\left[\log_e z\right] = \lim_{u \to 0}\left[\frac{-x\sin ux + ix\cos\phi\cos ux}{\cos ux + i\sin ux\cos\phi}\right]$$

$$= ix\cos\phi,$$

whence

$$\lim_{u \to 0} z = e^{ix\cos\phi}. \qquad \text{(V-40.8)}$$

By virtue of Eqs. (V-40.8) and (V-40.6) we find that Eq. (V-40.5) becomes

$$\lim_{n\to\infty} P_n\left(\cos\frac{x}{n}\right) = \frac{1}{\pi}\int_0^\pi e^{ix\cos\phi}d\phi \qquad \text{(V-40.9)}$$

$$= \frac{1}{\pi}\left[\int_0^\pi \cos(x\cos\phi)d\phi + i\int_0^\pi \sin(x\cos\phi)d\phi\right].$$
$$\text{(V-40.10)}$$

Let us write the first integral on the right in Eq. (V-40.10) with t as the variable of integration and write it as the sum of two integrals:

$$\int_0^\pi \cos(x\cos t)dt = \int_0^{\pi/2}\cos(x\cos t)dt + \int_{\pi/2}^\pi \cos(x\cos t)dt.$$
$$\text{(V-40.11)}$$

The transformations $\phi = t + \pi/2$, $\phi = t - \pi/2$ make the integrals on the right in Eq. (V-40.11) equivalent to

$$\int_{\pi/2}^\pi \cos(x\sin\phi)d\phi \quad \text{and} \quad \int_0^{\pi/2}\cos(x\sin\phi)d\phi.$$

Thus, we have

$$\int_0^\pi \cos(x\cos\phi)d\phi = \int_0^\pi \cos(x\sin\phi)d\phi. \qquad \text{(V-40.12)}$$

The result expressed in Eq. (V-40.12) can also be obtained directly from the definition of definite integral as limit of a sum, inasmuch as the cosine function is an odd function, so to speak, with respect to $\pi/2$ but is even with respect to zero. In similar manner, either by transformations of variable of integration or directly from the definition of definite integral as limit of a sum, one finds that the second integral on the right in Eq. (V-40.10) has the value zero. Thus, Eq. (V-40.10) becomes

$$\lim_{n\to\infty} P_n\left(\cos\frac{x}{n}\right) = \frac{1}{\pi}\int_0^\pi \cos(x\sin\phi)d\phi$$

$$= J_0(x) \qquad \text{by Eq. (V-40.3).}$$

V-41. If x and y are taken as rectangular coordinates with r and θ as the corresponding polar coordinates (the pole at the origin and the ray $\theta = 0$ along the positive x-axis), show that

$$\sum_{n=0}^{\infty} \frac{P_n(\cos \theta)}{n!} r^n = e^x J_0(y). \qquad \star$$

We start with the generating expansion given in Prob. V-23, namely

$$\sum_{n=0}^{\infty} \frac{P_n(t)}{n!} h^n = e^{th} J_0(\sqrt{1 - t^2}\, h), \qquad \text{(V-41.1)}$$

where we have written t in place of x. And now in Eq. (V-41.1) we replace t by x/r and h by r, obtaining

$$\sum_{n=0}^{\infty} \frac{P_n\left(\dfrac{x}{r}\right)}{n!} r^n = e^x J_0(\sqrt{1 - (x^2/r^2)}\, r), \qquad \text{(V-41.2)}$$

that is,

$$\sum_{n=0}^{\infty} \frac{P_n(\cos \theta)}{n!} r^n = e^x J_0(y). \qquad \text{(V-41.3)}$$

REMARKS. 1. Actually we have $y = \pm \sqrt{1 - (x^2/r^2)}\, r$, not just $\sqrt{1 - (x^2/r^2)}\, r$. But the function J_0 is an even function of its argument by Eq. (V-0.4), so that

$$J_0(- \sqrt{1 - (x^2/r^2)}\, r) = J_0(\sqrt{1 - (x^2/r^2)}\, r).$$

2. It is, perhaps, not too trite to make the observation that Eq. (V-41.1) reduces to the well-known Maclaurin series for e^h when we take $t = 1$ in Eq. (V-41.1) by virtue of the fact that every

Legendre polynomial $P_n(t) = 1$ at $t = 1$ together with the fact that $J_0(0) = 1$.

3. The right side of Eq. (V-41.3) can be written alternatively as

$$e^{r \cos \theta} J_0(r \sin \theta).$$

TABLE V-1
Formulas

$$\frac{d}{dx}\Big[x^p q_p(ax)\Big] = \begin{cases} ax^p q_{p-1}(ax) & q = I, J, Y, H^{(1)}, H^{(2)} \quad \text{(A)} \\ -ax^p q_{p-1}(ax) & q = K \quad \text{(B)} \end{cases}$$

$$\frac{d}{dx}\Big[x^{-p} q_p(ax)\Big] = \begin{cases} -ax^{-p} q_{p+1}(ax) & q = J, K, Y, H^{(1)}, H^{(2)} \quad \text{(C)} \\ ax^{-p} q_{p+1}(ax) & q = I \quad \text{(D)} \end{cases}$$

$$\frac{d}{dx}\Big[q_p(ax)\Big] = \begin{cases} aq_{p-1}(ax) - \dfrac{p}{x} q_p(ax) & q = I, J, Y, H^{(1)}, H^{(2)} \quad \text{(E)} \\ -aq_{p-1}(ax) - \dfrac{p}{x} q_p(ax) & q = K \quad \text{(F)} \end{cases}$$

$$\frac{d}{dx}\Big[q_p(ax)\Big] = \begin{cases} -aq_{p+1}(ax) + \dfrac{p}{x} q_p(ax) & q = J, K, Y, H^{(1)}, H^{(2)} \quad \text{(G)} \\ aq_{p+1}(ax) + \dfrac{p}{x} q_p(ax) & q = I \quad \text{(H)} \end{cases}$$

$$\frac{d}{dx}\Big[q_p(ax)\Big] = \begin{cases} \dfrac{a}{2}\Big[q_{p-1}(ax) - q_{p+1}(ax)\Big] & q = J, Y, H^{(1)}, H^{(2)} \quad \text{(I)} \\ \dfrac{a}{2}\Big[q_{p-1}(ax) + q_{p+1}(ax)\Big] & q = I \quad \text{(J)} \\ -\dfrac{a}{2}\Big[q_{p-1}(ax) + q_{p+1}(ax)\Big] & q = K \quad \text{(K)} \end{cases}$$

$$q_p(ax) = \begin{cases} \dfrac{ax}{2p}\Big[q_{p-1}(ax) + q_{p+1}(ax)\Big] & q = J, Y, H^{(1)}, H^{(2)} \quad \text{(L)} \\ \dfrac{ax}{2p}\Big[q_{p-1}(ax) - q_{p+1}(ax)\Big] & q = I \quad \text{(M)} \\ \dfrac{ax}{2p}\Big[-q_{p-1}(ax) + q_{p+1}(ax)\Big] & q = K \quad \text{(N)} \end{cases}$$

TABLE V-2
$J_0(x)$

x	0	.1	.2	.3	.4	.5	.6	.7	.8	.9
0	1.0000	.9975	.9900	.9776	.9604	.9385	.9120	.8812	.8463	.8075
1	.7652	.7196	.6711	.6201	.5669	.5118	.4554	.3980	.3400	.2818
2	.2239	.1666	.1104	.0555	.0025	−.0484	−.0968	−.1424	−.1850	−.2243
3	−.2601	−.2921	−.3202	−.3443	−.3643	−.3801	−.3918	−.3992	−.4026	−.4018
4	−.3971	−.3887	−.3766	−.3610	−.3423	−.3205	−.2961	−.2693	−.2404	−.2097
5	−.1776	−.1443	−.1103	−.0758	−.0412	−.0068	.0270	+.0599	.0917	.1220
6	.1506	.1773	.2017	.2238	.2433	.2601	.2740	.2851	.2931	.2981
7	.3001	.2991	.2951	.2882	.2786	.2663	.2516	.2346	.2154	.1944
8	.1717	.1475	.1222	.0960	.0692	.0419	.0146	−.0125	−.0392	−.0653
9	−.0903	−.1142	−.1367	−.1577	−.1768	−.1939	−.2090	−.2218	−.2323	−.2403
10	−.2459	−.2490	−.2496	−.2477	−.2434	−.2366	−.2276	−.2164	−.2032	−.1881
11	−.1712	−.1528	−.1330	−.1121	−.0902	−.0677	−.0446	−.0213	+.0020	.0250
12	.0477	.0697	.0908	.1108	.1296	.1469	.1626	.1766	.1887	.1988
13	.2069	.2129	.2167	.2183	.2177	.2150	.2101	.2032	.1943	.1836
14	.1711	.1570	.1414	.1245	.1065	.0875	.0679	.0476	.0271	.0064
15	−.0142	−.0346	−.0544	−.0736	−.0919	−.1092	−.1253	−.1401	−.1533	−.1650

When $x > 15.9$,

$$J_0(x) \simeq \sqrt{\left(\frac{2}{\pi x}\right)} \left\{ \sin(x + \tfrac{1}{4}\pi) + \frac{1}{8x} \sin(x - \tfrac{1}{4}\pi) \right\}$$

$$\simeq \frac{.7979}{\sqrt{x}} \left\{ \sin(57.296x + 45)^\circ + \frac{1}{8x} \sin(57.296x - 45)^\circ \right\}.$$

$J_1(x)$

x	0	.1	.2	.3	.4	.5	.6	.7	.8	.9
0	.0000	.0499	.0995	.1483	.1960	.2423	.2867	.3290	.3688	.4059
1	.4401	.4709	.4983	.5220	.5419	.5579	.5699	.5778	.5815	.5812
2	.5767	.5683	.5560	.5399	.5202	.4971	.4708	.4416	.4097	.3754
3	.3391	.3009	.2613	.2207	.1792	.1374	.0955	.0538	.0128	−.0272
4	−.0660	−.1033	−.1386	−.1719	−.2028	−.2311	−.2566	−.2791	−.2985	−.3147
5	−.3276	−.3371	−.3432	−.3460	−.3453	−.3414	−.3343	−.3241	−.3110	−.2951
6	−.2767	−.2559	−.2329	−.2081	−.1816	−.1538	−.1250	−.0953	−.0652	−.0349
7	−.0047	+.0252	.0543	.0826	.1096	.1352	.1592	.1813	.2014	.2192
8	.2346	.2476	.2580	.2657	.2708	.2731	.2728	.2697	.2641	.2559
9	.2453	.2324	.2174	.2004	.1816	.1613	.1395	.1166	.0928	.0684
10	.0435	.0184	−.0066	−.0313	−.0555	−.0789	−.1012	−.1224	−.1422	−.1603
11	−.1768	−.1913	−.2039	−.2143	−.2225	−.2284	−.2320	−.2333	−.2323	−.2290
12	−.2234	−.2157	−.2060	−.1943	−.1807	−.1655	−.1487	−.1307	−.1114	−.0912
13	−.0703	−.0489	−.0271	−.0052	+.0166	.0380	.0590	.0791	.0984	.1165
14	.1334	.1488	.1626	.1747	.1850	.1934	.1999	.2043	.2066	.2069
15	.2051	.2013	.1955	.1879	.1784	.1672	.1544	.1402	.1247	.1080

When $x > 15.9$,

$$J_1(x) \simeq \sqrt{\left(\frac{2}{\pi x}\right)} \left\{ \sin(x - \tfrac{1}{4}\pi) + \frac{3}{8x} \sin(x + \tfrac{1}{4}\pi) \right\}$$

$$\simeq \frac{.7979}{\sqrt{x}} \left\{ \sin(57.296x - 45)^\circ + \frac{3}{8x} \sin(57.296x + 45)^\circ \right\}.$$

TABLE V-3

$$J_2(x)$$

x	0	.1	.2	.3	.4	.5	.6	.7	.8	.9
0	.0000	.0012	.0050	.0112	.0197	.0306	.0437	.0588	.0758	.0946
1	.1149	.1366	.1593	.1830	.2074	.2321	.2570	.2817	.3061	.3299
2	.3528	.3746	.3951	.4139	.4310	.4461	.4590	.4696	.4777	.4832
3	.4861	.4862	.4835	.4780	.4697	.4586	.4448	.4283	.4093	.3879
4	.3641	.3383	.3105	.2811	.2501	.2178	.1846	.1506	.1161	.0813

When $0 \leq x < 1$, $\qquad J_2(x) \simeq \dfrac{x^2}{8}\left(1 - \dfrac{x^2}{12}\right).$

$$J_3(x)$$

x	0	.1	.2	.3	.4	.5	.6	.7	.8	.9
0	.0000	.0000	.0002	.0006	.0013	.0026	.0044	.0069	.0102	.0144
1	.0196	.0257	.0329	.0411	.0505	.0610	.0725	.0851	.0988	.1134
2	.1289	.1453	.1623	.1800	.1981	.2166	.2353	.2540	.2727	.2911
3	.3091	.3264	.3431	.3588	.3734	.3868	.3988	.4092	.4180	.4250
4	.4302	.4333	.4344	.4333	.4301	.4247	.4171	.4072	.3952	.3811

When $0 \leq x < 1$, $\qquad J_3(x) \simeq \dfrac{x^3}{48}\left(1 - \dfrac{x^2}{16}\right).$

$$J_4(x)$$

x	0	.1	.2	.3	.4	.5	.6	.7	.8	.9
0	.0000	.0000	.0000	.0000	.0001	.0002	.0003	.0006	.0010	.0016
1	.0025	.0036	.0050	.0068	.0091	.0118	.0150	.0188	.0232	.0283
2	.0340	.0405	.0476	.0556	.0643	.0738	.0840	.0950	.1067	.1190
3	.1320	.1456	.1597	.1743	.1891	.2044	.2198	.2353	.2507	.2661
4	.2811	.2958	.3100	.3236	.3365	.3484	.3594	.3693	.3780	.3853

When $0 \leq x < 1$, $\qquad J_4(x) \simeq \dfrac{x^4}{384}\left(1 - \dfrac{x^2}{20}\right).$

TABLE V-4

Zeros of $J_0(x)$, $J_1(x)$, $J_2(x)$, $J_3(x)$, $J_4(x)$, $J_5(x)$

m	$j_{0,m}$	$j_{1,m}$	$j_{2,m}$	$j_{3,m}$	$j_{4,m}$	$j_{5,m}$
1	2.4048	3.8317	5.1356	6.3802	7.5883	8.7715
2	5.5201	7.0156	8.4172	9.7610	11.0647	12.3386
3	8.6537	10.1735	11.6198	13.0152	14.3725	15.7002
4	11.7915	13.3237	14.7960	16.2235	17.6160	18.9801
5	14.9309	16.4706	17.9598	19.4094	20.8269	22.2178
6	18.0711	19.6159	21.1170	22.5827	24.0190	25.4303
7	21.2116	22.7601	24.2701	25.7482	27.1991	28.6266
8	24.3525	25.9037	27.4206	28.9084	30.3710	31.8117
9	27.4935	29.0468	30.5692	32.0649	33.5371	34.9888
10	30.6346	32.1897	33.7165	35.2187	36.6990	38.1599

TABLE V-5

$Y_0(x)$

x	0	.1	.2	.3	.4	.5	.6	.7	.8	.9
0	$-\infty$	−1.534	−1.081	−.8073	−.6060	−.4445	−.3085	−.1907	−.0868	+.0056
1	.0883	.1622	.2281	.2865	.3379	.3824	.4204	.4520	.4774	.4968
2	.5104	.5183	.5208	.5181	.5104	.4981	.4813	.4605	.4359	.4079
3	.3769	.3431	.3071	.2691	.2296	.1890	.1477	.1061	.0645	.0234
4	−.0169	−.0561	−.0938	−.1296	−.1633	−.1947	−.2235	−.2494	−.2723	−.2921
5	−.3085	−.3216	−.3313	−.3374	−.3402	−.3395	−.3354	−.3282	−.3177	−.3044
6	−.2882	−.2694	−.2483	−.2251	−.1999	−.1732	−.1452	−.1162	−.0864	−.0563
7	−.0259	+.0042	.0339	.0628	.0907	.1173	.1424	.1658	.1872	.2065
8	.2235	.2381	.2501	.2595	.2662	.2702	.2715	.2700	.2659	.2592
9	.2499	.2383	.2245	.2086	.1907	.1712	.1502	.1279	.1045	.0804
10	.0557	.0307	.0056	−.0193	−.0437	−.0675	−.0904	−.1122	−.1326	−.1516
11	−.1688	−.1843	−.1977	−.2091	−.2183	−.2252	−.2299	−.2322	−.2322	−.2298
12	−.2252	−.2184	−.2095	−.1986	−.1858	−.1712	−.1551	−.1375	−.1187	−.0989
13	−.0782	−.0569	−.0352	−.0134	+.0085	+.0301	+.0512	.0717	.0913	.1099
14	.1272	.1431	.1575	.1703	.1812	.1903	.1974	.2025	.2056	.2065
15	.2055	.2023	.1972	.1902	.1813	.1706	.1584	.1446	.1295	.1132

When $x > 15.9$,

$$Y_0(x) \simeq \sqrt{\left(\frac{2}{\pi x}\right)} \left\{ \sin(x - \tfrac{1}{4}\pi) - \frac{1}{8x} \sin(x + \tfrac{1}{4}\pi) \right\}$$

$$\simeq \frac{.7979}{\sqrt{x}} \left\{ \sin(57.296x - 45)° - \frac{1}{8x} \sin(57.296x + 45)° \right\}.$$

TABLE V-6

$$Y_1(x)$$

x	0	.1	.2	.3	.4	.5	.6	.7	.8	.9
0	$-\infty$	−6.459	−3.324	−2.293	−1.781	−1.471	−1.260	−1.103	−.9781	−.8731
1	−.7812	−.6981	−.6211	−.5485	−.4791	−.4123	−.3476	−.2847	−.2237	−.1644
2	−.1070	−.0517	+.0015	+.0523	.1005	.1459	.1884	.2276	.2635	.2959
3	.3247	.3496	.3707	.3879	.4010	.4102	.4154	.4167	.4141	.4078
4	.3979	.3846	3680	.3484	.3260	.3010	.2737	.2445	.2136	.1812
5	.1479	.1137	.0792	.0445	.0101	−.0238	−.0568	−.0887	−.1192	−.1481
6	−.1750	−.1998	−.2223	−.2422	−.2596	−.2741	−.2857	−.2945	−.3002	−.3029
7	−.3027	−.2995	−.2934	−.2846	−.2731	−.2591	−.2428	−.2243	−.2039	−.1817
8	−.1581	−.1331	−.1072	−.0806	−.0535	−.0262	+.0011	+.0280	.0544	.0799
9	+.1043	.1275	.1491	.1691	.1871	.2032	.2171	.2287	.2379	.2447
10	.2490	.2508	.2502	.2471	.2416	.2337	.2236	.2114	.1973	.1813
11	.1637	.1446	.1243	.1029	.0807	.0579	.0348	.0114	−.0118	−.0347
12	−.0571	−.0787	−.0994	−.1189	−.1371	−.1538	−.1689	−.1821	−.1935	−.2028
13	−.2101	−.2152	−.2182	−.2190	−.2176	−.2140	−.2084	−.2007	−.1912	−.1798
14	−.1666	−.1520	−.1359	−.1186	−.1003	−.0810	−.0612	−.0408	−.0202	+.0005
15	.0211	.0413	.0609	.0799	.0979	.1148	.1305	.1447	.1575	.1686

When $x > 15.9$,

$$Y_1(x) \simeq \sqrt{\left(\frac{2}{\pi x}\right)}\left\{\sin(x - \tfrac{3}{4}\pi) + \frac{3}{8x}\sin(x - \tfrac{1}{4}\pi)\right\}$$

$$\simeq \frac{.7979}{\sqrt{x}}\left\{\sin(57.296x - 135)^\circ + \frac{3}{8x}\sin(57.296x - 45)^\circ\right\}.$$

TABLE V-7

$$I_0(x)$$

	x	0	.1	.2	.3	.4	.5	.6	.7	.8	.9
	0	1.0000	1.0025	1.0100	1.0226	1.0404	1.0635	1.0920	1.1263	1.1665	1.2130
	1	1.2661	1.3262	1.3937	1.4693	1.5534	1.6467	1.7500	1.8640	1.9896	2.1277
	2	2.2796	2.4463	2.6291	2.8296	3.0493	3.2898	3.5533	3.8417	4.1573	4.5027
	3	4.8808	5.2945	5.7472	6.2426	6.7848	7.3782	8.0277	8.7386	9.5169	10.369
$10\times$	4	1.1302	1.2324	1.3442	1.4668	1.6010	1.7481	1.9093	2.0858	2.2794	2.4915
$10\times$	5	2.7240	2.9789	3.2584	3.5648	3.9009	4.2695	4.6738	5.1173	5.6038	6.1377
$10\times$	6	6.7234	7.3663	8.0718	8.8462	9.6962	10.629	11.654	12.779	14.014	15.370
$10^2\times$	7	1.6859	1.8495	2.0292	2.2266	2.4434	2.6816	2.9433	3.2309	3.5468	3.8941
$10^2\times$	8	4.2756	4.6950	5.1559	5.6626	6.2194	6.8316	7.5046	8.2445	9.0580	9.9524
$10^3\times$	9	1.0936	1.2017	1.3207	1.4514	1.5953	1.7535	1.9275	2.1189	2.3294	2.5610

When $x \geq 10$, $\quad I_0(x) \simeq \dfrac{.3989e^x}{x^{1/2}}\left\{1 + \dfrac{1}{8x} + \dfrac{9}{128x^2} + \dfrac{75}{1024x^3}\right\}.$

TABLE V-8

$$I_1(x)$$

	x	0	.1	.2	.3	.4	.5	.6	.7	.8	.9
	0	0	.0501	.1005	.1517	.2040	.2579	.3137	.3719	.4329	.4971
	1	.5652	.6375	.7147	.7973	.8861	.9817	1.0848	1.1963	1.3172	1.4482
	2	1.5906	1.7455	1.9141	2.0978	2.2981	2.5167	2.7554	3.0161	3.3011	3.6126
	3	3.9534	4.3262	4.7343	5.1810	5.6701	6.2058	6.7927	7.4357	8.1404	8.9128
10×	4	.97595	1.0688	1.1706	1.2822	1.4046	1.5389	1.6863	1.8479	2.0253	2.2199
10×	5	2.4336	2.6680	2.9254	3.2080	3.5182	3.8588	4.2328	4.6436	5.0946	5.5900
10×	6	6.1342	6.7319	7.3886	8.1100	8.9026	9.7735	10.730	11.782	12.938	14.208
10²×	7	1.5604	1.7138	1.8825	2.0679	2.2717	2.4958	2.7422	3.0131	3.3110	3.6385
10²×	8	3.9987	4.3948	4.8305	5.3096	5.8366	6.4162	7.0538	7.7551	8.5266	9.3754
10³×	9	1.0309	1.1336	1.2467	1.3710	1.5079	1.6585	1.8241	2.0065	2.2071	2.4280

When $x \geq 10$, $\qquad I_1(x) \simeq \dfrac{.3989e^x}{x^{1/2}}\left\{1 - \dfrac{3}{8x} - \dfrac{15}{128x^2} - \dfrac{105}{1024x^3}\right\}.$

TABLE V-9

Functions of equal order and argument

n	$J_n(n)$	$n^{1/3}J_n(n)$	$J_n'(n)$	$n^{2/3}J_n'(n)$	$-Y_n(n)$	$-n^{1/3}Y_n(n)$
1	.4401	.4401	.3251	.3251	.7812	.7812
2	.3528	.4445	.2239	.3554	.6174	.7779
3	.3091	.4457	.1770	.3682	.5385	.7767
4	.2811	.4463	.1490	.3756	.4889	.7761
5	.2611	.4465	.1301	.3804	.4537	.7758
6	.2458	.4467	.1163	.3838	.4268	.7756
7	.2336	.4468	.1056	.3865	.4054	.7754
8	.2235	.44691	.0971	.3885	.3877	.7753
9	.2149	.44697	.0902	.3902	.3727	.77526
10	.2075	.44702	.0844	.3916	.3598	.77520
11	.2010	.44705	.0794	.3928	.3485	.77515
12	.1953	.44708	.0751	.3938	.3386	.77510
13	.1901	.44710	.0714	.3947	.3296	.77507
14	.1855	.44712	.0681	.3955	.3216	.77504
15	.1813	.44714	.0651	.3962	.3143	.77502
16	.1775	.44715	.0625	.3968	.3076	.77500
17	.1739	.44717	.0601	.3973	.3014	.77498
18	.1706	.44718	.0579	.3978	.2957	.77496
19	.1676	.44719	.0559	.3983	.2904	.77495
20	.1647	.44719	.0541	.3987	.2855	.77494

TABLE V-10

$$K_0(x)$$

	x	0	.1	.2	.3	.4	.5	.6	.7	.8	.9
	0	∞	2.4271	1.7527	1.3725	1.1145	.9244	.7775	.6605	.5653	.4867
	1	.4210	.3656	.3185	.2782	.2437	.2138	.1880	.1655	.1459	.1288
$10^{-1}\times$	2	1.1389	1.0078	.8926	.7914	.7022	.6235	.5540	.4926	.4382	.3901
$10^{-1}\times$	3	.3474	.3095	.2759	.2461	.2196	.1960	.1750	.1563	.1397	.1248
$10^{-2}\times$	4	1.1160	.9980	.8927	.7988	.7149	.6400	.5730	.5132	.4597	.4119
$10^{-2}\times$	5	.3691	.3308	.2966	.2659	.2385	.2139	.1918	.1721	.1544	.1386
$10^{-3}\times$	6	1.2440	1.1167	1.0025	.9001	.8083	.7259	.6520	.5857	.5262	.4728
$10^{-3}\times$	7	.4248	.3817	.3431	.3084	.2772	.2492	.2240	.2014	.1811	.1629
$10^{-4}\times$	8	1.4647	1.3173	1.1849	1.0658	.9588	.8626	.7761	.6983	.6283	.5654
$10^{-4}\times$	9	.5088	.4579	.4121	.3710	.3339	.3006	.2706	.2436	.2193	.1975

When $x \geq 10$, $\qquad K_0(x) \simeq \dfrac{1.2533e^{-x}}{x^{1/2}}\left\{1 - \dfrac{1}{8x} + \dfrac{9}{128x^2} - \dfrac{75}{1024x^3}\right\}.$

TABLE V-11

$$K_1(x)$$

	x	0	.1	.2	.3	.4	.5	.6	.7	.8	.9
	0	∞	9.8538	4.7760	3.0560	2.1844	1.6564	1.3028	1.0503	.8618	.7165
	1	.6019	.5098	.4346	.3725	.3208	.2774	.2406	.2094	.1826	.1597
$10^{-1}\times$	2	1.3987	1.2275	1.0790	.9498	.8372	.7389	.6528	.5774	.5111	.4529
$10^{-1}\times$	3	.4016	.3563	.3164	.2812	.2500	.2224	.1979	.1763	.1571	.1400
$10^{-2}\times$	4	1.2484	1.1136	.9938	.8872	.7923	.7078	.6325	.5654	.5055	.4521
$10^{-2}\times$	5	.4045	.3619	.3239	.2900	.2597	.2326	.2083	.1866	.1673	.1499
$10^{-3}\times$	6	1.3439	1.2050	1.0805	.9691	.8693	.7799	.6998	.6280	.5636	.5059
$10^{-3}\times$	7	.4542	.4078	.3662	.3288	.2953	.2653	.2383	.2141	.1924	.1729
$10^{-4}\times$	8	1.5537	1.3964	1.2552	1.1283	1.0143	.9120	.8200	.7374	.6631	.5964
$10^{-4}\times$	9	.5364	.4825	.4340	.3904	.3512	.3160	.2843	.2559	.2302	.2072

When $x \geq 10$, $\qquad K_1(x) \simeq \dfrac{1.2533e^{-x}}{x^{1/2}}\left\{1 + \dfrac{3}{8x} - \dfrac{15}{128x^2} + \dfrac{105}{1024x^3}\right\}.$

TABLE V-12
Struve Function

$$\mathbf{H}_0(x)$$

x	0	.1	.2	.3	.4	.5	.6	.7	.8	.9
0	.0000	.0636	.1268	.1891	.2501	.3096	.3669	.4218	.4740	.5230
1	.5687	.6106	.6486	.6824	.7118	.7367	.7570	.7726	.7835	.7895
2	.7909	.7875	.7796	.7673	.7506	.7300	.7054	.6773	.6459	.6114
3	.5743	.5348	.4934	.4503	.4060	.3608	.3151	.2694	.2238	.1789
4	.1350	.0924	.0515	.0125	−.0243	−.0585	−.0901	−.1187	−.1442	−.1664
5	−.1852	−.2006	−.2124	−.2208	−.2256	−.2268	−.2247	−.2193	−.2107	−.1990
6	−.1846	−.1674	−.1479	−.1262	−.1025	−.0773	−.0507	−.0230	+.0054	.0343
7	.0634	.0923	.1208	.1485	.1753	.2009	.2249	.2472	.2677	.2860
8	.3020	.3156	.3267	.3352	.3410	.3442	.3446	.3423	.3374	.3299
9	.3199	.3075	.2929	.2763	.2578	.2375	.2158	.1929	.1689	.1441
10	.1187	.0931	.0674	.0420	.0169	−.0074	−.0309	−.0532	−.0742	−.0936
11	−.1114	−.1274	−.1413	−.1532	−.1629	−.1703	−.1754	−.1781	−.1786	−.1767
12	−.1725	−.1662	−.1577	−.1472	−.1348	−.1206	−.1048	−.0877	−.0693	−.0498
13	−.0295	−.0086	+.0127	+.0342	.0557	.0770	.0978	.1179	.1372	.1554
14	.1724	.1881	.2022	.2146	.2252	.2340	.2408	.2456	.2484	.2491
15	.2477	.2443	.2389	.2316	.2225	.2116	.1990	.1850	.1696	.1530

When $x > 15.9$,

$$\mathbf{H}_0(x) \simeq Y_0(x) + \frac{2}{\pi x}$$

$$\simeq \frac{.7979}{\sqrt{x}}\left[\sin(57.296x - 45)° - \frac{1}{8x}\sin(57.296x + 45)°\right] + \frac{.6366}{x}.$$

TABLE V-13
Struve Function
$$\mathbf{H}_1(x)$$

x	0	.1	.2	.3	.4	.5	.6	.7	.8	.9
0	.0000	.0021	.0085	.0190	.0336	.0522	.0746	.1006	.1301	.1628
1	.1985	.2368	.2774	.3201	.3645	.4103	.4570	.5044	.5521	.5997
2	.6468	.6930	.7381	.7817	.8235	.8632	.9004	.9349	.9665	.9950
3	1.020	1.042	1.060	1.074	1.085	1.092	1.095	1.094	1.089	1.081
4	1.070	1.055	1.037	1.016	.9921	.9660	.9376	.9073	.8754	.8421
5	.8078	.7728	.7375	.7021	.6670	.6324	.5987	.5661	.5350	.5056
6	.4782	.4529	.4299	.4095	.3917	.3768	.3647	.3556	.3495	.3464
7	.3463	.3492	.3549	.3634	.3746	.3883	.4044	.4226	.4428	.4647
8	.4881	.5128	.5385	.5649	.5918	.6190	.6460	.6728	.6989	.7243
9	.7485	.7715	.7930	.8128	.8307	.8466	.8604	.8719	.8810	.8876
10	.8918	.8935	.8928	.8895	.8839	.8760	.8658	.8535	.8392	.8232
11	.8055	.7863	.7659	.7444	.7222	.6993	.6760	.6526	.6293	.6063
12	.5839	.5621	.5414	.5218	.5035	.4868	.4717	.4584	.4470	.4376
13	.4302	.4251	.4220	.4212	.4226	.4260	.4316	.4392	.4488	.4601
14	.4732	.4878	.5038	.5211	.5394	.5586	.5784	.5987	.6193	.6400
15	.6605	.6807	.7003	.7192	.7371	.7540	.7697	.7839	.7966	.8077

When $x > 15.9$,

$$\mathbf{H}_1(x) \simeq Y_1(x) + \frac{2}{\pi}\left\{1 + \frac{1}{x^2}\right\}$$

$$\simeq \frac{.7979}{\sqrt{x}}\left\{\sin(57.296x - 135)° + \frac{3}{8x}\sin(57.296x - 45)°\right\}$$
$$+ .6366\left(1 + \frac{1}{x^2}\right).$$

TABLE V-14
Ber x

x	0	.1	.2	.3	.4	.5	.6	.7	.8	.9
0	1.0000	1.0000	1.0000	.9999	.9996	.9990	.9980	.9962	.9936	.9898
1	.9844	.9771	.9676	.9554	.9401	.9211	.8979	.8700	.8367	.7975
2	.7517	.6987	.6377	.5680	.4890	.4000	.3001	.1887	.0651	− .0714
3	− .2214	− .3855	− .5644	− .7584	− .9680	− 1.194	− 1.435	− 1.693	− 1.967	− 2.258
4	− 2.563	− 2.884	− 3.219	− 3.568	− 3.928	− 4.299	− 4.678	− 5.064	− 5.453	− 5.843
5	− 6.230	− 6.611	− 6.980	− 7.334	− 7.667	− 7.974	− 8.247	− 8.479	− 8.664	− 8.794
6	− 8.858	− 8.849	− 8.756	− 8.569	− 8.276	− 7.867	− 7.329	− 6.649	− 5.816	− 4.815
7	− 3.633	− 2.257	− .6737	+ 1.131	3.169	5.455	7.999	10.81	13.91	17.29
8	20.97	24.96	29.25	33.84	38.74	43.94	49.42	55.19	61.21	67.47
9	73.94	80.58	87.35	94.21	101.1	108.0	114.7	121.3	127.5	133.4
10	138.8									

When $x > 10$,

$$\text{ber } x \simeq \frac{.3989e^{x/\sqrt{2}}}{\sqrt{x}}\left\{\sin(40.514x + 67.5)° + \frac{1}{8x}\sin(40.514x + 22.5)°\right\}.$$

TABLE V-15

Bei x

x	0	.1	.2	.3	.4	.5	.6	.7	.8	.9
0	.0000	.0025	.0100	.0225	.0400	.0625	.0900	.1224	.1599	.2023
1	.2496	.3017	.3587	.4204	.4867	.5576	.6327	.7120	.7953	.8821
2	.9723	1.065	1.161	1.259	1.357	1.457	1.557	1.656	1.753	1.847
3	1.938	2.023	2.102	2.172	2.233	2.283	2.320	2.341	2.345	2.330
4	2.293	2.231	2.142	2.024	1.873	1.686	1.461	1.195	.8837	.5251
5	.1160	−.3467	−.8658	−1.444	−2.085	−2.789	−3.560	−4.399	−5.307	−6.285
6	−7.335	−8.454	−9.644	−10.90	−12.22	−13.61	−15.05	−16.54	−18.07	−19.64
7	−21.24	−22.85	−24.46	−26.05	−27.61	−29.12	−30.55	−31.88	−33.09	−34.15
8	−35.02	−35.67	−36.06	−36.16	−35.92	−35.30	−34.25	−32.71	−30.65	−28.00
9	−24.71	−20.72	−15.98	−10.41	−3.969	3.411	11.79	21.22	31.76	43.46
10	56.37									

When $x > 10$,

$$\text{bei } x \simeq \frac{.3989 e^{x/\sqrt{2}}}{\sqrt{x}} \left\{ \sin(40.514x - 22.5)° + \frac{1}{8x} \sin(40.514x - 67.5)° \right\}.$$

TABLE V-16

Ber$'x$

x	0	.1	.2	.3	.4	.5	.6	.7	.8	.9
0	.0000	.0000	−.0005	−.0017	−.0040	−.0078	−.0135	−.0214	−.0320	−.0455
1	−.0624	−.0831	−.1078	−.1370	−.1709	−.2100	−.2545	−.3048	−.3612	−.4238
2	−.4931	−.5691	−.6520	−.7420	−.8392	−.9436	−1.055	−1.174	−1.299	−1.431
3	−1.570	−1.714	−1.864	−2.018	−2.175	−2.336	−2.498	−2.661	−2.822	−2.981
4	−3.135	−3.282	−3.420	−3.547	−3.659	−3.754	−3.828	−3.878	−3.901	−3.891
5	−3.845	−3.759	−3.627	−3.445	−3.206	−2.907	−2.541	−2.102	−1.586	−.9844
6	−.2931	+.4943	1.384	2.380	3.490	4.717	6.067	7.544	9.151	10.89
7	12.76	14.77	16.92	19.19	21.60	24.13	26.78	29.53	32.38	35.31
8	38.31	41.35	44.42	47.47	50.49	53.44	56.28	58.97	61.45	63.68
9	65.60	67.14	68.25	68.83	68.82	68.13	66.67	64.35	61.07	56.72
10	51.20									

When $x > 10$,

$$\text{ber}' x \simeq \frac{.3989 e^{x/\sqrt{2}}}{\sqrt{x}} \left\{ \sin(40.514x + 112.5)° - \frac{3}{8x} \sin(40.514x + 67.5)° \right\}.$$

TABLE V-17

Bei′x

x	0	.1	.2	.3	.4	.5	.6	.7	.8	.9
0	.0000	.0500	.1000	.1500	.2000	.2499	.2998	.3496	.3991	.4485
1	.4974	.5458	.5935	.6403	.6860	.7303	.7727	.8131	.8509	.8857
2	.9170	.9442	.9666	.9836	.9944	.9983	.9943	.9815	.9590	.9257
3	.8805	.8223	.7499	.6621	.5577	.4353	.2937	.1315	−.0525	−.2597
4	−.4911	−.7481	−1.032	−1.343	−1.683	−2.053	−2.452	−2.882	−3.342	−3.833
5	−4.354	−4.905	−5.484	−6.089	−6.720	−7.373	−8.045	−8.734	−9.433	−10.14
6	−10.85	−11.55	−12.23	−12.90	−13.54	−14.13	−14.67	−15.15	−15.45	−15.85
7	−16.04	−16.11	−16.03	−15.79	−15.37	−14.74	−13.88	−12.76	−11.37	−9.681
8	−7.660	−5.285	−2.530	+.6341	4.232	8.290	12.83	17.88	23.47	29.60
9	36.30	43.58	51.46	59.94	69.01	78.68	88.94	99.76	111.1	123.0
10	135.3									

When $x > 10$,

$$\text{bei}' \, x \simeq \frac{.3989 e^{x/\sqrt{2}}}{\sqrt{x}} \left\{ \sin(40.514x + 22.5)° - \frac{3}{8x} \sin(40.514x - 22.5)° \right\}.$$

TABLE V-18

Ker x

	x	0	.1	.2	.3	.4	.5	.6	.7	.8	.9
	0	+∞	2.420	1.733	1.337	1.063	.8559	.6931	.5614	.4529	.3625
	1	.2867	.2228	.1689	.1235	.0851	.0529	.0260	.0037	−.0147	−.0297
	2	−.0417	−.0511	−.0583	−.0637	−.0674	−.0697	−.0708	−.0710	−.0703	−.0689
	3	−.0670	−.0647	−.0620	−.0590	−.0559	−.0526	−.0493	−.0460	−.0426	−.0394
$10^{-3} \times$	4	−3,618	−3,308	−3,011	−2,726	−2,456	−2,200	−1,960	−1,734	−1,525	−1,330
$10^{-3} \times$	5	−1,151	−986.5	−835.9	−698.9	−574.9	−463.2	−363.2	−274.0	−195.2	−125.8
$10^{-6} \times$	6	−653.0	−129.5	+319.1	699.1	1,017	1,278	1,488	1,653	1,777	1,866
$10^{-6} \times$	7	1,922	1,951	1,956	1,940	1,907	1,860	1,800	1,731	1,655	1,572
$10^{-6} \times$	8	1,486	1,397	1,306	1,216	1,126	1,037	951.1	867.5	787.1	710.2
$10^{-6} \times$	9	637.2	568.1	503.0	442.2	385.5	333.0	284.6	240.2	199.6	162.8
$10^{-6} \times$	10	129.5									

When $x > 10$,

$$\text{ker} \, x \simeq \frac{1.2533 e^{-x/\sqrt{2}}}{\sqrt{x}} \left\{ \sin(40.514x + 112.5)° + \frac{1}{8x} \sin(40.514x - 22.5)° \right\}.$$

TABLE V-19
Kei x

	x	0	.1	.2	.3	.4	.5	.6	.7	.8	.9
	0	−.7854	−.7769	−.7581	−.7331	−.7038	−.6716	−.6374	−.6022	−.5664	−.5305
	1	−.4950	−.4601	−.4262	−.3933	−.3617	−.3314	−.3026	−.2752	−.2494	−.2251
	2	−.2024	−.1812	−.1614	−.1431	−.1262	−.1107	−.0964	−.0834	−.0716	−.0608
$10^{-5}\times$	3	−5,112	−4,240	−3,458	−2,762	−2,145	−1,600	−1,123	−707.7	−348.7	−41.08
$10^{-6}\times$	4	+2,198	4,386	6,194	7,661	8,826	9,721	10,380	10,830	11,100	11,210
$10^{-6}\times$	5	11,190	11,050	10,820	10,510	10,140	9,716	9,255	8,766	8,258	7,739
$10^{-6}\times$	6	7,216	6,696	6,183	5,681	5,194	4,724	4,274	3,846	3,440	3,058
$10^{-6}\times$	7	2,700	2,366	2,057	1,770	1,507	1,267	1,048	849.8	671.4	511.7
$10^{-6}\times$	8	369.6	244.0	133.9	+38.09	−44.49	−114.9	−174.1	−223.3	−263.2	−294.9
$10^{-6}\times$	9	−319.2	−336.8	−348.6	−355.2	−357.4	−355.7	−350.8	−343.0	−332.9	−321.0
$10^{-6}\times$	10	−307.5									

When $x > 10$,

$$\operatorname{kei} x \simeq \frac{1.2533 e^{-x/\sqrt{2}}}{\sqrt{x}}\left\{ - \sin(40.514x + 22.5)^\circ + \frac{1}{8x}\sin(40.514x + 67.5)^\circ \right\}.$$

TABLE V-20
Ker'x

	x	0	.1	.2	.3	.4	.5	.6	.7	.8	.9
	0	−∞	−9.961	−4.923	−3.220	−2.352	−1.820	−1.457	−1.191	−.9873	−.8259
	1	−.6946	−.5859	−.4946	−.4172	−.3511	−.2942	−.2451	−.2027	−.1659	−.1341
$10^{-4}\times$	2	−1,066	−828.2	−623.4	−447.5	−297.1	−169.3	−61.36	+29.04	+104.0	165.3
$10^{-4}\times$	3	214.8	253.7	283.6	305.6	320.7	329.9	334.1	334.0	330.4	323.8
$10^{-4}\times$	4	314.8	303.7	291.3	277.7	263.2	248.1	232.8	217.3	201.9	186.8
$10^{-4}\times$	5	171.9	157.5	143.7	130.4	117.7	105.8	94.47	83.88	74.00	64.81
$10^{-4}\times$	6	56.32	48.50	41.33	34.79	28.85	23.49	18.67	14.36	10.54	7.164
$10^{-6}\times$	7	420.5	163.3	−58.39	−247.4	−406.6	−538.8	−646.5	−732.2	−798.2	−846.7
$10^{-6}\times$	8	−879.7	−899.2	−906.9	−904.4	−893.2	−874.7	−850.0	−820.4	−786.8	−750.2
$10^{-6}\times$	9	−711.2	−670.7	−629.3	−587.5	−545.8	−504.5	−464.1	−424.8	−386.8	−350.4
$10^{-6}\times$	10	−315.6									

When $x > 10$,

$$\operatorname{ker'} x \simeq - \frac{1.2533 e^{-x/\sqrt{2}}}{\sqrt{x}}\left\{ \sin(40.514x + 67.5)^\circ + \frac{3}{8x}\sin(40.514x + 112.5)^\circ \right\}.$$

TABLE V-21

Kei′x

	x	0	.1	.2	.3	.4	.5	.6	.7	.8	.9
	0	.0000	.1460	.2229	.2743	.3095	.3332	.3482	.3563	.3590	.3574
	1	.3524	.3445	.3345	.3227	.3096	.2956	.2809	.2658	.2504	.2351
	2	.2198	.2048	.1901	.1759	.1621	.1489	.1363	.1243	.1129	.1021
$10^{-5} \times$	3	9,204	8,259	7,378	6,558	5,799	5,098	4,454	3,864	3,325	2,835
$10^{-5} \times$	4	2,391	1,991	1,631	1,310	1,024	771.5	549.2	355.0	186.5	41.52
$10^{-6} \times$	5	−820.0	−1,861	−2,726	−3,433	−4,000	−4,440	−4,769	−5,000	−5,146	−5,217
$10^{-6} \times$	6	−5,224	−5,176	−5,082	−4,951	−4,788	−4,600	−4,393	−4,170	−3,939	−3,701
$10^{-6} \times$	7	−3,460	−3,218	−2,979	−2,745	−2,517	−2,296	−2,084	−1,881	−1,689	−1,507
$10^{-6} \times$	8	−1,336	−1,177	−1,028	−890.2	−763.2	−646.7	−540.4	−443.8	−356.5	−278.1
$10^{-6} \times$	9	−208.1	−145.9	−91.09	−43.15	−1.559	+34.16	64.49	89.89	110.8	127.7
$10^{-6} \times$	10	140.9									

When $x > 10$,

$$\text{kei}' \, x \simeq \frac{1.2533 e^{-x/\sqrt{2}}}{\sqrt{x}} \left\{ \sin(40.514x - 22.5)^\circ \right.$$

$$\left. + \frac{3}{8x} \sin(40.514x + 22.5)^\circ \right\}.$$

TABLE V-22

Ber$_n$ x, bei$_n$ x, ber$'_n$ x, and bei$'_n$ x, from $n = 1$ to 5

x	1	2	3	4	5	6	7	8	9	10
ber$_1$ x	−.3959	−.9971	−1.733	−1.869	.3598	7.462	20.37	32.51	20.72	−59.48
bei$_1$ x	+.3076	.2998	−.4875	−2.564	−5.798	−7.877	−2.317	21.67	72.05	131.9
ber$'_1$ x	−.4767	−.7205	−.6360	.6587	4.251	10.21	14.68	5.866	−37.11	−132.1
bei$'_1$ x	+.2120	−.3058	−1.364	−2.793	−3.328	.2355	12.78	36.88	61.75	45.13
ber$_2$ x	.0104	.1653	.8084	2.318	4.488	5.243	−.9504	−22.89	−65.87	−111.8
bei$_2$ x	−.1247	−.4792	−.8910	−.7254	1.422	7.432	17.59	25.44	10.13	−66.61
ber$'_2$ x	+.0416	.3278	1.031	1.976	2.050	−1.455	−12.49	−32.59	−50.96	−28.84
bei$'_2$ x	−.2480	−.4378	−.2865	.8538	3.785	8.369	11.02	1.301	−38.55	−122.0
ber$_3$ x	+.0138	.0856	.1304	−.2826	−2.094	−6.430	−12.88	−15.42	3.167	72.25
bei$_3$ x	.0156	.1442	.5654	1.438	2.454	+1.901	−4.407	−22.58	−54.54	−81.42
ber$'_3$ x	.0394	.0936	.0720	−.9141	−2.923	−5.748	−6.249	3.980	38.35	104.5
bei$'_3$ x	.0486	.2394	.6363	1.074	.6956	−2.499	−11.22	−25.71	−35.56	−7.513
ber$_4$ x	−.0026	−.0410	−.1933	−.4931	−.6287	.6483	6.084	19.09	38.67	46.58
bei$_4$ x	−.00013	−.0083	−.0930	−.4999	−1.7276	−4.230	−7.117	−5.289	14.08	70.50
ber$'_4$ x	−.0104	−.0806	−.2343	−.3237	.2483	2.770	8.745	17.32	19.14	−12.15
bei$'_4$ x	−.00078	−.0248	−.1835	−.7167	−1.834	−3.071	−1.922	7.704	34.55	80.47
ber$_5$ x	+.00019	.0068	.0586	.2731	.8510	1.831	2.209	−1.821	−18.62	−58.72
bei$_5$ x	−.00018	−.0048	−.0255	−.0335	.2114	1.476	5.242	12.81	21.38	15.19
ber$'_5$ x	.00097	+.0178	.1048	.3608	.8151	1.007	−.8472	−8.624	−26.96	−53.43
bei$'_5$ x	−.00087	−.0110	−.0283	.0467	.5656	2.220	5.590	9.234	+5.504	−24.51

TABLE V-23

Ker$_n$ x, kei$_n$ x, ker$'_n$ x, and kei$'_n$ x, from $n = 1$ to 5

x	1	2	3	4	5	6	7	8	9	10
r_1x	−7,403	−2,308	−499.0	53.51	127.4	76.76	27.44	3.229	−3.558	−3.228
i_1x	−2,420	800.5	802.7	391.7	115.8	2.884	−21.49	−15.67	−6.501	−1.235
r'_1x	8,876	2,880	1,002	226.9	−23.18	−59.20	−36.60	−13.52	−1.853	+1.582
i'_1x	7,947	736.3	−380.1	−369.3	−183.7	−56.13	1.566	9.852	7.485	3.214
					Multiply all values by 10⁻⁴					
r_2x	4,180	2,615	1,284	481.3	111.8	−10.88	−29.10	−18.20	−6.834	−1.013
i_2x	18,842	3,090	368.0	−179.4	−180.6	−90.94	−28.21	−1.497	4.772	3.706
r'_2x	−1,415	−1,549	−1,071	−555.5	−216.7	−52.69	4.111	13.35	8.531	3.359
i'_2x	−41,208	−5,288	−1,166	−149.4	80.46	82.55	46.65	13.74	1.020	−2.150
					Multiply all values by 10⁻⁴					
r_3x	48,873	2,980	−364.5	−520.7	−292.8	−114.5	−27.07	2.677	7.205	4.563
i_3x	−62,697	−8,868	−2,360	−605.2	−76.85	45.12	44.65	22.63	7.148	.473
r'_3x	−16,290	−850.4	−80.36	17.70	22.44	12.92	5.213	1.292	−.0944	−.3273
i'_3x	17,772	1,297	300.8	92.11	25.29	3.405	−1.977	−2.030	−1.059	−.3479
				Multiply all values of ker$_3$x and kei$_3$x by 10⁻⁴ and all those of ker$'_3$x and kei$'_3$x by 10⁻³						
r_4x	−47,753	−2,775	−410.6	−57.09	7.143	12.38	7.257	2.878	.6807	−.0722
i_4x	3,981	940.0	348.5	137.4	49.43	14.00	1.780	−1.193	−1.154	−.5843
r'_4x	192,000	5,966	740.2	136.7	20.43	−3.344	−5.361	−3.229	−1.318	−.3272
i'_4x	−8,035	−1,042	−323.6	−131.4	−54.82	−20.62	−6.088	−.8148	.5168	.5229
					Multiply all values by 10⁻³					
r_5x	287.8	10.21	1.468	327.1 ×	77.13 ×	12.98 ×	−1.719 ×	−3.146 ×	−1.874 ×	−.7460 ×
i_5x	253.9	6.077	.3531	−52.99 ×	−56.32 ×	−29.38 ×	−11.77 ×	−3.455 ×	−.4175 ×	.3241 ×
r'_5x	−1,408	−24.23	−2.403	−465.6 ×	−117.1 ×	−29.47 ×	−5.162 ×	.7744 ×	1.375 ×	.8372 ×
i'_5x	−1,306	−17.82	−1.125	−71.26 ×	26.42 ×	23.33 ×	11.28 ×	5.038 ×	1.529 ×	.2001 ×
					Multiply values marked × by 10⁻³					

TABLE V-24

$$J_0(xi^{\frac{3}{2}}) = M_0(x)e^{i\theta_0(x)} = \text{ber } x + i \text{ bei } x$$

x	$M_0(x)$	$\log_{10}\{\sqrt{x}M_0(x)\}$	$\theta_0(x)$	x	$M_0(x)$	$\log_{10}\{\sqrt{x}M_0(x)\}$	$\theta_0(x)$
.00	1.000	. .	.00°	2.0	1.229	.2401	52.29°
.05	1.000	$\bar{1}$.3995	.04	2.1	1.274	.2663	56.74
.10	1.000	$\bar{1}$.5000	.14	2.2	1.325	.2933	61.22
.15	1.000	$\bar{1}$.5880	.32	2.3	1.381	.3210	65.71
.20	1.000	$\bar{1}$.6505	.57	2.4	1.443	.3493	70.19
.25	1.000	$\bar{1}$.6990	.90°	2.5	1.511	.3783	74.65°
.30	1.000	$\bar{1}$.7386	1.29	2.6	1.586	.4077	79.09
.35	1.000	$\bar{1}$.7721	1.75	2.7	1.666	.4375	83.50
.40	1.000	$\bar{1}$.8012	2.29	2.8	1.754	.4676	87.87
.45	1.001	$\bar{1}$.8269	2.90	2.9	1.849	.4980	92.21
.50	1.001	$\bar{1}$.8499	3.58°	3.0	1.950	.5286	96.52°
.55	1.001	$\bar{1}$.8708	4.33	3.1	2.059	.5594	100.79
.60	1.002	$\bar{1}$.8900	5.15	3.2	2.176	.5902	105.03
.65	1.003	$\bar{1}$.9077	6.04	3.3	2.301	.6212	109.25
.70	1.004	$\bar{1}$.9242	7.01	3.4	2.434	.6521	113.43
.75	1.005	$\bar{1}$.9397	8.04°	3.5	2.576	.6830	117.60°
.80	1.006	$\bar{1}$.9543	9.14	3.6	2.728	.7140	121.75
.85	1.008	$\bar{1}$.9682	10.31	3.7	2.889	.7449	125.87
.90	1.010	$\bar{1}$.9815	11.55	3.8	3.061	.7758	129.99
.95	1.013	$\bar{1}$.9943	12.86	3.9	3.244	.8067	134.10
1.00	1.016	.0067	14.23°	4.0	3.439	.8375	138.19°
1.05	1.019	.0187	15.66	4.5	4.618	.9910	158.59
1.10	1.023	.0304	17.16	5.0	6.231	1.1441	178.93
1.15	1.027	.0419	18.72	5.5	8.447	1.2969	199.28
1.20	1.032	.0533	20.34	6.0	11.50	1.4498	219.62
1.25	1.038	.0645	22.02°	7.0	21.55	1.7560	260.29°
1.30	1.044	.0756	23.75	8.0	40.82	2.0624	300.92
1.35	1.051	.0867	25.54	9.0	77.96	2.3690	341.52
1.40	1.059	.0978	27.37	10.0	149.8	2.6756	382.10
1.45	1.067	.1089	29.26	11.0	289.5	2.9824	422.66
1.50	1.077	.1201	31.19°	12.0	561.8	3.2892	463.22°
1.55	1.087	.1314	33.16	14.0	2,137	3.9029	544.32
1.60	1.098	.1428	35.17	16.0	8,217	4.5168	625.40
1.65	1.111	.1544	37.22	18.0	3,185$_1$	5.1307	706.46
1.70	1.124	.1661	39.30	20.0	1,242$_2$	5.7447	787.52
1.75	1.139	.1779	41.41°	25.0	3,809$_3$	7.2798	990.15°
1.80	1.154	.1900	43.54	30.0	1,192$_5$	8.8150	1192.75
1.85	1.171	.2022	45.70	35.0	3,786$_6$	10.3502	1395.35
1.90	1.189	.2146	47.88	40.0	1,215$_8$	11.8856	1597.94
1.95	1.208	.2273	50.08	45.0	3,929$_9$	13.4209	1800.53

(1,215$_8$ represents $1,215 \times 10^8$)

When $x > 45$, $M_0(x)$ and $\theta_0(x)$ can be found to 4 decimal places and to the nearest .001°, respectively, from the formulae:

$$\log_{10}M_0(x) \simeq .307093x + \frac{.0384}{x} - .39909 - \tfrac{1}{2}\log_{10}x,$$

$$\theta_0(x) \simeq 40.51423x - \frac{5.06}{x} - 22.5 \text{ (degrees)}.$$

TABLE V-25
$$J_1(xi^{\frac{3}{2}}) = M_1(x)e^{i\theta_1(x)} = \text{ber}_1 x + i \text{ bei}_1 x$$

x	$M_1(x)$	$\log_{10}\{\sqrt{x}M_1(x)\}$	$\theta_1(x)$	x	$M_1(x)$	$\log_{10}\{\sqrt{x}M_1(x)\}$	$\theta_1(x)$
.00	.0000	..	135.00°	2.25	1.199	.2548	170.50°
.05	.0250	$\bar{3}$.7474	135.02	2.30	1.232	.2715	172.03
.10	.0500	$\bar{2}$.1990	135.07	2.35	1.266	.2881	173.58
.15	.0750	$\bar{2}$.4631	135.16	2.40	1.301	.3045	175.16
.20	.1000	$\bar{2}$.6505	135.29	2.45	1.337	.3207	176.76
.25	.1250	$\bar{2}$.7959	135.45°	2.50	1.374	.3368	178.39°
.30	.1500	$\bar{2}$.9147	135.64	2.55	1.411	.3529	180.03
.35	.1750	$\bar{1}$.0151	135.88	2.60	1.450	.3688	181.70
.40	.2000	$\bar{1}$.1021	136.15	2.65	1.489	.3846	183.39
.45	.2250	$\bar{1}$.1788	136.45	2.70	1.530	.4004	185.10
.50	.2500	$\bar{1}$.2475	136.79°	2.80	1.615	.4317	188.57°
.55	.2751	$\bar{1}$.3096	137.17	2.90	1.705	.4628	192.11
.60	.3001	$\bar{1}$.3663	137.58	3.00	1.800	.4938	195.71
.65	.3252	$\bar{1}$.4185	138.03	3.10	1.901	.5247	199.37
.70	.3502	$\bar{1}$.4669	138.51	3.20	2.009	.5555	203.08
.75	.3753	$\bar{1}$.5119	139.03°	3.30	2.124	.5863	206.83°
.80	.4004	$\bar{1}$.5541	139.58	3.40	2.246	.6171	210.62
.85	.4256	$\bar{1}$.5937	140.17	3.50	2.376	.6479	214.44
.90	.4508	$\bar{1}$.6311	140.80	3.60	2.515	.6788	218.30
.95	.4760	$\bar{1}$.6665	141.46	3.70	2.664	.7096	222.17
1.00	.5013	$\bar{1}$.7001	142.16°	3.80	2.823	.7405	226.07°
1.05	.5267	$\bar{1}$.7321	142.89	4.00	3.173	.8025	233.90
1.10	.5521	$\bar{1}$.7627	143.66	4.25	3.681	.8801	243.77
1.15	.5776	$\bar{1}$.7920	144.46	4.50	4.278	.9579	253.67
1.20	.6032	$\bar{1}$.8201	145.29	5.00	5.809	1.1136	273.55
1.25	.6290	$\bar{1}$.8471	146.17°	5.5	7.925	1.2692	293.48°
1.30	.6548	$\bar{1}$.8731	147.07	6.0	10.85	1.4245	313.45
1.35	.6808	$\bar{1}$.8982	148.02	6.5	14.90	1.5795	333.46
1.40	.7070	$\bar{1}$.9225	148.99	7.0	20.50	1.7343	353.51
1.45	.7333	$\bar{1}$.9460	150.00	7.5	28.27	1.8889	373.59
1.50	.7598	$\bar{1}$.9688	151.04°	8.0	39.07	2.0434	393.69°
1.55	.7866	$\bar{1}$.9909	152.12	9.0	74.97	2.3520	433.96
1.60	.8136	.0125	153.23	10.0	144.7	2.6604	474.28
1.65	.8408	.0335	154.38	11.0	280.4	2.9685	514.63
1.70	.8684	.0539	155.55	12.0	545.6	3.2765	555.02
1.75	.8962	.0739	156.76°	14.0	2,084	3.8920	635.84°
1.80	.9244	.0935	158.00	16.0	8,038	4.5072	716.72
1.85	.9530	.1127	159.27	18.0	$3,123_1$	5.1222	797.63
1.90	.9819	.1315	160.57	20.0	$1,220_2$	5.7370	878.57
1.95	1.011	.1499	161.90	25.0	$3,755_3$	7.2736	1080.98
2.00	1.041	.1680	163.27°	30.0	$1,178_5$	8.8099	1283.45°
2.05	1.072	.1859	164.66	35.0	$3,748_6$	10.3459	1485.94
2.10	1.102	.2035	166.08	40.0	$1,204_8$	11.8817	1688.46
2.15	1.134	.2208	167.53	45.0	$3,899_9$	13.4175	1890.98
2.20	1.166	.2379	169.00	50.0	$1,270_{11}$	14.9532	2093.52

$(1,178_5 \text{ represents } 1,178 \times 10^5)$

When $x > 50$, $M_1(x)$ and $\theta_1(x)$ can be found to 4 decimal places and to the nearest .001°, respectively, from the formulae:

$$\log_{10}M_1(x) \simeq .307093x - \frac{.1152}{x} - .39909 - \tfrac{1}{2}\log_{10}x,$$

$$\theta_1(x) \simeq 40.51423x + \frac{15.19}{x} + 67.5 \text{ (degrees)}.$$

TABLE V-26

$$K_0(xi^{\frac{1}{2}}) = N_0(x)e^{i\phi_0(x)} = \ker x + i \ker x$$

x	$N_0(x)$	$-\phi_0(x)$ (degrees)	x	$N_0(x)$	$-\phi_0(x)$ (degrees)
.0	∞	.000	5.0	.016 052	224.182
.1	2.542 1	17.794	5.1	.014 814	228.249
.2	1.891 7	23.626	5.2	.013 674	232.315
.3	1.525 0	28.733	5.3	.012 624	236.381
.4	1.274 6	33.517	5.4	.011 656	240.446
.5	1.087 94	38.119	5.5	.010 764 0	244.511
.6	.941 68	42.604	5.6	.009 942 0	248.575
.7	.823 26	47.008	5.7	.009 184 1	252.639
.8	.725 17	51.353	5.8	.008 485 2	256.703
.9	.642 54	55.654	5.9	.007 840 6	260.766
1.0	.572 03	59.920	6.0	.007 246 0	264.829
1.1	.511 25	64.159	6.1	.006 697 3	268.892
1.2	.458 43	68.375	6.2	.006 191 0	272.954
1.3	.412 21	72.573	6.3	.005 723 6	277.016
1.4	.371 55	76.755	6.4	.005 292 2	281.078
1.5	.335 60	80.925	6.5	.004 893 8	285.139
1.6	.303 68	85.083	6.6	.004 526 0	289.200
1.7	.275 25	89.232	6.7	.004 186 2	293.261
1.8	.249 85	93.372	6.8	.003 872 4	297.321
1.9	.227 09	97.505	6.9	.003 582 4	301.382
2.0	.206 64	101.632	7.0	.003 314 5	305.442
2.1	.188 24	105.753	7.1	.003 067 0	309.502
2.2	.171 65	109.869	7.2	.002 838 2	313.561
2.3	.156 66	113.981	7.3	.002 626 7	317.621
2.4	.143 09	118.088	7.4	.002 431 1	321.680
2.5	.130 81	122.192	7.5	.002 250 4	325.739
2.6	.119 66	126.293	7.6	.002 083 2	329.798
2.7	.109 53	130.390	7.7	.001 928 6	333.857
2.8	.100 319	134.485	7.8	.001 785 7	337.915
2.9	.091 936	138.578	7.9	.001 653 4	341.974
3.0	.084 299	142.668	8.0	.001 531 1	346.032
3.1	.077 335	146.756	8.1	.001 417 9	350.090
3.2	.070 979	150.842	8.2	.001 313 2	354.148
3.3	.065 175	154.927	8.3	.001 216 3	358.205
3.4	.059 870	159.009	8.4	.001 126 68	362.263
3.5	.055 018	163.090	8.5	.001 043 69	366.321
3.6	.050 578	167.170	8.6	.000 966 89	370.378
3.7	.046 513	171.249	8.7	.000 895 79	374.435
3.8	.042 789	175.326	8.8	.000 829 97	378.492
3.9	.039 376	179.402	8.9	.000 769 04	382.549
4.0	.036 246	183.477	9.0	.000 712 63	386.606
4.1	.033 374	187.551	9.1	.000 660 39	390.663
4.2	.030 738	191.624	9.2	.000 612 02	394.720
4.3	.028 318	195.697	9.3	.000 567 22	398.776
4.4	.026 095	199.768	9.4	.000 525 73	402.832
4.5	.024 052	203.839	9.5	.000 487 30	406.889
4.6	.022 174	207.909	9.6	.000 451 71	410.945
4.7	.020 447	211.978	9.7	.000 418 73	415.001
4.8	.018 859	216.047	9.8	.000 388 19	419.057
4.9	.017 398	220.115	9.9	.000 359 89	423.113
			10.0	.000 333 67	427.169

TABLE V-27

$$i^{-1}K_1(xi^{\frac{1}{2}}) = N_1(x)e^{i\phi_1(x)} = \ker_1 x + i\kei_1 x$$

x	$N_1(x)$	$-\phi_1(x)$ (degrees)	x	$N_1(x)$	$-\phi_1(x)$ (degrees)
.0	∞	135.000	5.0	.017 213	317.730
.1	9.962 0	135.840	5.1	.015 864	321.736
.2	4.928 0	137.593	5.2	.014 624	325.743
.3	3.231 5	139.869	5.3	.013 484	329.752
.4	2.372 3	142.497	5.4	.012 435	333.762
.5	1.850 1	145.376	5.5	.011 470 6	337.774
.6	1.497 6	148.443	5.6	.010 582 8	341.786
.7	1.243 10	151.656	5.7	.009 765 5	345.800
.8	1.050 59	154.984	5.8	.009 013 0	349.815
.9	.899 90	158.403	5.9	.008 320 0	353.831
1.0	.778 87	161.899	6.0	.007 681 5	357.849
1.1	.679 69	165.496	6.1	.007 093 2	361.867
1.2	.597 11	169.066	6.2	.006 550 9	365.886
1.3	.527 47	172.721	6.3	.006 051 1	369.906
1.4	.468 10	176.413	6.4	.005 590 2	373.927
1.5	.417 05	180.139	6.5	.005 165 1	377.949
1.6	.372 81	183.892	6.6	.004 773 0	381.972
1.7	.334 24	187.671	6.7	.004 411 2	385.995
1.8	.300 43	191.471	6.8	.004 077 4	390.019
1.9	.270 64	195.291	6.9	.003 769 4	394.044
2.0	.244 29	199.128	7.0	.003 485 0	398.070
2.1	.220 90	202.980	7.1	.003 222 4	402.096
2.2	.200 07	206.846	7.2	.002 980 0	406.123
2.3	.181 47	210.724	7.3	.002 756 1	410.150
2.4	.164 81	214.614	7.4	.002 549 3	414.178
2.5	.149 85	218.513	7.5	.002 358 3	418.207
2.6	.136 41	222.422	7.6	.002 181 8	422.236
2.7	.124 29	226.339	7.7	.002 018 7	426.266
2.8	.113 353	230.264	7.8	.001 868 0	430.296
2.9	.103 466	234.195	7.9	.001 728 7	434.327
3.0	.094 515	238.134	8.0	.001 599 9	438.358
3.1	.086 402	242.078	8.1	.001 480 8	442.389
3.2	.079 039	246.028	8.2	.001 370 8	446.421
3.3	.072 348	249.982	8.3	.001 269 0	450.454
3.4	.066 264	253.942	8.4	.001 174 85	454.487
3.5	.060 724	257.905	8.5	.001 087 79	458.520
3.6	.055 677	261.873	8.6	.001 007 26	462.554
3.7	.051 074	265.845	8.7	.000 932 76	466.588
3.8	.046 873	269.820	8.8	.000 863 83	470.623
3.9	.043 036	273.799	8.9	.000 800 06	474.657
4.0	.039 530	277.780	9.0	.000 741 04	478.693
4.1	.036 323	281.765	9.1	.000 686 43	482.728
4.2	.033 389	285.752	9.2	.000 635 88	486.764
4.3	.030 703	289.742	9.3	.000 589 10	490.800
4.4	.028 242	293.734	9.4	.000 545 79	494.836
4.5	.025 986	297.728	9.5	.000 505 70	498.873
4.6	.023 918	301.725	9.6	.000 468 58	502.910
4.7	.022 021	305.724	9.7	.000 434 21	506.947
4.8	.020 280	309.724	9.8	.000 402 39	510.985
4.9	.018 681	313.726	9.9	.000 372 92	515.023
			10.0	.000 345 63	519.061

6 APPLICATIONS

OF BESSEL FUNCTIONS

INTRODUCTION

In this chapter we present a selection of problems (with solutions) involving applications of Bessel functions. The selection is a modest one. But we trust that it gives a fair indication of the wide range of applications which can be made of the Bessel functions.

The first ten problems are purely mathematical applications in solving certain differential equations and in the expansion of simple functions in series of Bessel functions. The remaining two-thirds of the problems (in which several of the first ten are turned to account) comprise problems of application in mechanics, physics, hydrodynamics, heat-transfer, and electrical engineering. In the majority of these problems we have deemed it worthwhile to begin at the beginning, that is, to formulate the differential equations involved. These problems include the problem in which Bessel functions were first brought to light, namely, Daniel Bernouilli's problem (in the year 1732) concerning small oscillations in a hanging chain. They also include a problem of the modern era—the problem of flux distribution in a nuclear reactor.

Problems: Solutions of Equations Reducible to Bessel's Equation

VI-1. Find the general solution of

$$x^2y'' + xy' + (D^2x^2 - p^2)y = 0 \qquad \text{(VI-1.1)}$$

where the D and p are constants and the primes denote first and second derivatives with respect to x. ★

Equation (VI-1.1) differs from Bessel's equation (V-0.1) only in that D^2x^2 takes the place of x^2. Accordingly, we should be able to get the general solution of Eq. (VI-1.1) by making the change of variable $u = Dx$. Then $du/dx = D$ and $x = (1/D)u$. So, for the derivatives in Eq. (VI-1.1) we have

$$y' = \frac{dy}{dx} = \frac{dy}{du}\frac{du}{dx} = D\frac{dy}{du}, \qquad y'' = D^2\frac{d^2y}{du^2}. \qquad \text{(VI-1.2)}$$

Thus Eq. (VI-1.1) becomes

$$\frac{1}{D^2}u^2D^2\frac{d^2y}{du^2} + \frac{1}{D}uD\frac{dy}{du} + (u^2 - p^2)y = 0, \qquad \text{(VI-1.3)}$$

that is,

$$u^2\frac{d^2y}{du^2} + u\frac{dy}{du} + (u^2 - p^2)y = 0. \qquad \text{(VI-1.4)}$$

Equation (VI-1.4) is Bessel's equation whose general solution by Eq. (V-0.8) is $y = Z_p(u)$. This means that

$$y = Z_p(Dx) \qquad \text{(VI-1.5)}$$

is the general solution of

$$x^2y'' + xy' + (D^2x^2 - p^2)y = 0. \qquad \text{(VI-1.6)}$$

VI-2. Show that the general solution of

$$x^2y'' + (1 - 2A)xy' + \left[D^2E^2x^{2E} + (A^2 - E^2p^2) \right]y = 0, \qquad \text{(VI-2.1)}$$

is

$$y = x^A Z_p(Dx^E). \qquad ★ \qquad \text{(VI-2.2)}$$

We let $u = Dx^E$. By Eq. (V-0.8) the function $Z_p(u)$ is a linear combination of $J_p(u)$ and $J_{-p}(u)$ or of $J_p(u)$ and $Y_p(u)$. Let us show that any one of the three functions

$$y = x^A J_p(u), \quad y = x^A J_{-p}(u), \quad y = x^A Y_p(u), \quad \text{where} \quad u = Dx^E$$

is a solution of Eq. (VI-2.1). Then it will follow that Eq. (VI-2.2) is the general solution thereof.

Let us take $y = x^A J_p(u)$. For simplicity of notation we will omit the argument u and write J_p for $J_p(u)$. Primes on J will denote derivatives with respect to u. Primes on y will denote derivatives with respect to x. Then we have

$$
\begin{aligned}
y' &= \frac{d}{dx}\left[x^A J_p \right] \\
&= x^A \frac{d}{dx} J_p + J_p \frac{d}{dx} x^A \\
&= x^A J_p' DE x^{E-1} + J_p A x^{A-1} \\
&= DE x^{A+E-1} J_p' + A x^{A-1} J_p.
\end{aligned}
\tag{VI-2.3}
$$

By differentiating Eq. (VI-2.3) with respect to x we get

$$y'' = D^2 E^2 x^{A+2E-2} J_p'' + DE\left[(A + E - 1)x^{A+E-2} + A x^{A+E-2} \right] J_p' + A(A - 1)x^{A-2} J_p. \tag{VI-2.4}$$

Substitution of the right sides of Eqs. (VI-2.3) and (VI-2.4) for y' and y'' in the left side of Eq. (VI-2.1) makes the left side thereof become

$$E^2 x^A \left[D^2 x^{2E} J_p'' + D x^E J_p' + (D^2 x^{2E} - p^2) J_p \right].$$

Since $Dx^E = u$, we see that the left side of Eq. (VI-2.1) thus has become

$$E^2 x^A \left[u^2 \frac{d^2 J_p}{du^2} + u \frac{dJ_p}{du} + (u^2 - p^2) J_p \right].$$

The expression in brackets is now none other than the left side of Bessel's equation (V-0.1) with u as independent variable and J_p as function thereof. We conclude that Eq. (VI-2.1) is satisfied by $y = x^A J_p(u)$. Moreover, J_p may be replaced throughout by J_{-p} and Y_p in the preceding demonstration. It follows that Eq. (VI-2.1) is satisfied by an arbitrary linear combination of $x^A J_p(Dx^E)$ and $Y_p(Dx^E)$ or of $x^A J_p(Dx^E)$ and $x^A J_{-p}(Dx^E)$. Thus the general solution of Eq. (VI-2.1) is

$$y = x^A Z_p(u)$$
$$= x^A Z_p(Dx^E).$$

REMARK. One finds in similar manner that the general solution of

$$x^2 y'' + \left[(1 - 2A)x - 2BCx^{C+1} \right] y'$$
$$+ \left[D^2 E^2 x^{2E} + B^2 C^2 x^{2C} + BC(2A - C)x^C \right.$$
$$\left. + (A^2 - E^2 p^2) \right] y = 0 \qquad \text{(VI-2.5)}$$

is

$$y = x^A e^{Bx^C} Z_p(Dx^E). \qquad \text{(VI-2.6)}$$

VI-3. Obtain the general solution of

$$4y'' + 9xy = 0$$

in terms of Bessel Functions. ⋆

Multiplying by x^2 and dividing by 4, we have

$$x^2 y'' + \frac{9}{4} x^3 y = 0.$$

We can now identify with Eq. (VI-2.1), namely

$$x^2 y'' + (1 - 2A)xy' + \left[D^2 E^2 x^{2E} + (A^2 - E^2 p^2) \right] y = 0,$$

as follows:

$$1- 2A = 0, \quad 2E = 3, \quad D^2E^2 = \tfrac{9}{4}, \quad A^2 - E^2p^2 = 0,$$

whence

$$A = \tfrac{1}{2}, \quad E = \tfrac{3}{2}, \quad D = 1, \quad p = \tfrac{1}{3}.$$

Thus, by Eq. (VI-2.2), the general solution is

$$\begin{aligned}
y &= x^A Z_p(Dx^E) \\
&= x^{1/2} Z_{1/3}(x^{3/2}) \\
&= x^{1/2}[C_1 J_{1/3}(x^{3/2}) + C_2 J_{-1/3}(x^{3/2})] \qquad \text{by Eq. (V-0.8).}
\end{aligned}$$

VI-4. Express the general solution of

$$y'' + \frac{1}{x}y' - \left(1 + \frac{4}{x^2}\right)y = 0$$

in terms of pertinent Bessel functions. ★

As in Prob. VI-3 we make identification with Eq. (VI-2.1). This requires multiplication of the given equation by x^2:

$$x^2 y'' + xy' + (- x^2 - 4)y = 0,$$

which is now Eq. (VI-2.1) with

$$1 - 2A = 1, \quad 2E = 2, \quad D^2E^2 = -1, \quad A^2 - E^2p^2 = -4.$$

Thus we have

$$A = 0, \quad E = 1, \quad D = \sqrt{-1} = i, \quad p = 2,$$

so that, by Eq. (VI-2.2), the general solution of the given equation is

$$\begin{aligned}
y &= x^A Z_p(Dx^E) \\
&= Z_2(ix) \\
&= C_1 I_2(x) + C_2 K_2(x) \qquad \text{by Eq. (V-0.12).}
\end{aligned}$$

VI-5. Tell the general solution of

$$x^2 y'' + x^2 y' + \frac{1}{4} y = 0$$

in terms of appropriate Bessel Functions.   ★

Since the first derivative y' has the coefficient x^2, the given equation requires application of the more general Eq. (VI-2.5), so that we have

$$1 - 2A = 0, \quad C + 1 = 2, \quad -2BC = 1, \quad A^2 - E^2 p^2 = \tfrac{1}{4},$$
$$D^2 E^2 x^{2E} + B^2 C^2 x^{2C} = 0,$$

whence

$$A = \tfrac{1}{2}, \quad C = 1, \quad B = -\tfrac{1}{2}, \quad E^2 p^2 = 0, \quad D^2 E^2 x^{2E} + \tfrac{1}{4} x^2 = 0.$$

We must still determine D, E and p from the last two equations. The first of these, namely, $E^2 p^2 = 0$ is satisfied by either $E = 0$ or $p = 0$. If we try $E = 0$, then the other equation $D^2 E^2 x^{2E} + \tfrac{1}{4} x^2 = 0$ requires x to be identically zero, which is unsuitable. So, we must take $p = 0$ with $E \neq 0$. Then we require that

$$D^2 E^2 x^{2E} = -\tfrac{1}{4} x^2,$$

whence $2E = 2$, $D^2 E^2 = -\dfrac{1}{4}$ so that $E = 1$, $D = \dfrac{\sqrt{-1}}{2} = \dfrac{i}{2}$.

Thus the general solution of the given equation, by Eq. (VI-2.6), is

$$y = x^A e^{Bx^C} Z_p(Dx^E)$$
$$= x^{1/2} e^{-x/2} Z_0(\tfrac{1}{2} i x)$$
$$= x^{1/2} e^{-x/2} \Big[C_1 I_0(\tfrac{1}{2} x) + K_0(\tfrac{1}{2} x) \Big] \qquad \text{by Eq. (V-0.12).}$$

VI-6. Solve

$$x\frac{d^2y}{dx^2} + \frac{dy}{dx} - ixy = 0, \qquad i = \sqrt{-1}. \qquad \star \qquad \text{(VI-6.1)}$$

Multiplying both sides of Eq. (VI-6.1) by x and using primes to denote derivatives of y by x, we get

$$x^2y'' + xy' - ix^2y = 0,$$

which can now be solved by Eq. (VI-2.1). We have

$$1 - 2A = 1, \quad 2E = 2, \quad D^2E^2 = -i, \quad A^2 - E^2p^2 = 0,$$

whence

$$A = 0, \quad E = 1, \quad D = \sqrt{-i} = i \cdot i^{1/2} = i^{3/2}, \quad p = 0.$$

Then by Eqs. (VI-2.2) and (V-0.8), the general solution of Eq. (VI-6.1) is

$$y = C_1 J_0(i^{3/2}x) + C_2 Y_0(i^{3/2}x), \qquad \text{(VI-6.2)}$$

where C_1 and C_2 are arbitrary constants.

REMARKS.   In problems requiring solution of an equation of type (VI-6.1), the function Y_0 is often found to be unsuitable in view of imposed boundary conditions. The solution is then taken to be $C_1 J_0(i^{3/2}x)$ or a composite of such functions. In particular, the function $y = J_0(i^{3/2}x)$ is a solution of Eq. (VI-6.1). As mentioned in the introduction to Chapter V, the complex function $J_0(i^{3/2}x)$ is often written

$$J_0(i^{3/2}x) = \text{ber } x + i \text{ bei } x. \qquad \text{(VI-6.3)}$$

Thus, we may say that

$$y = \text{ber } x + i \text{ bei } x \qquad \text{(VI-6.4)}$$

is a particular solution of

$$xy'' + y' - ixy = 0. \qquad \text{(VI-6.5)}$$

Problems: Specific Expansions in Bessel Functions

VI-7. Express the function $f(x) = 1$ on the open interval $0 < x < L$ as an infinite series of Bessel functions of order zero. *

This problem is clearly an exercise in application of Prob. V-27. So, we should have $f(x)$ defined at every point on the closed interval $0 \leq x \leq L$. We take $f(0) = C_1$ and $f(L) = C_2$. The choice of the numbers C_1 and C_2 is immaterial; for, by the remarks at the end of Prob. V-26 (which apply to Prob. V-27 as well), the series expansion to be obtained via Prob. V-27 for the $f(x)$ of the present problem will converge to zero at $x = L$ and to 1 at $x = 0$ regardless of choice of C_1 and C_2.

By Eq. (V-27.2) we have

$$
\begin{aligned}
A_n &= \frac{2}{[LJ_1(\alpha_n)]^2} \int_0^L x \cdot 1 \cdot J_0\left(\frac{\alpha_n}{L}x\right) dx \\
&= \frac{2}{[LJ_1(\alpha_n)]^2} \int_0^L \frac{L}{\alpha_n} \frac{d}{dx}\left[x J_1\left(\frac{\alpha_n}{L}x\right)\right] dx \qquad \text{by Prob. V-8} \\
&= \frac{2}{\alpha_n L [J_1(\alpha_n)]^2} \left[x J_1\left(\frac{\alpha_n}{L}x\right)\right]_{x=0}^{x=L} \\
&= \frac{2}{\alpha_n J_1(\alpha_n)}.
\end{aligned}
$$

Thus the desired expansion is

$$
1 = 2 \sum_{n=1}^{\infty} \frac{J_0(\alpha_n x/L)}{\alpha_n J_1(\alpha_n)}, \qquad 0 < x < L. \tag{VI-7.1}
$$

Using Table V-4 to get the zeros α_1, α_2, α_3, and Table V-2 to obtain the values for $J_1(\alpha_n)$, we have

$$
\begin{aligned}
1 = 1.602 J_0\left(2.405\frac{x}{L}\right) &- 1.065 J_0\left(5.520\frac{x}{L}\right) \\
&+ 0.8512 J_0\left(8.654\frac{x}{L}\right) + \cdots .
\end{aligned}
$$

VI-8.   Expand the function $f(x) = x^2$, $0 \leqq x \leqq 1$ in a series of the form $A_1 J_0(\alpha_1 x) + A_2 J_0(\alpha_2 x) + A_3 J_0(\alpha_3 x) + \cdots$ where α_n denotes the nth positive zero of $J_0(x)$.   ★

By Eq. (V-26.2) we have

$$A_n = \frac{2}{[J_1(\alpha_n)]^2} \int_0^1 x^3 J_0(\alpha_n x) dx$$

$$= \frac{2}{[J_1(\alpha_n)]^2} \left[\frac{1}{\alpha_n^4} \left\{ \alpha_n^3 J_1(\alpha_n) - 2\alpha_n^2 J_2(\alpha_n) \right\} \right] \qquad \text{by Prob. V-30}$$

$$= \frac{2}{[J_1(\alpha_n)]^2} \left[\frac{J_1(\alpha_n)}{\alpha_n} - \frac{2 J_2(\alpha_n)}{\alpha_n^2} \right].$$

But, by Prob. V-7 with $p = 1$,

$$J_1(\alpha_n) = \frac{\alpha_n}{2} \left[J_2(\alpha_n) + J_0(\alpha_n) \right],$$

whence

$$J_2(\alpha_n) = \frac{2 J_1(\alpha_n)}{\alpha_n},$$

since $J_0(\alpha_n) = 0$. Accordingly,

$$A_n = \frac{2}{[J_1(\alpha_n)]^2} \left[\frac{J_1(\alpha_n)}{\alpha_n} - \frac{4 J_1(\alpha_n)}{\alpha_n^3} \right]$$

$$= \frac{2(\alpha_n^2 - 4)}{\alpha_n^3 J_1(\alpha_n)};$$

and the expression requested is

$$x^2 = 2 \sum_{n=1}^{\infty} \frac{(\alpha_n^2 - 4) J_0(\alpha_n x)}{\alpha_n^3 J_1(\alpha_n)}, \qquad 0 \leqq x \leqq 1.$$

VI-9.   Express as a series of the form

$$A_1 J_0(\alpha_1 x) + A_2 J_0(\alpha_2 x) + A_3 J_0(\alpha_3 x) + \cdots$$

the function $f(x) = J_0(Cx)$, $0 \leqq x \leqq 1$ where α_1, α_2, α_3, $\cdots$ are the positive zeros of $J_0(x)$ and C is any constant other than one of the numbers α_1, α_2, α_3, $\cdots$.   ★

By Eq. (V-26.2) we have

$$A_n = \frac{2}{[J_1(\alpha_n)]^2} \int_0^1 x J_0(Cx) J_0(\alpha_n x) dx$$

$$= \frac{2}{[J_1(\alpha_n)]^2} \frac{\alpha_n J_0(C) J_1(\alpha_n)}{\alpha_n^2 - C^2} \qquad \text{by Eq. (V-25.2)}$$

$$= \frac{2\alpha_n J_0(C)}{(\alpha_n^2 - C^2) J_1(\alpha_n)};$$

and the desired expansion is

$$J_0(Cx) = 2J_0(C) \sum_{n=1}^{\infty} \frac{\alpha_n J_0(\alpha_n x)}{(\alpha_n^2 - C^2) J_1(\alpha_n)}.$$

VI-10. Show that the Fourier series for the ordinate y of the cycloid

$$x = R(\phi - \sin \phi), \quad y = R(1 - \cos \phi) \qquad \text{(VI-10.1)}$$

is

$$y = R\left[\frac{3}{2} + 2\sum_{n=1}^{\infty} J_n''(n) \cos \frac{nx}{R}\right]. \qquad \star \qquad \text{(VI-10.2)}$$

Since the cycloid given by Eq. (VI-10.1) is symmetric with respect to the y-axis, which makes y an even function of x, the Fourier series for y will be a Fourier cosine series. Moreover, y is a periodic function of x of period $2\pi R$. Consequently, the Fourier expansion will be of the form

$$y = a_0 + \sum_{n=1}^{\infty} a_n \cos \frac{nx}{R}, \qquad \text{(VI-10.3)}$$

here

$$a_0 = \frac{1}{\pi R} \int_0^{\pi R} y \, dx \quad \text{and} \quad a_n = \frac{2}{\pi R} \int_0^{\pi R} y \cos \frac{nx}{R} \, dx, \qquad n > 0.$$

$$\text{(VI-10.4)}$$

To determine the coefficients required by Eq. (VI-10.4) we make use of the parametric representation Eq. (VI-10.1). The period for y with respect to ϕ is now 2π. Moreover, when ϕ varies from 0 to π, then x varies monotonically and continuously and differentiably from 0 to πR with $dx = R(1 - \cos \phi)d\phi$. Accordingly, the coefficient formulas (VI-10.4) when expressed in terms of ϕ become

$$a_0 = \frac{R}{\pi} \int_0^\pi (1 - \cos \phi)^2 d\phi,$$

$$a_n = \frac{2R}{\pi} \int_0^\pi \cos (n\phi - n \sin \phi)(1 - \cos \phi)^2 d\phi, \qquad n = 1, 2, 3, \cdots.$$

$$(\text{VI-10.5})$$

The coefficient a_0 is readily found by elementary integration to be $3R/2$. The remaining coefficients are not so readily determined. However, a clue to their determination is furnished by Eq. (V-37.7), namely

$$J_n(x) = \frac{1}{\pi} \int_0^\pi \cos (n\phi - x \sin \phi)d\phi. \qquad (\text{VI-10.6})$$

Indeed, if the factor $(1 - \cos \phi)^2$ in the integrand of the integral for a_n be written out as $1 - 2 \cos \phi + \cos^2 \phi = 2 - \cos \phi - \sin^2 \phi$ and then the integral be written as the sum of three integrals, the first of the three integrals with its coefficient $2R/\pi$ will be none other than $2RJ_n(x)$. But what of the other two integrals? Answer: apply Leibniz's rule for differentiation under the integral sign. This technique we have found useful in problems elsewhere in this book. See, for example, Prob. I-20. Differentiation of Eq. (VI-10.6) yields

$$J_n'(x) = \frac{1}{\pi} \int_0^\pi \sin (n\phi - x \sin \phi) \sin \phi \, d\phi,$$

$$J_n''(x) = \frac{-1}{\pi} \int_0^\pi \cos (n\phi - x \sin \phi) \sin^2 \phi \, d\phi.$$

Moreover,

$$2\int_0^\pi \cos(n\phi - x\sin\phi)\cos\phi\,d\phi = \int_0^\pi \cos\left[(n+1)\phi - x\sin\phi\right]d\phi$$
$$+ \int_0^\pi \cos\left[(n-1)\phi - x\sin\phi\right]d\phi$$
$$= \pi J_{n+1}(x) + \pi J_{n-1}(x)$$

by Eq. (VI-10.6)

$$= \pi\frac{2n}{x}J_n(x) \qquad \text{by Eq. (V-7.1).}$$

So, we have from Eq. (VI-10.5), wherein the x in Eq. (VI-10.6) is now n,

$$a_n = 2R\left[J_n(n) - \frac{n}{n}J_n(n) + J_n''(n)\right] = 2RJ_n''(n).$$

Thus we find that the Fourier series for the ordinate y of the cycloid (VI-10.1) is given by Eq. (VI-10.2).

REMARK. Equation (VI-10.2) is of interest in that it exhibits y explicitly as a function of x. It is worth noting here that, on the other hand, determination of x as an explicit function of y is readily effected as a closed form in terms of elementary functions. One has only to solve the second of Eqs. (VI-10.1) for ϕ in terms of y, obtaining $\phi = \arccos[(R - y)/R]$, and then to substitute this value for ϕ in the first of Eqs. (VI-10.1). The result is readily seen to be

$$x = R\left[\cos^{-1}\left(\frac{R-y}{R}\right) \pm \frac{\sqrt{2R-y^2}}{R}\right]. \qquad \text{(VI-10.7)}$$

This form of the equation of the cycloid is encountered, for instance, in the calculus of variations problem of determining the equation of the brachystochrone.

Problems in Dynamics

VI-11. A coil spring is such that a force of 8 lbs. will stretch the spring 6 inches. If the spring is suspended in a vertical position with a 4-lb. weight attached to the lower end (hanging free), and then the lower end is pushed up to a point 2 inches above the point of equilibrium and released, determine the equation of motion. Use 32 ft./sec². for g.   ⋆

The differential equation is formulated as follows.

$$\text{Force} = (\text{mass})(\text{acceleration})$$

$$= \frac{d}{dt}\left(\frac{W}{g}\frac{dy}{dt}\right).$$

Since W/g is a constant in this problem, we get by carrying out the indicated differentiation

$$F = \frac{W}{g}\frac{d^2y}{dt^2}.$$

This force is equated to the restorative force, which is the product of the spring constant k and the displacement y of the lower end from the point of equilibrium. We have

$$8 \text{ lbs.} = k \text{ (stretch-distance in feet)}$$

$$= k\tfrac{6}{12},$$

whence $k = 16$. Thus the restorative force is $-16y$. So, upon equating these two forces, we have the differential equation of motion (ignoring friction):

$$\frac{4}{32}\frac{d^2y}{dt^2} + 16\,y = 0$$

or

$$\frac{d^2y}{dt^2} + 128y = 0. \qquad\qquad (\text{VI-11.1})$$

Although Eq. (VI-11.1) is immediately recognized as the well known equation of simple harmonic motion and is quickly and easily solved in terms of trigonometric functions, we will solve it here via Bessel functions as an exercise in application of Eq. (VI-2.1). We now identify Eq. (VI-11.1) with Eq. (VI-2.1) by multiplying both sides of Eq. (VI-11.2) by t^2. The variable t in the present problem plays the role of x in Eq. (VI-2.1). We have

$$t^2 y'' + 128 t^2 y = 0,$$

so that

$$1 - 2A = 0, \quad 2E = 2, \quad D^2 E^2 = 128, \quad A^2 - E^2 p^2 = 0,$$

whence

$$A = 1/2, \quad E = 1, \quad D = \sqrt{128}, \quad p = 1/2.$$

Then by Eq. (V-0.8) we have

$$\begin{aligned}
y &= t^A Z_p(Dt^E) \\
&= t^{1/2} Z_{1/2}(\sqrt{128}\, t) \\
&= t^{1/2}\Big[C_1 J_{1/2}(\sqrt{128}\, t) + C_2 J_{-1/2}(\sqrt{128}\, t) \Big]. \quad \text{(VI-11.2)}
\end{aligned}$$

By Eqs. (V-18.4 and 17.4) we may write Eq. (VI-11.2) as

$$y = A \sin(\sqrt{128}\, t) + B \cos(\sqrt{128}\, t), \quad \text{(VI-11.3)}$$

where $A = \sqrt{2/\pi}\, C_1$ and $B = \sqrt{2/\pi}\, C_2$.

The constants A and B in Eq. (VI-11.3) are determined by the initial conditions taking positive y downward

$$y = -2 \text{ inches} = -\tfrac{1}{6} \text{ ft. at } t = 0, \, dy/dt = 0 \text{ at } t = 0,$$

whence $A = 0$ and $B = -1/6$. Thus the desired equation of motion is

$$y = -\tfrac{1}{6} \cos(\sqrt{128}\, t).$$

VI-12. Starting at rest at distance L from the origin O a mass-particle P of varying mass m is attracted to the origin by a force directed always toward the origin and having magnitude proportional to the product my, where y is the distance of P from the origin. The mass m of P decreases with the time t according to the formula

$$m = \frac{1}{a + bt},$$ (VI-12.1)

where a and b are constants. The problem is to find the time required for P to reach 0.   ★

Our starting point is the Newtonian equation

$$\frac{d\vec{M}}{dt} = \vec{F}$$ (VI-12.2)

where $\vec{M}$ is the vector momentum and $\vec{F}$ is the acting force. For the problem at hand Eq. (VI-12.2) yields

$$\frac{d}{dt}\left(m\frac{dy}{dt}\right) = -k^2 my,$$

that is,

$$m\frac{d^2y}{dt^2} + \frac{dm}{dt}\frac{dy}{dt} + k^2 my = 0,$$ (VI-12.3)

where k^2 is the constant of proportionality involved in the magnitude of $\vec{F}$.

If we make the change of variable

$$a + bt = bx,$$ (VI-12.4)

so that $m = 1/bx$ and $dy/dt = dy/dx$, then Eq. (VI-12.3) becomes

$$\frac{d^2y}{dx^2} - \frac{1}{x}\frac{dy}{dx} + k^2 y = 0,$$

that is,

$$x^2 y'' - xy' + k^2 x^2 y = 0.$$ (VI-12.5)

Identifying Eq. (VI-12.5) with Eq. (VI-2.1), we have

$$1 - 2A = -1, \quad 2E = 2, \quad D^2E^2 = k^2, \quad A^2 - E^2p^2 = 0,$$

whence $A = 1$, $E = 1$, $D = k$, $p = 1$.
Thus, by Eq. (VI-2.1), the general solution of Eq. (VI-12.5) is

$$y = C_1 x J_1(kx) + C_2 x Y_1(kx). \tag{VI-12.6}$$

To determine the constants C_1 and C_2 we have two conditions:

(a) $y = L$ when mass is at rest at $t = 0$, that is, when $x = a/b$,
(b) the velocity $dy/dt = dy/dx = 0$ when $x = a/b$.

If we put $q = ka/b$ and make use of formula (A) in Table V-1, these equations of condition are

$$\left. \begin{array}{l} C_1 J_1(q) + C_2 Y_1(q) = \dfrac{bL}{a} \\[2mm] C_1 J_0(q) + C_2 Y_0(q) = 0. \end{array} \right\} \tag{VI-12.7}$$

By Prob. V-16 the determinant of Eqs. (VI-12.7), namely $J_1(q) Y_0(q) - J_0(q) Y_1(q)$ has the value of $2/\pi q = 2b/\pi ka$.

Accordingly, the solution of Eqs. (VI-12.7) is

$$C_1 = \tfrac{1}{2}\pi kL\, Y_0(ka/b), \quad C_2 = -\tfrac{1}{2}\pi kL J_0(ka/b).$$

To find the time required for P to reach O, we set $y = 0$ in Eq. (VI-12.6) and solve for x. This we can do by locating two successive entries in Tables V-2, V-5, V-6 which make the difference

$$Y_0(ka/b)J_1(kx) - J_0(ka/b) Y_1(kx) \tag{VI-12.8}$$

respectively positive and negative. Once x has been found with desired accuracy, the corresponding value of the time t is given by Eq. (VI-12.4).

VI-13. A mass-particle of varying mass m is repelled from the origin by a force $\vec{F}$ of varying magnitude but always directed away from the origin. The mass of the particle increases with the time t as follows:

$$m = m(t) = a + bt, \tag{VI-13.1}$$

where a and b are positive constants. The magnitude of $\vec{F}$ is proportional to the product of the mass m and the distance x of the particle from the origin:

$$F = \left|\vec{F}\right| = k^2mx, \tag{VI-13.2}$$

where $k^2 \neq 0$ and is the constant of proportionality. Since $F = 0$ at $x = 0$, in order to get the particle going we assume an initial speed V_0 greater than zero at the origin at time $t = 0$. Express the distance x and the speed V each as function of time t. It is to be understood that the motion is to be stopped somehow before t becomes very large, so that there will be no conflict with the theory of relativity.   *

This problem is clearly a variant of Prob. VI-12, where the mass decreases and the force attracts to the origin. Accordingly, we will omit those details of the solution which are essentially the same as in the previous problem. It will be seen, however, that the present problem involves the functions I_0, I_1, K_0, K_1 instead of their counterparts J_0, J_1, Y_0, Y_1.

Instead of $(d/dt)[m\,dy/dt] = -k^2my$, as in the previous problem, we now have

$$\frac{d}{dt}\left(m\frac{dx}{dt}\right) = k^2mx, \tag{VI-13.3}$$

that is,

$$m\frac{d^2x}{dt^2} + \frac{dm}{dt}\frac{dx}{dt} - k^2mx = 0. \tag{VI-13.4}$$

Making the same sort of change of time variable as in Prob. VI-12, namely $a + bt = u$, we find that Eq. (VI-13.4) becomes

$$mb^2\frac{d^2x}{du^2} + b^2\frac{dx}{du} - k^2mx = 0. \qquad \text{(VI-13.5)}$$

Multiplying Eq. (VI-13.5) through by m/b^2 and using primes to denote first and second derivative by u and observing that now $m = u$, we get

$$u^2x'' + ux' - h^2u^2x = 0, \qquad \text{(VI-13.6)}$$

where

$$h = \frac{k}{b}. \qquad \text{(VI-13.7)}$$

Identifying Eq. (VI-13.6) with Eq. (VI-2.1) as we did for the corresponding equation in Prob. VI-12, we find that the general solution of Eq. (VI-13.6) is

$$x = C_1I_0(hu) + C_2K_0(hu). \qquad \text{(VI-13.8)}$$

We have yet to determine the constants C_1 and C_2 by application of the two initial conditions. At time $t = 0$, that is, when $u = a$, we have given that $x = 0$:

$$0 = C_1I_0(ha) + C_2K_0(ha). \qquad \text{(VI-13.9)}$$

We also have given that the velocity $V = V_0$ at time $u = a$. Now,

$$\begin{aligned} V = \frac{dx}{dt} &= C_1\frac{d}{dt}\Big[I_0(hu)\Big] + C_2\frac{d}{dt}\Big[K_0(hu)\Big] \\ &= \left\{C_1\frac{d}{du}\Big[I_0(hu)\Big] + C_2\frac{d}{du}\Big[K_0(hu)\Big]\right\}\frac{du}{dt} \\ &= \Big[C_1hI_1(hu) - C_2hK_1(hu)\Big]b. \end{aligned}$$

Thus,

$$\frac{V_0}{bh} = C_1 I_1(ha) - C_2 K_1(ha). \qquad \text{(VI-13.10)}$$

In taking derivatives with respect to u we have applied formulas (D) and (C) in Table V-1.

Equations (VI-13.9) and (VI-13.10) determine C_1 and C_2 in terms of the given constants V_0, k, a, b. If we write these equations together as

$$\left.\begin{aligned}
I_1(ha)C_1 - K_1(ha)C_2 &= \frac{V_0}{bh} \\
I_0(ha)C_1 + K_0(ha)C_2 &= 0
\end{aligned}\right\}, \qquad \text{(VI-13.11)}$$

we notice that the determinant of the coefficients, namely $I_1(ha)K_0(ha) + I_0(ha)K_1(ha)$, is equal to $1/ha$ by Eq. (V-16.2). Accordingly, solution of Eqs. (VI-13.11) by Cramer's Rule (by determinants) yields

$$C_1 = \frac{aV_0}{b}K_0(ha), \qquad C_2 = -\frac{aV_0}{b}I_0(ha). \qquad \text{(VI-13.12)}$$

Recalling that $u = a + bt$ and that $h = k/b$, we find by substitution from Eq. (VI-13.12) in Eq. (VI-13.8) that

$$x = \frac{aV_0}{b}\left[K_0\left(\frac{ka}{b}\right)I_0\left(kt + \frac{ka}{b}\right) - I_0\left(\frac{ka}{b}\right)K_0\left(kt + \frac{ka}{b}\right)\right]. \qquad \text{(VI-13.13)}$$

Differentiation of Eq. (VI-13.13) together with application of formulas (D) and (C) in Table V-1 will provide the requested expression for the speed V as a function of t. In applying the formulas just indicated we put $x = kt + ka/b$ and remember that

$$\frac{d}{dt}I_0(x) = \left[\frac{d}{dx}I_0(x)\right]\frac{dx}{dt} = k\frac{d}{dx}I_0(x). \qquad \text{(VI-13.14)}$$

We find that

$$V = \frac{dx}{dt} = \frac{aV_0k}{b}\left[K_0\left(\frac{ka}{b}\right)I_1\left(kt + \frac{ka}{b}\right) + I_0\left(\frac{ka}{b}\right)K_1\left(kt + \frac{ka}{b}\right)\right].$$
(VI-13.15)

VI-14. A simple pendulum consisting of a string (considered weightless) with constant mass m attached at lower end is so arranged that, as the pendulum oscillates, the string is continuously lengthened at a constant rate from the support so that the length r at time t is $a + bt$, where the initial length a and velocity b are constants. The oscillations are quite small, so that θ may be taken as sufficient approximation for $\sin \theta$, where θ is the angle which the string makes with the vertical. θ_0 is the angle at $t = 0$. Determine θ as a function of t.   ★

By Newton's second law of motion applied to angular momentum, the time rate of change of angular momentum (with respect to the point of support) is equal in magnitude to the moment (about the point of support) of the restoring force of gravity.

The angular momentum A, sometimes called the moment of momentum, is the product of the rotational inertia I_r about the point of the support and the angular velocity $d\theta/dt$. Since $I_r = mr^2$, we have for the angular momentum

$$A = mr^2\frac{d\theta}{dt},$$
(VI-14.1)

whence it follows that the time rate of change of angular momentum is

$$\frac{dA}{dt} = \frac{d}{dt}\left(mr^2\frac{d\theta}{dt}\right).$$
(VI-14.1a)

The restoring force is the component of gravity normal to the string. Its magnitude is, therefore, $mg \sin \theta$, which we may replace with sufficient approximation by $mg\theta$. Its moment arm about the point of support is r. Then by Newton's second law we have

$$\frac{dA}{dt} = -(mg\theta)r. \tag{VI-14.2}$$

The minus sign on the right in Eq. (VI-14.2) is required because A as defined by Eq. (VI-14.1) is a signed quantity and θ is signed and because dA/dt is negative when the pendulum is slowing down on an upswing when θ is positive. Similarly, the other cases—such as speeding up on a downswing when θ (taken clockwise) is negative—will be seen to require the minus sign on the right in Eq. (VI-14.2).

Now we equate the right side of Eq. (VI-14.1a) with the right side of Eq. (VI-14.2). In carrying out the indicated differentiation in Eq. (VI-14.1a) we observe that r is a function of t. So we have

$$m\left(r^2\frac{d^2\theta}{dt^2} + 2r\frac{dr}{dt}\frac{d\theta}{dt}\right) + mgr\theta = 0. \tag{VI-14.3}$$

Replacing r by $a + bt$, dr/dt by b and dividing out m in Eq. (VI-14.3), we have

$$\frac{d^2\theta}{dt^2} + \frac{2b}{a + bt}\frac{d\theta}{dt} + \frac{g}{a + bt}\theta = 0. \tag{VI-14.4}$$

For the sake of simplicity we introduce a new time variable z defined by $a + bt = bz$. Then $d\theta/dt = d\theta/dz$; and Eq. (VI-14.4) becomes, upon multiplying through by z^2,

$$z^2\theta'' + 2z\theta' + k^2z\theta = 0, \tag{VI-14.5}$$

where the primes denote first and second derivative with respect to z and $k^2 = g/b$.

Equation (VI-14.5) can be solved by application of Eq. (VI-2.1) by identifying z with x and θ with y. We have

$$1 - 2A = 2, \quad 2E = 1, \quad D^2E^2 = k^2, \quad A^2 - E^2p^2 = 0,$$

whence $A = -\frac{1}{2}$, $E = \frac{1}{2}$, $D = 2k$, $p = 1$.

Thus, by Eqs. (VI-2.2) and (V-0.8), the general solution of Eq. (VI-14.5) is

$$\theta = \frac{C_1}{\sqrt{z}} J_1(2k\sqrt{z}) + \frac{C_2}{\sqrt{z}} Y_1(2k\sqrt{z}). \qquad \text{(VI-14.6)}$$

It remains only to determine the constants C_1 and C_2. We count time from an instant of maximum angular displacement θ_0, when the velocity $d\theta/dt = 0$. At that instant $t = 0$ we have $r = a$ and $z = a/b$. Differentiating Eq. (VI-14.6) and applying formula (C) in Table V-1 with $x = \sqrt{z}$, we get

$$\frac{d\theta}{dt} = \frac{d\theta}{dz} = -\frac{C_1 k}{z} J_2(2k\sqrt{z}) - \frac{C_2 k}{z} Y_2(2k\sqrt{z}). \qquad \text{(VI-14.7)}$$

At time $t = 0$, Eq. (VI-14.7) yields

$$0 = C_1 J_2(2k\sqrt{a/b}) + C_2 Y_2(2k\sqrt{a/b}). \qquad \text{(VI-14.8)}$$

At time $t = 0$, Eq. (VI-14.6) yields

$$\theta_0 \sqrt{a/b} = C_1 J_1(2k\sqrt{a/b}) + C_2 Y_1(2k\sqrt{a/b}). \qquad \text{(VI-14.9)}$$

Solving Eqs. (VI-14.8) and (VI-14.9), we find by application of Eq. (V-16.1) that

$$C_1 = -\frac{\pi\theta_0 a}{b}\left(\frac{g}{b}\right)^{1/2} Y_2\left(\frac{2\sqrt{ag}}{b}\right),$$

$$C_2 = \frac{\pi\theta_0 a}{b}\left(\frac{g}{b}\right)^{1/2} J_2\left(\frac{2\sqrt{ag}}{b}\right).$$

Substituting these values for C_1 and C_2 in Eq. (VI-14.6) and noting that $z = (a + bt)/b$, we have for the solution to our problem

$$\theta = \frac{\pi\theta_0 a}{b}\sqrt{\frac{g}{a+bt}}\left[J_2\left(\frac{2\sqrt{ag}}{b}\right)Y_1\left(\frac{2}{b}\sqrt{g(a+bt)}\right)\right.$$
$$\left. - J_1\left(\frac{2}{b}\sqrt{g(a+bt)}\right)Y_2\left(\frac{2\sqrt{ag}}{b}\right)\right].$$

VI-15. Uniform radial pressure P is applied to the rim of a homogeneous circular disc of thickness $2h$ and radius R. The pressure P is continuously increased from $P = 0$. Determine the value of P at which the disc will buckle, that is, be bent permanently so that there is no longer any restorative (elastic) force. The rim of the disc is not clamped. The following are given.

(a) If cylindrical coordinates (r, θ, z) are taken with origin at center of the disc and with z-axis perpendicular to the plane faces of the disc and if it be assumed that displacements of points of the disc are parallel to the z-axis and are the same for all points at the same distance from the z-axis, then the equation of equilibrium is

$$\left\{\frac{1}{r}\frac{d}{dr}\left(r\frac{d}{dr}\right)\right\}\left[\left\{\frac{1}{r}\frac{d}{dr}\left(r\frac{d}{dr}\right) + k^2\right\}z\right] = 0, \qquad \text{(VI-15.1)}$$

where z denotes displacement of points at distance r from the z-axis: $z = z(r)$.

(b) There is no displacement at the rim:

$$z(R) = 0. \qquad \text{(VI-15.2)}$$

(c) When the disc buckles the bending moment at the rim vanishes. This is equivalent to the condition

$$\frac{d^2z}{dr^2} + \frac{\sigma}{r}\frac{dz}{dr} = 0, \qquad r = R, \qquad \text{(VI-15.3)}$$

where σ is a positive constant less than unity and known as Poisson's ratio. (See, for instance, Seely and Smith, *Advanced Mechanics of Materials*, Wiley, 1952.)

(d) The constant k in Eq. (VI-15.1) is given by

$$k^2 = \frac{3P(1 - \sigma)}{h^2E}, \qquad \text{(VI-15.4)}$$

where E is Young's modulus.   ★

Since the operators are permutable we can write

$$\left[\frac{1}{r}\frac{d}{dr}\left(r\frac{d}{dr}\right)\right]z = 0, \qquad \left[\frac{1}{r}\frac{d}{dr}\left(r\frac{d}{dr} + k^2\right)\right]z = 0. \tag{VI-15.5}$$

The first of Eqs. (VI-15.5) becomes

$$\frac{d^2z}{dr^2} + \frac{1}{r}\frac{dz}{dr} = 0, \tag{VI-15.6}$$

whose general solution is

$$z = A \log_e r + B, \tag{VI-15.7}$$

where A and B are arbitrary constants. The second of Eqs. (VI-15.5) becomes

$$\frac{d^2z}{dr^2} + \frac{1}{r}\frac{dz}{dr} + k^2z = 0, \tag{VI-15.8}$$

that is,

$$r^2z'' + rz' + k^2r^2z = 0, \tag{VI-15.9}$$

where the primes denote first and second derivative with respect to r. Equation (VI-15.9) is readily solved by making identification with Eq. (VI-1.6) in which we take $r = x$, $z = y$, $D = k$. Thus, by Eqs. (VI-1.5) and (V-0.8), the general solution of Eq. (VI-15.7) is

$$z = CJ_0(kr) + DY_0(kr). \tag{VI-15.10}$$

The general solution for Eq. (VI-15.1) is the sum of the solutions given by Eqs. (VI-15.7) and (VI-15.10):

$$z = A \log_e r + B + CJ_0(kr) + DY_0(kr). \tag{VI-15.11}$$

For, the four terms on the right in Eq. (VI-15.11) are not linearly dependent and the solution (VI-15.11) contains four arbitrary constants as required for the general solution of a differential equation of fourth order. [Each operator in (VI-15.1) involves second order derivatives; and when the operations are performed in tandem, the result involves derivatives of order four and less.]

The conditions of the problem prohibit z from becoming infinite at any point of the disc. So, we must take $A = 0$ and $D = 0$, since $\log_e r$ and $Y_0(kr)$ both become infinite at $r = 0$. We have, then,

$$z = B + CJ_0(kr). \tag{VI-15.12}$$

The boundary condition (b) requires the additive constant B to be related to the multiplicative constant C:

$$B = - CJ_0(kR).$$

This leaves C undetermined, unless a further displacement condition be empirically established for a point of the disc for which $r < R$. However, we can determine the pressure P with C undetermined.

To determine the value of P at which the disc buckles we now proceed as follows. We put z, as given by Eq. (VI-15.12), in Eq. (VI-15.3) and solve the resulting equation (called the critical equation) for k. Then we put this value for k in Eq. (VI-15.4) and solve for P.

From Eq. (VI-15.12) we have

$$\frac{dz}{dr} = C \frac{d}{dr}\left[J_0(kr) \right];$$

and application of formula (C) in Table V-1 yields

$$\frac{dz}{dr} = - kCJ_1(kr). \tag{VI-15.13}$$

Similarly, differentiation of Eq. (VI-15.13) with application of formula (E) in Table V-1 yields

$$\frac{d^2z}{dr^2} = - kC\left[kJ_0(kr) - \frac{J_1(kr)}{r} \right]. \tag{VI-15.14}$$

Multiplication of Eq. (VI-15.13) by σ/r and addition of the resulting equation to Eq. (VI-15.14) makes the condition for buckling, namely Eq. (VI-15.3), to be

$$(- kC)\, kJ_0(kr) + \frac{\sigma - 1}{r}J_1(kr) = 0 \text{ at } r = R,$$

which means that

$$kRJ_0(kR) = (1 - \sigma)J_1(kR). \qquad \text{(VI-15.15)}$$

As the pressure P is continuously increased from $P = 0$, k as given by Eq. (VI-15.4) will be continuously increased from $k = 0$. Buckling will, therefore, occur at the least positive k for which the critical Eq. (VI-15.15) is satisfied. This value of k will in turn determine by Eq. (VI-15.4) the value of P at which the buckling occurs.

VI-16. A homogeneous straight steel wire having circular cross-section is in a nearly vertical position with its lower end clamped at a small angle θ_0 to the vertical and with the upper end free. The wire will be stable in this position if its length will not be too great and if θ_0 is sufficiently small. Determine the critical length, that is, the upper limit of lengths for which the wire will be stable in the vertical position. It is given that, if x denotes height (above the level of the clamped end) of our arbitrary point P of a wire in a stable position for small θ_0 and y denotes small horizontal displacement of P from the vertical line through the clamped end, then y satisfies the equation

$$\frac{d^3y}{dx^3} + \frac{W}{EI}\frac{dy}{dx} = 0, \qquad \text{(VI-16.1)}$$

where W is the weight of that portion of the wire which is above P and EI is the flexural rigidity of the wire, E denoting Young's modulus and I denoting the moment of inertia of the cross-sectional area A with respect to a diameter. $\qquad \star$

We let L denote the length of the wire and let w denote the weight per unit of length. Then

$$W = w(L - x), \tag{VI-16.2}$$

so that Eq. (VI-16.1) becomes

$$\frac{d^3y}{dx^3} + \frac{w}{EI}(L - x)\frac{dy}{dx} = 0. \tag{VI-16.3}$$

In order to solve Eq. (VI-16.3) we will let $p = dy/dx$. And for convenience we will let $k^2 = w/EI$; and we will let $z = L - x$. Then in place of Eq. (VI-16.3) we have

$$\frac{d^2p}{dz^2} + k^2zp = 0,$$

that is,

$$z^2p'' + k^2z^3p = 0. \tag{VI-16.4}$$

We now have our differential equation for small displacements in a form where we can easily solve it by identification with Eq. (VI-2.1) where p plays the role of y and z plays the role of x. We have

$$1 - 2A = 0, \quad 2E = 3, \quad D^2E^2 = k^2, \quad A^2 - E^2p^2 = 0,$$

whence

$$A = \tfrac{1}{2}, \quad E = \tfrac{3}{2}, \quad D = \tfrac{2}{3}k, \quad p = \tfrac{1}{3}.$$

Thus, the general solution of Eq. (VI-16.4) via Eqs. (VI-2.2) and (V-0.8) is

$$p = z^{1/2}\Big[C_1J_{1/3}(\tfrac{2}{3}kz^{3/2}) + C_2J_{-1/3}(\tfrac{2}{3}kz^{3/2}) \Big],$$

where C_1 and C_2 are arbitrary constants. In order to simplify the notation let us put $a = 2k/3$ and $r = z^{3/2}$, so that the general solution of Eq. (VI-16.4) now reads

$$p = r^{1/3}\Big[C_1J_{1/3}(ar) + C_2J_{-1/3}(ar) \Big]. \tag{VI-16.5}$$

Evaluation of the constants C_1 and C_2 is determined by the boundary conditions at the ends of the wire. At the upper end, which is free, the bending moment is zero. This is equivalent to requiring

$$\frac{d^2y}{dx^2} = 0 \qquad \text{at } x = L,$$

which in turn is equivalent to

$$\frac{dp}{dz} = 0 \qquad \text{at } z = 0, \tag{VI-16.6}$$

since $d^2y/dx^2 = dp/dx = (dp/dz)(-1)$. But $r = z^{3/2}$. Thus, Eq. (VI-16.6) requires that

$$\left[\frac{dp}{dr}\right]\left[\frac{3}{2}z^{1/2}\right] = 0 \text{ at } r = 0. \tag{VI-16.7}$$

Removing the nonvanishing factor $\frac{3}{2}$ and replacing $z^{1/2}$ by $r^{1/3}$, we find that Eq. (VI-16.7) requires

$$r^{1/3}\frac{dp}{dr} = 0 \qquad \text{at } r = 0. \tag{VI-16.8}$$

Let us now apply the left side of Eq. (VI-16.8) to p as given by Eq. (VI-16.5). Then Eq. (VI-16.8) becomes

$$C_1 r^{1/3}\frac{d}{dr}\left[r^{1/3}J_{1/3}(ar)\right] + C_2 r^{1/3}\frac{d}{dr}\left[r^{1/3}J_{-1/3}(ar)\right] = 0 \qquad \text{at } r = 0,$$

which, by application of formulas (A) and (C) in Table V-1, is the same as

$$C_1 r^{1/3}\left[ar^{1/3}J_{-2/3}(ar)\right] + C_2 r^{1/3}\left[-ar^{1/3}J_{2/3}(ar)\right] = 0 \qquad \text{at } r = 0$$

or, since $a \neq 0$,

$$C_1 r^{2/3}J_{-2/3}(ar) - C_2 r^{2/3}J_{2/3}(ar) = 0 \qquad \text{at } r = 0. \tag{VI-16.9}$$

Since the leading term in the series for $J_{-2/3}(ar)$ as given in Eq. (V-0.2) is a term in $r^{-2/3}$, it follows that the first term on the left in Eq. (VI-16.9) is a constant which does not vanish when $C_1 \neq 0$. On the other hand, since the leading term in the series expansion for $J_{2/3}(ar)$ is a term in $r^{2/3}$, the second term on the left in Eq. (VI-16.9) vanishes at $r = 0$ regardless of what constant we take for C_2. Consequently, in order for Eq. (VI-16.9) to hold we must take $C_1 = 0$. Thus, Eq. (VI-16.5) reduces to

$$p = C_2 r^{1/3} J_{-1/3}(ar). \qquad \text{(VI-16.10)}$$

At the lower end, where $x = 0$ and $r = L^{3/2}$, $\quad p = p_0 = \tan \theta_0$. This boundary condition requires by Eq. (VI-16.10) that

$$C_2 = \frac{p_0}{\sqrt{L} J_{-1/3}(aL^{3/2})},$$

which makes

$$p = \frac{p_0 r^{1/3} J_{-1/3}(ar)}{\sqrt{L} J_{-1/3}(aL^{3/2})}. \qquad \text{(VI-16.11)}$$

Since $p = dy/dx$, one might integrate Eq. (VI-16.11), remembering that $r = (L - x)^{3/2}$, and thus obtain formula for vertical displacement (due to loading) in terms of distance above level of clamped end. But, happily, that is not needed for the determination of the critical length. Such determination can be made at once via Eq. (VI-16.11) as follows. Equation (VI-16.11) expresses the slope of each point of wire in stable position at small p_0. Every such slope will be finite. This means that Eq. (VI-16.11) holds (for sufficiently small p_0) for positive values of L less than the least positive L such that the denominator in Eq. (VI-16.11) vanishes, which means the least positive L such that

$$J_{-1/3}(aL^{3/2}) = 0. \qquad \text{(VI-16.12)}$$

Equation (VI-16.12) is the critical equation, that is, the equation which determines the critical length L_c, namely the length beyond which the wire will not be stable in vertical position.

The least positive zero of the function $J_{-1/3}$ can be found in tables to be approximately 1.87. Since we have not listed values for $J_{-1/3}$ in this book, let us see what we can do on our own to get an approximation for the least positive zero of $J_{-1/3}$. Taking only the first three terms of the series expansion given in Eq. (V-0.2) with $p = -\frac{1}{3}$, we want

$$\frac{x^{-1/3}}{2^{-1/3}\Gamma(\frac{2}{3})} - \frac{x^{5/3}}{2^{5/3}\Gamma(\frac{5}{3})} + \frac{x^{11/3}}{2^{11/3}2!\Gamma(\frac{8}{3})} = 0,$$

that is,

$$\frac{32}{\Gamma(\frac{2}{3})} - \frac{8x^2}{\Gamma(\frac{5}{3})} + \frac{x^4}{\Gamma(\frac{8}{3})} = 0. \qquad \text{(VI-16.13)}$$

Now, by Eq. (I-4.1) we have $\Gamma(\frac{5}{3}) = \frac{2}{3}\Gamma(\frac{2}{3})$ and $\Gamma(\frac{8}{3}) = (\frac{5}{3})(\frac{2}{3})\Gamma(\frac{2}{3})$. This makes Eq. (VI-16.13) become

$$9x^4 - 120x^2 + 320 = 0,$$

whose solutions are

$$x^2 = \frac{120 \pm \sqrt{2880}}{18}. \qquad \text{(VI-16.14)}$$

The positive values of x given by Eq. (VI-16.14) are approximate values (probably not very close) for the least and next-to-least positive zero of $J_{-1/3}(x)$. To get the least positive zero, we take the negative sign in Eq. (VI-16.14), obtaining

$$x \cong 1.92$$

as against 1.87 from the tables. Let us, however, use 1.87.

Returning to the equation for critical length, namely Eq. (VI-16.12), we find the critical length to be given by

$$aL^{3/2} \cong 1.87. \qquad \text{(VI-16.15)}$$

Recalling that $a = 2k/3 = (2/3)\sqrt{w/EI}$, we find from Eq. (VI-16.15) that the critical length L_c is given by

$$L_c \cong 1.99\left(\frac{EI}{w}\right)^{1/3}. \qquad \text{(VI-16.16)}$$

VI-17. Apply the result obtained in Eq. (VI-16.16) in Prob. VI-16 to find the critical length L_c for vertical stability of steel wire having the following specifications:

    (a)   cross-section circular, diameter .06 inches;

    (b)   density 489 pounds per cubic foot;

    (c) $E = 3.20 \times 10^7$ pounds per square inch.   *

The formula to be applied is

$$L_c = 1.99 \left(\frac{EI}{w}\right)^{1/3}, \tag{VI-17.1}$$

where E is Young's modulus and I is the moment of inertia of the cross-sectional area A about a diameter and w is weight per unit of length.

One finds in an appropriate textbook that the moment of inertia of a circular area about a diameter is

$$I = \tfrac{1}{4}\pi r^4. \tag{VI-17.2}$$

Since $r = .03$ by specification **(a)**, we find by Eq. (VI-17.1) that

$$I \cong 6.4 \times 10^{-7}. \tag{VI-17.3}$$

In Eq. (VI-17.3) we have implicitly committed ourselves to taking the inch as unit of length. We must, therefore, compute the linear density w in pounds per inch. Accordingly, we have

$$w = \left(\frac{489}{1728}\right)\left[\pi(.03)^2\right] \cong 8.00 \times 10^{-4}. \tag{VI-17.4}$$

Taking I from Eq. (VI-17.3) and w from Eq. (VI-17.4) and E from specification **(c)**, we find by Eq. (VI-17.1) that the critical length is

$$L_c \cong 1.99 \left[\frac{(3.20 \times 10^7)(6.4 \times 10^{-7})}{8.00 \times 10^{-4}}\right]^{1/3}$$

$$\cong 58.7 \text{ inches.}$$

Problem: Flux Distribution in a Nuclear Reactor

VI-18. Determine the radial flux distribution ϕ in a bare nuclear reactor in the shape of a long (compared to radius) right circular cylinder of radius R, being given that

(a) the flux ϕ satisfies the equation

$$\nabla^2\phi + B^2\phi = 0 \qquad \text{(VI-18.1)}$$

where $\nabla^2\phi$ is the Laplacian of ϕ and B^2 is a positive constant to be determined by boundary conditions,

(b) ϕ is a function of one variable only, namely distance from longitudinal axis of reactor, that is, ϕ is symmetric with respect to this axis,

(c) if cylindrical coordinates (r, θ, z) be taken with the z-axis in the longitudinal axis of the reactor, then the flux ϕ satisfies the conditions

$$\phi(R) = 0, \qquad \text{(VI-18.2)}$$

$$\phi(r) > 0, \quad 0 < r < R. \qquad \star \qquad \text{(VI-18.3)}$$

In terms of cylindrical coordinates Eq. (VI-18.1) is

$$\frac{\partial^2\phi}{\partial r^2} + \frac{\partial^2\phi}{\partial z^2} + \frac{1}{r^2}\frac{\partial^2\phi}{\partial\theta^2} + \frac{1}{r}\frac{\partial\phi}{\partial r} + B^2\phi = 0. \qquad \text{(VI-18.4)}$$

But, the given condition (b) means that ϕ is independent of both θ and z. This makes Eq. (VI-18.4) reduce to

$$\frac{d^2\phi}{dr^2} + \frac{1}{r}\frac{d\phi}{dr} + B^2\phi = 0. \qquad \text{(VI-18.5)}$$

If, now, we multiply both sides of Eq. (VI-18.5) by r^2 and use primes to denote first and second derivative of ϕ with respect to r, then Eq. (VI-18.5) becomes

$$r^2\phi'' + r\phi' + B^2r^2\phi = 0. \qquad \text{(VI-18.6)}$$

Solution of Eq. (VI-18.6) is immediately obtained by way of Eq. (VI-1.6) if we identify r with x and take $D = B$ and $p = 0$. Thus, by Eqs. (VI-1.5) and (V-0.8) the general solution of Eq. (VI-18.6) is

$$\phi = C_1 J_0(Br) + C_2 Y_0(Br) \qquad \text{(VI-18.7)}$$

where C_1 and C_2 are arbitrary constants.

Now in the use of cylindrical coordinates, negative values are permissible for r as well as for θ and z. This means, by condition (**b**), that

$$\phi(-r, \theta, z) = \phi(r, \theta, z).$$

Consequently, we must take $C_2 = 0$ in Eq. (VI-18.7), since $Y_0(x)$ is not an even function while $J_0(x)$ is an even function. Thus, we have

$$\phi = C_1 J_0(Br). \qquad \text{(VI-18.8)}$$

The constant C_1 in Eq. (VI-18.8) is determined by the power level at which the reactor is operating. C_1 is not zero.

The constant B is determined by the given condition (**c**). Eq. (VI-18.2) requires by Eq. (VI-18.8) that

$$J_0(BR) = 0. \qquad \text{(VI-18.9)}$$

Now, the function $J_0(x)$ has infinitely many zeros. To which of them shall we equate BR? The answer to this question is found by consideration of the condition stated in Eq. (VI-18.3), which requires that ϕ be positive for $0 < r < R$. This means that we must take BR equal to the least positive zero α_1 of $J_0(x)$ in order that $J_0(Br)$ will be positive for $0 \leqq Br < \alpha_1$. In Table V-4 we find that $\alpha_1 \cong 2.4048$ whence

$$B \cong \frac{2.4048}{R}.$$

So, we have

$$\phi \cong C_1 J_0 \left(\frac{2.4048}{R} r \right),$$

where, as mentioned above, C_1 is to be determined by the power level.

Problems: Heat-Flow Temperature Distribution

VI-19. Determine the steady-state temperature distribution T in a cooling fin (sometimes called spine) on an engine if the fin is a homogeneous solid in the shape of a right circular cone of length L and radius R at its base where it meets the body of the engine. Let $A(x)$ denote the area of the cross-section of the fin at right angles to its axis at distance x along the axis from the vertex. Let $C(x)$ denote the circumference of such cross-section. Let it be assumed that, for the purposes of this problem, sufficient accuracy will be had by supposing that the isothermal surfaces in the fin are the plane circular cross-sections at right angles to the axis. This makes T to be a function of x alone. Let it be given that, under this assumption (see Eckert and Drake, *Heat and Mass Transfer*, McGraw-Hill, 1959), the equation to be satisfied by T is

$$\frac{d}{dx}\left[A(x)\frac{dT}{dx} \right] = \frac{h}{k}\left[C(x) \right]\left[T - T_f \right], \qquad \text{(VI-19.1)}$$

where T_f is the temperature of the surrounding fluid, k is the thermal conductivity of the metal of which the fin is made, and h is a heat transfer coefficient. Let T_b denote the temperature (assumed constant) of the engine at the base of the fin.

Determine also the value, at the base of the fin, of the rate of heat flow Q given by

$$Q = Q(x) = kA(x)\frac{dT}{dx}. \qquad \star \qquad \text{(VI-19.2)}$$

By simple geometry involving similar triangles one finds that

$$A(x) = \frac{\pi R^2 x^2}{L^2}, \qquad C(x) = \frac{2\pi R x}{L}. \qquad \text{(VI-19.3)}$$

Let us denote the temperature excess by ϕ:

$$\phi = T - T_f. \qquad \text{(VI-19.4)}$$

Let us also put

$$m = \frac{2hL}{kR}. \qquad \text{(VI-19.5)}$$

Then, observing that $dT/dx = d\phi/dx$, substituting for $A(x)$ and for $C(x)$ from Eq. (VI-19.3) in Eq. (VI-19.1) and carrying out the differentiation indicated in Eq. (VI-19.1), we find that Eq. (VI-19.1) applied to the present problem becomes

$$x^2\phi'' + 2x\,\phi' - mx\,\phi = 0, \qquad \text{(VI-19.6)}$$

where the primes denote the first and second derivatives with respect to x.

We obtain solution for Eq. (VI-19.6) by identification with Eq. (VI-2.1) in which we take $y = \phi$. We have

$$1 - 2A = 2, \quad 2E = 1, \quad D^2E^2 = -m, \quad A^2 - E^2p^2 = 0,$$

whence

$$A = -\tfrac{1}{2}, \quad E = \tfrac{1}{2}, \quad D = 2i\sqrt{m}, \quad p = 1,$$

where $i = \sqrt{-1}$. Thus, by Eqs. (VI-2.2) and (V-0.12) the general solution of Eq. (VI-19.6) is

$$\phi = \frac{1}{\sqrt{x}}\left[C_1 I_1(2\sqrt{mx}) + C_2 K_1(2\sqrt{mx}) \right]. \qquad \text{(VI-19.7)}$$

The factor $1/\sqrt{x}$ in Eq. (VI-19.7) becomes infinite at the vertex of the cone where $x = 0$. But the ratio $C_1 I_1(2\sqrt{mx})/\sqrt{x}$ does *not* become infinite at $x = 0$. This ratio approaches $C_1\sqrt{m}$ as limit when $x \to 0$, as may be seen from Eqs. (V-0.9) and (V-0.2) whereby

$$I_1(2\sqrt{mx}) = \sqrt{mx} + \frac{(mx)^{3/2}}{2!} + \frac{(mx)^{5/2}}{2!3!} + \cdots.$$

On the other hand the ratio $C_2 K_1(2\sqrt{mx})/\sqrt{x}$ *does* become infinite as $x \to 0$ if C_2 is taken $\neq 0$, as may be seen in similar manner via Eqs. (V-0.11), (V-0.7), (V-0.2). Accordingly, we must take $C_2 = 0$, since no infinite temperatures are involved in our problem. So, we have

$$\phi = C_1 \frac{I_1(2\sqrt{mx})}{\sqrt{x}}. \qquad \text{(VI-19.8)}$$

And we shall take $\phi(x)$ at $x = 0$ to be

$$\phi(0) = \lim_{x \to 0} \phi(x) = C_1 \sqrt{m}.$$

It remains to determine C_1. This constant is determined by the boundary condition $T(L) = T_b$, which by Eq. (VI-19.4) makes $\phi(L) = T_b - T_f$. Thus,

$$C_1 = \frac{\sqrt{L}(T_b - T_f)}{I_1(2\sqrt{mL})};$$ (VI-19.9)

and the temperature T in the fin is

$$T = T_f + \phi = T_f + C_1 \frac{I_1(2\sqrt{mx})}{\sqrt{x}},$$ (VI-19.10)

where m is given by Eq. (VI-19.5) and C_1 is given by Eq. (VI-19.9).

To determine the rate Q of heat flow at the base of the fin as given by Eq. (VI-19.2) we have only to differentiate T as given by Eq. (VI-19.9) and then set $x = L$. Putting $u = \sqrt{x}$, we get

$$\frac{dT}{dx} = C_1 \frac{d}{dx} \left\{ x^{-1/2} I_1(2u\sqrt{m}) \right\}$$

$$= C_1 \frac{d}{dx} \left\{ u^{-1} I_1(2u\sqrt{m}) \right\}$$

$$= C_1 \frac{d}{du} \left\{ u^{-1} I_1(2u\sqrt{m}) \right\} \frac{du}{dx}.$$

We apply formula (D) in Table V-1, obtaining

$$\frac{dT}{dx} = C_1 \left\{ 2\sqrt{m} \, u^{-1} I_2(2u\sqrt{m}) \right\} \frac{1}{2\sqrt{x}}$$

$$= \sqrt{m} \, C_1 \frac{I_2(2\sqrt{mx})}{x}.$$ (VI-19.11)

Substituting from Eq. (VI-19.11) in Eq. (VI-19.2), we get

$$Q(x) = kA(x)\sqrt{m}\,C_1\frac{I_2(2\sqrt{mx})}{x},$$

so that the rate of heat flow at the base is

$$Q(L) = k\pi R^2 L^{-1}\sqrt{m}\,C_1 I_2(2\sqrt{mL}), \qquad \text{(VI-19.12)}$$

where m and C_1 are given by Eqs. (VI-19.5) and (VI-19.9) respectively.

REMARK.   In the usual steady-state heat-flow problem, where no presumption is taken regarding isothermal surfaces, the temperature T must satisfy Laplace's equation $\nabla^2 T = 0$ interior to the region of flow. In the present problem it was considered sufficiently accurate to assume the isothermal surfaces to be those portions of the surfaces $x = $ a constant contained in the region occupied by the fin, as was mentioned in the statement of the problem. That is why the differential equation to be satisfied by T is different from Laplace's equation.

VI-20.   Determine the temperature distribution T in a homogeneous right circular cylinder solid of height L and radius R, given the following conditions and assumptions. Heat is being produced in the cylinder. The rate Q' of heat production per unit volume per unit time at each point of the cylinder is a linear function of the temperature: $Q' = a + bT$, $a \neq 0$, $b \neq 0$. The surface of the cylinder is kept at a uniform temperature T_s. Assume that the production and conduction of the heat is such that   **(a)** T does not vary with time,   **(b)** T is a function only of distance r from the axis of the cylinder: $T = T(r)$,   **(c)** T decreases with increasing r, that is, heat flow is radially outward,   **(d)** T is continuous for $0 \leqq r \leqq R$ and differentiable for $0 \leqq r < R$.   ★

The amount of heat Q_H being produced per unit time at a given instant in a coaxial subcylinder H of radius r and height L is given by the integral of the rate of heat production Q' over the volume H:

$$Q_H = \iiint_H Q'dV = \int_0^r 2\pi rLQ'dr$$

$$= \int_0^r 2\pi rL(a + bT)dr.$$

$$\text{(VI-20.1)}$$

Since we are assuming T to be that of a steady-state heat conduction problem, the amount of heat being lost per unit time through the surface of the subcylinder H by conduction radially outward equals the amount of heat produced per unit time in H. Since the lateral surface area of the subcylinder H is $2\pi rL$, we have

$$Q_H = - 2\pi rLk\frac{dT}{dr}, \qquad \text{(VI-20.2)}$$

where k is the thermal conductivity of the material comprising the solid. The minus sign is needed here to make Q_H a positive quantity since the temperature gradient dT/dr is negative. Equating the two expressions for Q_H, we have

$$\int_0^r 2\pi rL(a + bT)dr = - 2\pi rLk\frac{dT}{dr}.$$

Differentiating both sides of this equation, we find that

$$2\pi rL(a + bT) = - 2\pi Lk\left(r\frac{d^2T}{dr^2} + \frac{dT}{dr}\right),$$

whence

$$\frac{d^2T}{dr^2} + \frac{1}{r}\frac{dT}{dr} + \frac{b}{k}T = - \frac{a}{k}. \qquad \text{(VI-20.3)}$$

Equation (VI-20.3) is a linear differential equation whose solution will be made up of a particular solution plus the complementary function. The complementary function is obtained by setting the left side of Eq. (VI-20.3) to zero:

$$\frac{d^2T}{dr^2} + \frac{1}{r}\frac{dT}{dr} + \frac{b}{k}T = 0,$$

whence, by multiplying through by r^2 we have

$$r^2T'' + rT' + \frac{b}{k}r^2T = 0. \qquad \text{(VI-20.4)}$$

The general solution of Eq. (VI-20.4) by Eqs. (VI-1.5) and (V-0.8) is

$$T = C_1 J_0\left(\sqrt{\frac{b}{k}}\,r\right) + C_2 Y_0\left(\sqrt{\frac{b}{k}}\,r\right). \qquad \text{(VI-20.5)}$$

For the particular solution, we observe that the right side of Eq. (VI-20.3) is a constant; so we take $T = $ a constant:

$$T = -\frac{a}{b}. \qquad \text{(VI-20.6)}$$

Thus, the general solution of Eq. (VI-20.3) is

$$T = -\frac{a}{b} + C_1 J_0\left(\sqrt{\frac{b}{k}}\,r\right) + C_2 Y_0\left(\sqrt{\frac{b}{k}}\,r\right). \qquad \text{(VI-20.7)}$$

To determine C_1 and C_2 we have the conditions

$$\left.\begin{array}{r} T \text{ is finite at } r = 0 \\ T(R) = T_s \end{array}\right\} \qquad \text{(VI-20.8)}$$

The first of the conditions in Eq. (VI-20.8) makes Y_0 unsuitable in Eq. (VI-20.7) since Y_0 becomes infinite at $r = 0$. So we must take $C_2 = 0$. Then the second of conditions (VI-20.8) requires that

$$C_1 = \frac{\left(\dfrac{a}{b}\right) + T_s}{J_0\left(\sqrt{\dfrac{b}{k}}\,R\right)}. \qquad \text{(VI-20.9)}$$

The solution to our problem is

$$T = \frac{\dfrac{a}{b} + T_s}{J_0\left(\sqrt{\dfrac{b}{k}}\,R\right)} J_0\left(\sqrt{\frac{b}{k}}\,r\right) - \frac{a}{b}. \qquad \text{(VI-20.10)}$$

REMARKS. 1. If there is heat absorption (instead of production) in the cylinder at a linear rate, all other conditions being the same, we have $Q' = a - bT$; and the solution Eq. (VI-20.10) is replaced by

$$T = \frac{T_s - \dfrac{a}{b}}{I_0\left(\sqrt{\dfrac{b}{k}}\,R\right)} I_0\left(\sqrt{\dfrac{b}{k}}\,r\right) + \frac{a}{b} \qquad \text{(VI-20.11)}$$

by Eq. (V-0.12).

2. The solutions obtained for T in Eqs. (VI-20.10) and (VI-20.11) are valid except when the trio of constants b, k, and R happen to be such that $\sqrt{b/k}\,R$ equals a zero of J_0.

Problems in Dynamics

VI-21. A perfectly flexible chain of length L and constant linear density ρ is fastened at one end to a fixed point from which it hangs vertically at rest in the positive x-axis with the origin $x = 0$ at the lower end of the chain. The chain is then caused to oscillate slightly in a fixed vertical plane by imparting to each of its points at time $t = 0$ an initial horizontal velocity v given by $v = F(x)$, where $F(x)$ is continuous and differentiable for $0 \leq x \leq L$ and where, of course, $F(L) = 0$.

Determine, as function of x and t, the horizontal displacement y of a point of the chain, making the following assumptions:

(a) the motion of each point of the chain is to be considered as taking place in a horizontal straight line,

(b) the magnitude of the tension $\vec{T}$ at each point P of the chain is given with sufficient approximation by the weight of that portion of the chain which is below P,

(c) the oscillations are so small that, if β denotes the acute angle made with the vertical by the line tangent to the chain at any point thereof, then $\sin \beta$ may be replaced, with sufficient approximation, by $\tan \beta$,

(d)   in setting up the differential equation of motion, an equation of sufficient approximation will be obtained by neglecting positive powers higher than the first power of infinitesimal quantities.   *

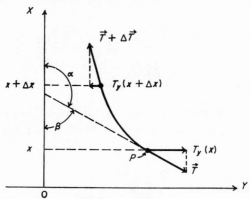

Figure VI-1.
Short piece of hanging chain.

Let us compute (approximately) first the net instantaneous horizontal component of tension for a short piece of chain from an arbitrary x to $x + \Delta x$. Since the x-axis points upward, the tension at $x + \Delta x$ will, by assumption **(b)**, be of greater magnitude than that of the tension at x. We take the y-axis horizontal through the point $x = 0$ at the bottom of the chain. We let $T_y(x)$ denote horizontal component of tension at x. We take our short piece of chain to be in the first quadrant and curving outward as x decreases. We let α denote the inclination to the positive x-direction of the line tangent to the chain. We let T denote the magnitude of the tension $\vec{T}$. Then we have

$$T_y(x) = T \sin \beta$$

$$= T \sin \alpha$$

$$= T(- \tan \alpha) \text{ by assumption (c).}$$

But $\tan \alpha$ is the slope of the chain, namely $\partial y / \partial x$. The partial derivative is required here, since y is a function of two independent variables x and t, and since we are holding t fixed while we analyze the momentary situation in a short piece of chain. Thus, at a given moment, we have

$$T_y(x) \cong - T \frac{\partial y}{\partial x}. \tag{VI-21.1}$$

The horizontal component of tension at the upper end of our short piece is given by Eq. (VI-21.1) with change of sign on the right and then evaluated at $x + \Delta x$, namely

$$T_y(x + \Delta x) \cong (T + \Delta T)\left(\frac{\partial y}{\partial x} + \Delta \frac{\partial y}{\partial x}\right). \tag{VI-21.2}$$

The change in sign is required by the fact that, for a short piece of chain, $T_y(x)$ and $T_y(x + \Delta x)$ are in opposite directions. Now, $T + \Delta T$ is given by the Taylor's series expansion for T, namely

$$T + \Delta T = T + \frac{\partial T}{\partial x}\Delta x + \frac{1}{2!}\frac{\partial^2 T}{\partial x^2}\Delta x^2 + \cdots \tag{VI-21.3}$$

Likewise, we have

$$\frac{\partial y}{\partial x} + \Delta \frac{\partial y}{\partial x} = \frac{\partial y}{\partial x} + \frac{\partial}{\partial x}\left(\frac{\partial y}{\partial x}\right)\Delta x + \frac{1}{2!}\frac{\partial^2}{\partial x^2}\left(\frac{\partial y}{\partial x}\right)\Delta x^2 + \cdots$$

$$= \frac{\partial y}{\partial x} + \frac{\partial^2 y}{\partial x^2}\Delta x + \frac{1}{2!}\frac{\partial^3 y}{\partial x^3}\Delta x^2 + \cdots \tag{VI-21.4}$$

In accordance with assumption (d) we shall retain only the first two terms on the right in Eqs. (VI-21.3) and (VI-21.4), since further terms involve powers of Δx higher than the first. By virtue of these curtailed versions of Eqs. (VI-21.3) and (VI-21.4) we may write Eq. (VI-21.2) as

$$T_y(x + \Delta x) \cong \left(T + \frac{\partial T}{\partial x}\Delta x\right)\left(\frac{\partial y}{\partial x} + \frac{\partial^2 y}{\partial x^2}\right)\Delta x. \tag{VI-21.5}$$

Carrying out the multiplication indicated on the right in Eq. (VI-21.5), we shall once more make use of assumption (d) in that we shall not retain the term involving the square of Δx. Thus, in place of Eq. (VI-21.5) we write

$$T_y(x + \Delta x) \cong T\frac{\partial y}{\partial x} + \left(T\frac{\partial^2 y}{\partial x^2} + \frac{\partial T}{\partial x}\frac{\partial y}{\partial x}\right)\Delta x,$$

that is,

$$T_y(x + \Delta x) \cong T\frac{\partial y}{\partial x} + \left[\frac{\partial}{\partial x}\left(T\frac{\partial y}{\partial x}\right)\right]\Delta x. \qquad (VI\text{-}21.6)$$

We take the net instantaneous horizontal component Y of tension on the short piece of chain to be the algebraic sum of oppositely signed horizontal forces acting at the ends of the piece, namely

$$Y = T_y(x) + T_y(x + \Delta x),$$

which, by Eqs. (VI-21.6) and (VI-21.1), is

$$Y = \left[\frac{\partial}{\partial x}\left(T\frac{\partial y}{\partial x}\right)\right]\Delta x. \qquad (VI\text{-}21.7)$$

The mass of m our short piece of chain is $\rho\Delta x$. And the magnitude T of tension at x is $g\rho x$, where g is gravitational acceleration. Thus, by Newton's law of motion $\vec{F} = m\vec{a}$, we have

$$\rho\Delta x \frac{\partial^2 y}{\partial t^2} = \left[\frac{\partial}{\partial x}\left(g\rho x\frac{\partial y}{\partial x}\right)\right]\Delta x. \qquad (VII\text{-}21.8)$$

In Eq. (VI-21.8) the acceleration factor $\partial^2 y/\partial t^2$ should really be considered as being evaluated at an appropriate point between x and $x + \Delta x$. But, if we assume continuity of this derivative and let $\Delta x \to 0$, then with the factor Δx removed from both sides of Eq. (VI-21.8), we may take our (approximate) equation of motion at time t and point x to be

$$\rho \frac{\partial^2 y}{\partial t^2} = \frac{\partial}{\partial x}\left(g\rho x\frac{\partial y}{\partial x}\right),$$

that is,

$$x\frac{\partial^2 y}{\partial x^2} + \frac{\partial y}{\partial x} = \frac{1}{g}\frac{\partial^2 y}{\partial t^2}. \qquad (VI\text{-}21.9)$$

It is well-known that partial differential equations can often be satisfied by product functions. Let us see if Eq. (VI-21.9) can be satisfied by a product function of the form

$$y = G(x)H(t) = GH. \tag{VI-21.10}$$

Using primes to denote first and second derivatives of G and H, each with respect to its own argument, we find that Eq. (VI-21.9) requires y as taken in Eq. (VI-21.10) to be such that

$$H(xG'' + G') = \frac{1}{g} GH'', \tag{VI-21.11}$$

that is,

$$\frac{\frac{d}{dx}(xG')}{G} = \frac{1}{g}\frac{H''}{H}. \tag{VI-21.12}$$

The left side of Eq. (VI-21.12) is independent of t while the right side is independent of x. In other words, Eq. (VI-21.12) requires a function of x alone to be identically equal to a function of t alone. This can only be so if each function is identically a constant. Thus, Eq. (VI-21.12) requires that

$$\frac{1}{g}\frac{H''}{H} = C \tag{VI-21.13}$$

that is,

$$H'' = CgH, \tag{VI-21.14}$$

where C is a constant. Now the statement of our problem implies that the motion of each point of the chain is assumed to be a periodic function of the time t. And solutions of Eq. (VI-21.14) are periodic when Cg is negative. So we take $C = -\lambda^2$. Then the general solution of Eq. (VI-21.14) is

$$H = C_1 \cos (\lambda \sqrt{g}\, t) + C_2 \sin (\lambda \sqrt{g}\, t) \tag{VI-21.15}$$

where C_1, C_2 and λ are arbitrary constants. Since λ is arbitrary, let us put $k^2 = \lambda^2 g$ as a convenience. This makes

$$H = C_1 \cos kt + C_2 \sin kt, \qquad k = \lambda \sqrt{g}. \quad \text{(VI-21.16)}$$

Reverting to Eq. (VI-21.12), we find that Eq. (VI-21.12) now requires

$$xG'' + G + \lambda^2 G = 0,$$

that is,

$$x^2 G'' + xG' + \lambda^2 xG = 0. \qquad \text{(VI-21.17)}$$

We can solve Eq. (VI-21.17) by application of Eq. (VI-2.1). We have

$$1 - 2A = 1, \quad 2E = 1, \quad D^2 E^2 = \lambda^2, \quad A^2 - E^2 p^2 = 0,$$

whence

$$A = 0, \quad E = 1/2, \quad D = \lambda, \quad p = 0.$$

Thus, by Eqs. (VI-2.2) and (V-0.8), the general solution of Eq. (VI-21.17) is

$$G = C_3 J_0(\lambda \sqrt{x}) + C_4 Y_0(\lambda \sqrt{x}), \qquad \text{(VI-21.18)}$$

where C_3 and C_4 are arbitrary constants.

Putting our results together, we find that Eq. (VI-21.9) can be satisfied by a solution of the kind indicated in Eq. (VI-21.10), where G and H are given respectively by Eqs. (VI-21.16) and (VI-21.18).

The Bessel function Y_0 is unsuitable in the problem at hand, since $Y_0(u)$ becomes infinite at $u = 0$. We must, therefore, take $C_4 = 0$. Since the upper end of the chain is fixed, we have $y = 0$ at $x = L$ for all t. This requires

$$J_0(\lambda \sqrt{L}) = 0.$$

Thus, we can satisfy Eq. (VI-21.9) by any function of the type

$$y = C_3 J_0(\lambda \sqrt{x})[C_1 \cos kt + C_2 \sin kt], \qquad \text{(VI-21.19)}$$

provided $\lambda\sqrt{L}$ is a zero of the function J_0 and provided $k = \lambda\sqrt{g}$. We must, of course, have $C_3 \neq 0$, for otherwise the chain would hang still without motion. It certainly will not hang still when an initial velocity (not identically zero) is imparted to its points.

At time $t = 0$ we have $y = 0$ at all x. Thus, Eq. (VI-21.19) requires that

$$0 = C_3 J_0(\lambda\sqrt{x})[C_1], \qquad 0 \leqq x \leqq L.$$

Now, $J_0(\lambda\sqrt{x})$ is certainly not identically zero. And we have $C_3 \neq 0$ as remarked in the preceding paragraph. Consequently, we must take $C_1 = 0$. Taking $C_1 = 0$ in Eq. (VI-21.19) and putting $A = C_3 C_2$, we can now say that any function of the type

$$y = A J_0(\lambda\sqrt{x}) \sin kt, \qquad \text{(VI-21.20)}$$

where λ is such that $J_0(\lambda\sqrt{L}) = 0$ and $k = \lambda\sqrt{g}$, will satisfy Eq. (VI-21.9) and will also satisfy two of the boundary conditions, namely initial zero displacement at all points and continued zero displacement for all t at the topmost point.

So far, so good. But now comes the difficulty. We have yet to fulfill the initial condition of prescribed imparted velocities at time $t = 0$, namely,

$$\left(\frac{\partial y}{\partial t}\right)_{t=0} = k A J_0(\lambda\sqrt{x}) \cos(0) = k A J_0(\lambda\sqrt{x}) = F(x), \qquad 0 \leqq x \leqq L.$$

But no single function of the type $k A J_0(\lambda\sqrt{x})$ equals a prescribed $F(x)$ at every x on the interval $0 \leqq x \leqq L$, unless perchance $F(x)$ itself happens to be given as such a function. How do we finish the solution of the problem when $F(x)$ is *not* $J_0(\lambda\sqrt{x})$ multiplied by a constant? Answer: construct a composite solution comprised of an infinite series of solutions of type Eq. (VI-21.20), namely

$$y = \sum_{j=1}^{\infty} A_j J_0(\lambda_j\sqrt{x}) \sin(\lambda_j\sqrt{g}\, t), \qquad \text{(VI-21.21)}$$

where the numbers λ_j are the positive zeros of the function J_0 and where the coefficients A_j are such that $(\partial y/\partial t)_{t=0} = F(x)$, that is,

$$\sum_{j=1}^{\infty} \lambda_j \sqrt{g}\, A_j J_0(\lambda_j \sqrt{x}) = F(x). \qquad \text{(VI-21.22)}$$

The expansion called for in Eq. (VI-21.22) can be effected by appropriate applications of Prob. V-27 as follows. First we set $\sqrt{x} = u$. Then Eq. (VI-21.22) becomes

$$f(u) = \sum_{j=1}^{\infty} \lambda_j \sqrt{g}\, A_j J_0(\lambda_j u), \qquad 0 \leq u \leq \sqrt{L}, \qquad \text{(VI-21.23)}$$

where $f(u)$ denotes $F(u^2)$. And now Eq. (VI-21.23) is an expansion of the type considered in Prob. V-27 where the independent variable is denoted by u instead of x and where $\lambda_j = \alpha_j, j = 1, 2, 3, \cdots$. Thus, the numbers A_j in Eq. (VI-21.23) are given by

$$\alpha_j \sqrt{g}\, A_j = \frac{2}{L[J_1(\alpha_j)]^2} \int_0^{\sqrt{L}} u f(u) J_0\left(\frac{\alpha_j}{\sqrt{L}} u\right) du, \qquad \text{(VI-21.24)}$$

that is,

$$A_j = \frac{1}{L\sqrt{g}\, \alpha_j [J_1(\alpha_j)]^2} \int_0^L F(x) J_0\left(\alpha_j \sqrt{\frac{x}{L}}\right) dx. \qquad \text{(VI-21.25)}$$

The solution of our problem is given by Eq. (VI-21.21) in which $\lambda_j = \alpha_j, j = 1, 2, 3, \cdots$ and in which the coefficients A_j are given by Eq. (VI-21.25).

REMARK. The problem of small oscillations of a uniform hanging chain was first studied by Daniel Bernoulli in 1732 and later by Euler in 1781. It is the problem whereby the functions now known as Bessel functions were first encountered in mathematics and its applications. The practical value of a formula for the displacements of points on an oscillating chain is most likely limited; however, the techniques that this problem caused to be developed occupy a position of importance in modern mathematical physics.

VI-22. A homogeneous, slightly tapered rod of length L is such that when properly placed with respect to the x-axis the cross-sectional area A perpendicular to the x-axis is given by

$$A = A(x) = \frac{a}{b + x} \qquad \text{(VI-22.1)}$$

where a and b are positive constants. The larger end is fixed and is in the plane $x = 0$. The smaller end is free and lies in the plane $x = a/(b + L)$. As result of an axial blow at the free end, the rod is vibrating longitudinally. Assuming that, for each cross-section, all the points in the cross-section undergo the same displacement u parallel to the x-axis, namely, $u = u(x, t)$ where t denotes time, determine the formula for u. $\star$

As is customary in many problems of applied mathematics, we shall be content to set up an approximate differential equation of motion, considering the physical and geometrical conditions to be such that we may, with satisfactory approximation, neglect powers higher than the first of infinitesimal quantities. Let us compute (approximately) the instantaneous net axial stress in a very thin cross-sectional slice whose plane bases are in planes cutting the x-axis at x and $x + \Delta x$. The corresponding areas of the bases are A and $A + \Delta A$. The corresponding displacements are u and $u + \Delta u$.

The stress at x is $EA[\partial u / \partial x]$, where E denotes Young's modulus. The area $A + \Delta A$ is given exactly by Taylor's series:

$A + \Delta A = A(x + \Delta x)$

$$= A + \frac{dA}{dx}\Delta x + \frac{1}{2!}\frac{d^2A}{dx^2}\Delta x^2 + \frac{1}{3!}\frac{d^3A}{dx^3}\Delta x^3 + \cdots . \qquad \text{(VI-22.1a)}$$

We have agreed to neglect powers higher than the first of small quantities. Accordingly, we have

$$A + \Delta A \cong A + \frac{dA}{dx}\Delta x.$$

Similarly, one finds via Taylor's Series expansion that a first approximation to the first derivative of u by x, with t held constant, at $x + \Delta x$ is

$$\frac{\partial u}{\partial x} + \left[\frac{\partial}{\partial x}\left(\frac{\partial u}{\partial x}\right)\right]\Delta x = \frac{\partial u}{\partial x} + \frac{\partial^2 u}{\partial x^2}\Delta x.$$

Thus, the stress at $x + \Delta x$ is approximately

$$E\left(A + \frac{dA}{dx}\Delta x\right)\left(\frac{\partial u}{\partial x} + \frac{\partial^2 u}{\partial x^2}\Delta x\right),$$

which in turn is approximated by

$$E\left[A\frac{\partial u}{\partial x} + \left(A\frac{\partial^2 u}{\partial x^2} + \frac{dA}{dx}\frac{\partial u}{\partial x}\right)\right]\Delta x,$$

where we have discarded the term involving the square of Δx.

The instantaneous net axial stress in the slice we take to be the excess of the stress at $x + \Delta x$ over the stress at x, namely

$$E\left(A\frac{\partial^2 u}{\partial x^2} + \frac{dA}{dx}\frac{\partial u}{\partial x}\right)\Delta x = E\left[\frac{\partial}{\partial x}\left(A\frac{\partial u}{\partial x}\right)\right]\Delta x. \quad \text{(VI-22.2)}$$

This is (approximately) the (signed) magnitude of the force acting on the slice, which by Newton's first law of motion equals the (signed) acceleration of the slice multiplied by its (approximate) mass.

Accordingly, we have

$$E\left[\frac{\partial}{\partial x}\left(A\frac{\partial u}{\partial x}\right)\right]\Delta x = (A\rho\Delta x)\frac{\partial^2 u}{\partial t^2} \quad \text{(VI-22.3)}$$

where ρ is the density of the material. Thus the (approximate) equation of motion is

$$\frac{1}{A}\frac{\partial}{\partial x}\left(A\frac{\partial u}{\partial x}\right) = k^2\frac{\partial^2 u}{\partial t^2}, \quad \text{(VI-22.4)}$$

where $k^2 = \rho/E$. $\quad$ (VI-22.5)

If the rod were of constant cross-sectional area, Eq. (VI-22.4) would be

$$\frac{\partial^2 u}{\partial t^2} = c^2 \frac{\partial^2 u}{\partial x^2}, \tag{VI-22.6}$$

c^2 denoting $1/k^2$. Equation (VI-22.6) is the *one-dimensional wave equation*, whose general solution is known to be

$$u = f(x - ct) + \phi(x + ct), \tag{VI-22.7}$$

f and ϕ being arbitrary functions of their respective compound arguments $x - ct$ and $x + ct$. If we assume that the longitudinal vibrations of the rod (of constant cross-section) are periodic in time, then it is known that Eq. (VI-22.6) is satisfied by functions of the form

$$u = B \sin (qx + cqt) \text{ or } u = B \cos (qx + cqt),$$

where B and q are arbitrary constants. This is equivalent to saying that Eq. (VI-22.6) is satisfied by functions of the form

$$u = \Big[B \sin qx \Big] \Big[\sin cqt \Big]$$

or any one of the three other forms obtained by replacing either sine or both by cosine. Indeed, one may even include a lead (or lag) additive term in the argument of the sine (or cosine) in either bracket. Thus, Eq. (VI-22.6) is satisfied by a function of the form

$$u = \Big[B \sin qx \Big] \Big[\sin (cqt + s) \Big], \tag{VI-22.8}$$

where B, q, and s are arbitrary constants.

The function u in Eq. (VI-22.8) is a product function of the form

$$u = g(x)h(t). \tag{VI-22.9}$$

Let us now try a product function of this type as solution of the equation of the present problem of the *tapered* rod, namely Eq. (VI-22.4). We take $h(t)$ to be of the same character as in Eq. (VI-22.8), namely $h = \sin (\lambda t + \mu)$, where λ and μ are constants to be determined by initial conditions. But since Eq. (VI-22.4) differs from its particular case, namely Eq. (VI-22.6), in that the area $A(x)$ is not constant, we shall not commit ourselves as to the nature of the function $g(x)$ in Eq. (VI-22.9), trusting that the differential Eq. (VI-22.4) will tell us what kind of function $g(x)$ shall be.

We, therefore, try

$$u = g(x) \sin(\lambda t + \mu) \qquad \text{(VI-22.10)}$$

in Eq. (VI-22.4). First we write Eq. (VI-22.4) as

$$\frac{\partial^2 u}{\partial x^2} + \frac{1}{A}\frac{dA}{dx}\frac{\partial u}{\partial x} - k^2\frac{\partial^2 u}{\partial t^2} = 0. \qquad \text{(VI-22.11)}$$

Substitution from (VI-22.10) in (VI-22.11) yields

$$\frac{d^2 g}{dx^2}\sin(\lambda t + \mu) + \frac{1}{A}\frac{dA}{dx}\frac{dg}{dx}\sin(\lambda t + \mu)$$
$$+ k^2\lambda^2\sin(\lambda t + \mu)g = 0. \qquad \text{(VI-22.12)}$$

The common factor $\sin(\lambda t + \mu)$ may be removed, since it will not be identically zero. From Eq. (VI-22.1) we have $A = a/(b + x)$. So, Eq. (VI-22.12) becomes

$$\frac{d^2 g}{dx^2} - \frac{1}{b + x}\frac{dg}{dx} + k^2\lambda^2 g = 0. \qquad \text{(VI-22.13)}$$

Let us put $b + x = z$. Then $g(x) = g(z - b) = h(z)$. And $dg/dx = dh/dz$, $d^2g/dx^2 = d^2h/dz^2$. Thus, Eq. (VI-22.13) multiplied through by z^2 becomes

$$z^2 h'' - zh' + k^2\lambda^2 z^2 h = 0, \qquad \text{(VI-22.14)}$$

where the primes denote first and second derivative with respect to z. Equation (VI-22.14) is an equation of the type solved in Prob. VI-2. Identifying Eq. (VI-22.14) with Eq. (VI-2.1), with z playing the role of x therein, we have

$$1 - 2A = -1, \quad 2E = 2, \quad D^2E^2 = k^2\lambda^2, \quad A^2E^2 - p^2 = 0,$$

whence

$$A = 1, \quad E = 1, \quad D = k\lambda, \quad p = 1.$$

Thus, we find by Eqs. (VI-2.2) and (V-0.8) that the function $h(z)$ shall be

$$h(z) = C_1 z J_1(k\lambda z) + C_2 z Y_1(k\lambda z) \qquad \text{(VI-22.15)}$$

We have, then, by Eq. (VI-22.15) and Eq. (VI-22.10) the solution of our problem:

$$u = \left[C_1 z J_1(k\lambda z) + C_2 z Y_1(k\lambda z) \right] \sin(\lambda t + \mu),$$
$$\text{(VI-22.16)}$$

where $z = b + x$, $k = \sqrt{\rho/E}$ by Eq. (VI-22.5), and where C_1, C_2, λ, and μ are constants to be determined by initial conditions. There remains only the determination of the four constants C_1, C_2, λ, μ.

At $x = 0$ there is no displacement, since the end of the rod at the origin is fixed and immovable. Thus, the displacement $u = 0$ when $z = b$ for all values of t. Consequently, the expression in brackets in Eq. (VI-22.16) must vanish when $z = b$, since the other factor $\sin(\lambda t + \mu)$ does not vanish identically. Recalling that $b \neq 0$, we have

$$0 = C_1 J_1(k\lambda b) + C_2 Y_1(k\lambda b). \qquad \text{(VI-22.17)}$$

At the free end of the rod we assume, that once the blow has been struck, there is no stress there: $\partial u/\partial x = 0$ at $x = L$ for all t. To require $\partial u/\partial x = 0$ at $x = L$ for all t is equivalent to requiring $dg/dx = 0$ at $x = L$, since the other factor in u does not vanish identically; and then to require $dg/dx = 0$ at $x = L$ is equivalent to requiring $dh/dz = 0$ at $z = b + L$. From Eq. (VI-22.15) we have

$$\frac{dh}{dz} = C_1\left[z\frac{d}{dz}J_1(k\lambda z) + J_1(k\lambda z) \right] + C_2\left[z\frac{d}{dz}Y_1(k\lambda z) + Y_1(k\lambda z) \right].$$
$$\text{(VI-22.18)}$$

Applying formula (E) in Table V-1 to the derivatives called for on the right in Eq. (VI-22.18), setting $z = b + L$ and collecting like terms we find that Eq. (VI-22.18) requires

$$0 = C_1 J_0(k[b + L]\lambda) + C_2 Y_0(k[b + L]\lambda). \qquad \text{(VI-22.19)}$$

In addition to Eqs. (VI-22.17) and (VI-22.19) we need two more equations to determine the four constants C_1, C_2, λ, μ. If, for example, the displacement $u(x, t) = h(z) \sin (\lambda t + \mu)$ where $h(z)$ is given by Eq. (VI-22.15) and $z = b + x$, be known for (x_1, t_1) and for (x_2, t_2), then Eq. (VI-22.16) will yield two additional equations in C_1, C_2, λ, μ. These two equations together with Eq. (VI-22.17) and Eq. (VI-22.19) will determine C_1, C_2, λ, μ.

REMARK. The one-dimensional wave equation Eq. (VI-22.6) is an important one in mathematical physics. It occurs also in such problems as the vibrations of a stretched elastic string and small oscillations in a thin tube of air as in an organ pipe. It is sometimes called D'Alembert's equation in honor of Jean-le-Rond D'Alembert who first solved it in 1747. Derivation of its solution Eq. (VI-22.7) will be found in any good text dealing with partial differential equations.

Problem: Fluid Velocity Imparted by Radially Pulsating Cylinder

VI-23. A circular cylinder of radius R and of great length (taken as infinite for the purposes of the problem) is surrounded laterally by an ideal compressible fluid. The cylinder is pulsating radially with frequency ω in such a manner that the radial velocity of each point of the lateral boundary of the cylinder is a sinusoidal function (the same for all such points) of the time t, the minimum absolute velocity being zero. Determine the velocities imparted by the pulsation to the points of the surrounding fluid. It is to be assumed that such velocities will also be sinusoidal in character. And it is given (from hydrodynamic theory) that the velocity vector field $\vec{V}$ of the imparted velocities is such that $\vec{V}$ is the gradient of a velocity potential function ϕ satisfying the *wave equation*

$$\nabla^2 \phi = \frac{1}{c^2} \frac{\partial^2 \phi}{\partial t^2}. \qquad \text{(VI-23.1)}$$

(Here the constant c is the speed of sound in the surrounding fluid.)   *

Cylindrical coordinates (r, θ, z) are certainly the most convenient to use in this problem, with the z-axis taken in the axis of the cylinder. Since $\vec{V}$ is always and everywhere radial with respect to the z-axis, it follows that ϕ is a function of r and t only. Accordingly, the Laplacian $\nabla^2\phi$ in cylindrical coordinates, namely

$$\frac{\partial^2\phi}{\partial r^2} + \frac{1}{r^2}\frac{\partial^2\phi}{\partial \theta^2} + \frac{\partial^2\phi}{\partial z^2} + \frac{1}{r}\frac{\partial\phi}{\partial r},$$

reduces to two terms, so that Eq. (VI-23.1) becomes

$$\frac{\partial^2\phi}{\partial r^2} + \frac{1}{r}\frac{\partial\phi}{\partial r} = \frac{1}{c^2}\frac{\partial^2\phi}{\partial t^2}. \tag{VI-23.2}$$

The character of the imparted velocities is such as to suggest that in this problem the use of complex quantities will be found convenient. Moreover, as in many problems calling for solution of such equations as Laplace's equation or the wave equation (see, for instance, Prob. V-22), we shall assume a product solution of the form

$$\phi = \phi(r, t) = F(r)G(t).$$

If we can determine a complex function ϕ which will satisfy Eq. (VI-23.2), it will follow by separation of reals and pure imaginaries that the real part of ϕ will satisfy Eq. (VI-23.2). So, we take

$$\phi = F(r)e^{i\omega t}, \tag{VI-23.3}$$

where $i = \sqrt{-1}$.

Putting ϕ from Eq. (VI-23.3) into Eq. (VI-23.2), we get

$$F''e^{i\omega t} + \frac{1}{r}F'e^{i\omega t} = \frac{1}{c^2}Fe^{i\omega t}(i\omega)^2, \tag{VI-23.4}$$

where the primes on F denote first and second derivative with respect to r. Suppressing the nonvanishing factor $e^{i\omega t}$, multiplying through by r^2, and transposing the term on the right, we find that Eq. (VI-23.4) becomes

$$r^2F'' + rF' + \frac{\omega^2}{c^2}r^2F = 0. \tag{VI-23.5}$$

By Prob. VI-2 together with Eq. (V-0.8) the general solution is seen to be

$$F = C_1 J_0\left(\frac{\omega r}{c}\right) + C_2 Y_0\left(\frac{\omega r}{c}\right). \qquad \text{(VI-23.6)}$$

Since we are employing a complex ϕ, it seems that it might be convenient even to take F as a complex function, that is, to express F in terms of Hankel functions (see Eqs. (V-0.19) and (V-0.20)). So we take

$$F = C_1 H_0^{(1)}\left(\frac{\omega r}{c}\right) + C_2 H_0^{(2)}\left(\frac{\omega r}{c}\right). \qquad \text{(VI-23.7)}$$

We now have

$$\phi = \left[C_1 H_0^{(1)}\left(\frac{\omega r}{c}\right) + C_2 H_0^{(2)}\left(\frac{\omega r}{c}\right)\right]e^{i\omega t}. \qquad \text{(VI-23.8)}$$

Our problem is to determine $\vec{V}$. Now, $\vec{V} = \text{grad}\left\{Re[\phi]\right\} = \nabla\left\{Re[\phi]\right\}$, where $Re[\phi]$ denotes the real part of ϕ. But, since $\vec{V}$ is everywhere directed radially with respect to the z-axis, it will suffice to determine the radial scalar component of $\vec{V}$, namely V_r, since all other components vanish. We then have

$$V_r = \frac{\partial}{\partial r}\left\{Re[\phi]\right\}$$

which is seen to be the same as

$$V_r = Re\left[\frac{\partial\phi}{\partial r}\right]. \qquad \text{(VI-23.9)}$$

Accordingly, we differentiate ϕ as given by Eq. (VI-23.8) with respect to r and take the real part of the result.

Applying formula (C) in Table V-1 with $x = \omega r/c$ and remembering that $(d/dr)f(x) = [(d/dx)f(x)](dx/dr)$, we get from Eqs. (VI-23.8) and (VI-23.9)

$$V_r = Re\left[-\frac{\omega}{c}\left\{C_1 H_1^{(1)}\left(\frac{\omega r}{c}\right) + C_2 H_1^{(2)}\left(\frac{\omega r}{c}\right)\right\}e^{i\omega t}\right].$$
$$\text{(VI-23.10}$$

We have not as yet determined C_1 and C_2. We must take $C_1 = 0$. For it can be shown from the relationship of the solution in terms of Hankel functions to the general time-periodic solution that solutions for ϕ of the form $e^{i\omega t}H_0^{(1)}(kr)$ represent inward bound waves, while solutions of the form $e^{i\omega t}H_0^{(2)}(kr)$ represent outward bound waves. Thus, we could have seen already in Eq. (VI-23.8) the need for taking $C_1 = 0$.

To determine C_2 we suppose the pulsating motion at the boundary of the cylinder to be known and such that the radial velocity of each point thereof is given by

$$\frac{dr}{dt} = \text{Re}\left[V_0 e^{i\omega t}\right]. \tag{VI-23.11}$$

It follows from Eqs. (VI-23.9) and (VI-23.11) that

$$(V_r)_{r=R} = \text{Re}\left[\left(\frac{\partial\phi}{\partial r}\right)_{r=R}\right],$$

which is equivalent to requiring that

$$V_0 e^{i\omega t} = \left(\frac{\partial\phi}{\partial r}\right)_{r=R}.$$

Thus, we require, by virtue of Eq. (VI-23.10) with $C_1 = 0$, that

$$V e^{i\omega t} = -\frac{\omega}{c}\, C_2 H_1^{(2)}\left(\frac{\omega R}{c}\right)e^{i\omega t},$$

whence

$$C_2 = -\frac{c}{\omega}\frac{V_0}{H_1^{(2)}\left(\dfrac{\omega R}{c}\right)}. \tag{VI-23.12}$$

Putting this value for C_2 in Eq. (VI-23.10) with $C_1 = 0$ and using the definitions of the Hankel functions from Eqs. (V-0.19) and (V-0.20) and separating out the real part of the result (remembering that $e^{i\omega t} = \cos \omega t + i \sin \omega t$), we find that

$$V_r = \frac{V_0}{\left[J_1\!\left(\dfrac{\omega R}{c}\right)\right]^2 + \left[Y_1\!\left(\dfrac{\omega R}{c}\right)\right]^2} \left\{ J_1\!\left(\frac{\omega R}{c}\right) \right.$$

$$\left[J_1\!\left(\frac{\omega r}{c}\right)\cos \omega t + Y_1\!\left(\frac{\omega r}{c}\right)\sin \omega t \right]$$

$$\left. - Y_1\!\left(\frac{\omega R}{c}\right)\!\left[J_1\!\left(\frac{\omega r}{c}\right)\sin \omega t - Y_1\!\left(\frac{\omega r}{c}\right)\cos \omega t \right] \right\}.$$

$$(\text{VI-23.13})$$

REMARK. It can be shown by virtue of the remark in Prob. V-34 that, for values of r and R large compared to the ratio c/ω, the solution for the radial velocities of the particles of the surrounding fluid as given by Eq. (VI-23.13) is "like" an elementary functions as follows:

$$V_r \sim V_0 \sqrt{\frac{R}{r}} \cos\!\left[\omega\!\left(t - \frac{r - R}{c} \right) \right], \qquad r \to \infty.$$

Problems: Heat-Flow Temperature Distribution

VI-24. A homogeneous circular lamina of radius 20 and diffusivity $\alpha = .71$ is provided with an initial temperature distribution $T_0 = 100 + r^2$, where r denotes distance from center of the lamina. Then the faces of the lamina are made thermally isolated while the temperature all along the rim is maintained at its initial value 500. Determine the subsequent temperature at time $t = 8$ at a point 10 from the center of the lamina, given that the equation to be satisfied by the temperature function T is

$$\frac{\partial T}{\partial t} = \alpha^2 \nabla^2 T. \qquad \star \qquad (\text{VI-24.1})$$

It will be convenient to use an adjusted temperature function U denoting excess of rim temperature over temperature at interior points, namely $U = 500 - T$, so that $U = 0$ along the rim. Equation (VI-24.1) is seen to hold for U as well as for T:

$$\frac{\partial U}{\partial t} = \alpha^2 \nabla^2 U. \tag{VI-24.2}$$

We naturally choose polar coordinates (r, θ) with pole at center of lamina. However, we see by symmetry of conditions that U is independent of θ: $U = U(r, t)$. The Laplacian of U then is

$$\nabla^2 U = \frac{\partial^2 U}{\partial r^2} + \frac{1}{r}\frac{\partial U}{\partial r}. \tag{VI-24.3}$$

Let us see if Eq. (VI-24.2) can be satisfied by a product function of the form

$$U = g(r)e^{-mt}. \tag{VI-24.4}$$

Then Eq. (VI-24.2) in conjunction with Eq. (VI-24.3) requires that

$$- me^{-mt}g(r) = \alpha^2 e^{-mt}\left[\frac{d^2g}{dr^2} + \frac{1}{r}\frac{dg}{dr}\right]. \tag{VI-24.5}$$

Dividing both sides of Eq. (VI-24.5) by the nonvanishing factor e^{-mt}, letting $k^2 = m/\alpha^2$, and multiplying through by r^2, we get

$$r^2\frac{d^2g}{dr^2} + r\frac{dg}{dr} + k^2r^2g = 0. \tag{VI-24.6}$$

By Prob. VI-1 and Eq. (V-0.8) the general solution of Eq. (VI-24.6) is

$$g = AJ_0(kr) + BY_0(kr) \tag{VI-24.7}$$

where A and B are constants. We cannot use $Y_0(kr)$ which becomes infinite as $r \to 0$. So we take $B = 0$.

Our trial solution is now

$$U = AJ_0(kr)e^{-k^2\alpha^2 t} \tag{VI-24.8}$$

Since the rim $r = 20$ is to be maintained at $U = 0$ for all t, Eq. (VI-24.8) requires that

$$A J_0(20k) = 0. \tag{VI-24.9}$$

Choice of $A = 0$ is certainly unsuitable. We may choose k so that $20k$ equals any one of the zeros of $J_0(x)$. But then the formula for U at time $t = 0$ would be a multiple of $J_0(kr)$ which certainly could not be identical with the initially prescribed value $U_0 = 500 - T_0 = 400 - r^2$.

The procedure for overcoming this impasse is the procedure used in Probs. IV-9, VI-21 and others. We express $U_0 = f(r) = 400 - r^2$, by Prob. V-27, as an infinite series of J_0-functions:

$$f(r) = \sum_{n=1}^{\infty} A_n J_0(k_n r), \qquad k_n = \frac{\alpha_n}{20}, \tag{VI-24.10}$$

where

$$A_n = \frac{2}{[20 J_1(\alpha_n)]^2} \int_0^{20} r f(r) J_0\left(\frac{\alpha_n}{20} r\right) dr. \tag{VI-24.11}$$

Since $U = 500 - T$, it follows by Eq. (VI-24.8) that the temperature at any point of the lamina at time t is given by

$$T = 500 - \sum_{n=1}^{\infty} A_n J_0\left(\frac{\alpha_n}{20} r\right) e^{-\alpha^2 \alpha_n^2 t / 400} \tag{VI-24.12}$$

where the coefficients A_1, A_2, $\cdots$ are given by Eq. (VI-24.11) with $f(r) = 400 - r^2$. The integral for each A_n in Eq. (VI-24.11) breaks up into two integrals, one of which may be evaluated by Eq. (V-28.3), the other by Eq. (V-30.1).

As an example of the determination of the values of the coefficients let us evaluate (approximately) the first coefficient A_1. From Eq. (VI-24.11) with $f(r) = 400 - r^2$ we have

$$A_1 = \frac{2}{[20 J_1(\alpha_1)]^2} \left[\int_0^{20} 400 r J_0\left(\frac{\alpha_1}{20} r\right) dr - \int_0^{20} r^3 J_0\left(\frac{\alpha_1}{20} r\right) dr \right]. \tag{VI-24.13}$$

In Table V-4 we find that $\alpha_1 \cong 2.4048 \cong 2.40$. Thus,

$$A_1 \cong \frac{1}{200[J_1(2.40)]^2}\left[400 \int_0^{20} rJ_0([.12]r)dr - \int_0^{20} r^3 J_0([.12]r)dr \right].$$
(VI-24.14)

Applying Eqs. (V-28.3) and (V-30.1), we get

$$A_1 \cong (.019)\left[\frac{(400)(20)}{.12}J_1(2.40) - \frac{8000}{.12}J_1(2.40) + \frac{800}{.144}J_2(2.40) \right].$$
(VI-24.15)

From Tables V-2 and V-3 we have $J_1(2.40) \cong .52$ and $J_2(2.40) \cong .43$. Putting these values in Eq. (VI-24.15) we obtain

$$A_1 \cong 4.56.$$

The remaining coefficients $A_2, A_3, A_4, \cdots$ are computed just as in Eq. (VI-24.13) through Eq. (VI-24.16), the only changes being that α_2 replaces α_1 in computing A_2, α_3 replaces α_1 in computing A_3, and so on.

Thus, we may indicate the solution of our problem at $t = 8$ and $r = 10$ by

$$T(r, t) = T(10, 8) = 500 - (4.56)J_0\left(\frac{2.40}{20}[10]\right)e^{-(0.71)^2(2.40)^2(8)/400}$$

$$- \sum_{n=2}^{\infty} A_n J_0\left(\frac{\alpha_n}{20}[10]\right)e^{-(0.71)^2\alpha_n^2(8)/400}.$$
(VI-24.16)

REMARKS. 1. The solution for T as given by Eq. (VI-24.12) indicates that, at any chosen interior point of the lamina, the temperature T will approach 500 as limit as $t \to \infty$.

2. Taking only the first two terms of Eq. (VI-24.16) as rough approximation for the solution of our problem at $r = 10$ with $t = 8$, we have

$$T(r, t) = T(10, 8) \cong 500 - (4.56)(.67)e^{-.06}$$

$$\cong 500 - 3 = 497.$$

VI-25. A homogeneous solid occupies the region R consisting of one-half of a right circular cylinder of radius 10 and height 20. Taking the axis of the cylinder in the z-axis with lower base in the xy-plane, we take R to be that half of the cylinder for which $y \geq 0$. The curved surface R, the lower base and the plane face containing the z-axis shall all be maintained at constant temperature $T_1 = 40$. This temperature $T_1 = 40$ shall be maintained also on the rim of the upper base. On the rest of the upper base shall be maintained a temperature distribution $T_2 = r^2\theta$, where r and θ together with z are cylindrical space coordinates with θ confined to the interval $0 < \theta < \pi$. Determine the steady state temperature distribution $T = T(r, \theta, z)$ in R. ✶

We shall find it convenient here, as in Prob. VI-24, to consider the function $U = T - 40$, which is such that $U = 0$ on the three faces of R other than the face $z = 20$. The temperature T must satisfy Laplace's equation $\nabla^2 T = 0$ at all interior points of R; the same must also be true for U, namely, $\nabla^2 U = 0$, since U differs from T only by an additive constant. Thus, we require

$$\frac{\partial^2 U}{\partial r^2} + \frac{1}{r^2}\frac{\partial^2 U}{\partial \theta^2} + \frac{\partial^2 U}{\partial z^2} + \frac{1}{r}\frac{\partial U}{\partial r} = 0 \qquad \text{(VI-25.1)}$$

at every interior point of R. And, as in other similar problems, we shall try to construct by composition of particular solutions a solution of Eq. (VI-25.1) which will cause T to take on the prescribed boundary temperature distributions. Moreover, we shall seek to obtain the particular solutions each as a product function of the form

$$U = F(r)G(\theta)H(z). \qquad \text{(VI-25.2)}$$

It will be convenient to remember that F is a function of r alone, G a function of θ alone, and H a function of z alone and to omit their respective arguments and to write simply

$$U = FGH. \qquad \text{(VI-25.2a)}$$

Using primes and double primes to denote first and second derivatives of F and G and H each with respect to its own argument, we get

$$\frac{\partial U}{\partial F} = F'GH, \qquad \frac{\partial^2 U}{\partial F^2} = F''GH,$$

$$\frac{\partial U}{\partial G} = G'FH, \qquad \frac{\partial^2 U}{\partial G^2} = G''FH,$$

$$\frac{\partial U}{\partial H} = H'FG, \qquad \frac{\partial^2 U}{\partial H^2} = H''FG.$$

Substituting these results in Eq. (VI-25.1) we get

$$F''GH + \frac{1}{r^2}FG''H + FGH'' + \frac{1}{r}F'GH = 0,$$

which may be written

$$\frac{r^2F''}{F} + \frac{rF'}{F} + \frac{r^2H''}{H} = -\frac{G''}{G}. \qquad \text{(VI-25.3)}$$

The right and left sides of Eq. (VI-25.3) must both equal one and the same constant C (sometimes called the separation constant), since one side is independent of θ while the other side is independent of r and z. We proceed to see what sort of constant C shall be, positive or negative or zero. Our procedure will be to make some assumptions regarding the constant C. If an assumption or premise leads to an absurd conclusion, we shall reject the assumption. First we shall try $C = 0$.

Then Eq. (VI-25.3) requires that $G'' = 0$. Integrating we get $G = A\theta + B$, where A and B are constants. Now on the plane lateral surface we need to have $U \equiv 0$. If we want the factor G to do the vanishing there so that Eq. (VI-25.2) would be $0 = F(r)(0)H(z)$, then $G(0) = G(\pi) = 0$. This would require A and B to equal zero, making U vanish everywhere in R. This absurd conclusion makes us reject the assumption that $C = 0$.

Now we try C negative, say $-k^2$ with $k \neq 0$. Then Eq. (VI-25.3) requires that $G'' = k^2 G$. The solution to this equation is

$$G = A \cosh k\theta + B \sinh k\theta.$$

And if G is to be the vanishing factor on the plane lateral surface, we need $A = B = 0$, making U identically zero in R. Let us then reject the assumption that C is negative.

There remains the possibility that C be positive: $C = k^2$ with $k \neq 0$. This time Eq. (VI-25.3) demands that

$$G = A \cos k\theta + B \sin k\theta.$$

And now we can have $G(0) = G(\pi) = 0$ by having $A = 0$ and $\sin k\pi = 0$. Thus, we can have the factor $G = 0$ everywhere on the plane lateral surface by taking for G any function of the family

$$G_n = B_n \sin n\theta, \qquad n = 1, 2, 3, \cdots \qquad \text{(VI-25.4)}$$

where B_n is an arbitrary constant $\neq 0$. So far we have made some progress. In the two previous assumptions on the constant C, where we let it equal zero or a negative number, we were able to satisfy the boundary condition $U = 0$ on the plane lateral face. But we had the untenable consequence that $U = 0$ at all interior points of R. However, when we took the separation constant as greater than zero we were able to get the boundary condition satisfied without having zero temperature interior to R at the same time.

And so, taking $C = k^2 = n^2$ where n is any positive integer, we go back to Eq. (VI-25.3) and set its left side equal to n^2. The resulting equation may be written

$$\frac{F''}{F} + \frac{F'}{rF} - \frac{n^2}{r^2} = -\frac{H''}{H}. \qquad \text{(VI-25.5)}$$

By the same argument as we applied to Eq. (VI-25.3) we conclude that both sides of Eq. (VI-25.5) must equal one and the same constant.

Let us call this separation constant M. We proceed as before to make assumptions regarding M and as before we will reject those assumptions which lead to contradictions. Let us try $M = 0$.

When $M = 0$, then Eq. (VI-25.5) requires $r^2F'' + rF' - n^2F = 0$. The solution to this equation is

$$F = Ar^n + \frac{B}{r^n}.$$

Since the term B/r^n becomes unsuitable when $r = 0$, we take $B = 0$. Now at each point of the lateral surface $r = 10$ at least one of the factors F, G, H shall vanish. The simplest way of accomplishing the vanishing there of U is to have one and the same factor of U vanish everywhere on that surface. No member of Eq. (VI-25.4) will vanish identically there. So it is up to F or H to be the vanishing factor for the curved surface. If we require $F = Ar^n$ to vanish when $r = 10$, then $A = 0$ and we have F identically zero with the impossible result that U vanishes identically.

If we try $M = -k^2$ with $k \neq 0$, then Eq. (VI-25.5) requires us to have $r^2F'' + rF' - (k^2r^2 + n^2)F = 0$, whose general solution by Prob. VI-2 and Eq. (V-0.12) is

$$F = AI_n(kr) + BK_n(kr).$$

Here again we take $B = 0$, since the term $BK_n(kr)$ becomes infinite when $r = 0$. And if F shall be the vanishing factor on the curved surface, we require $AI_n(10k) = 0$. So, with each n we may take any number k such that $10k$ is a zero of I_n.

If we try $M = k^2$ with $k \neq 0$, then Eq. (VI-25.5) demands that $r^2F'' + rF' + (k^2r^2 - n^2F) = 0$, whose general solution by Prob. VI-2 and Eq. (V-0.8) is

$$F = AJ_n(kr) + BY_n(kr).$$

Again we must take $B = 0$ because $Y_n(kr)$ becomes infinite at $r = 0$. And if F shall be the vanishing factor on the curved surface

$r = 10$, we require $AJ_n(10k) = 0$. Thus with each n we may take any number k such that $10k$ is a zero of J_n. We cannot know a priori whether to follow through with $M = -k^2$ or $M = k^2$. However, most of the developments of expansions and integrals in the theory of Bessel functions are with the J-functions; so we will follow through with $M = k^2$ in the hope we can develop a solution.

With each function G of the family in Eq. (VI-25.4) we may take any function F of the family

$$F_{nj} = A_{nj}J_n(k_{nj}r), \qquad j = 1, 2, 3, \cdots, \quad A_{nj} \neq 0, \quad \text{(VI-25.6)}$$

where

$$k_{nj} = \text{one-tenth of the } j^{\text{th}} \text{ positive zero of } J_n. \qquad \text{(VI-25.6a)}$$

Reverting to Eq. (VI-25.5) with $M = k_{nj}^2$, we require $H'' = k_{nj}^2 H$, whose general solution is

$$H = D \cosh(k_{nj}z) + E \sinh(k_{nj}z).$$

Let us call upon H to do the vanishing on the lower base of R, since no member of Eq. (VI-25.4) vanishes there nor does any member of Eq. (VI-25.6). Thus if we require $H(0) = 0$, we find $D = 0$, making

$$H = E \sinh(k_{nj}z). \qquad \text{(VI-25.7)}$$

Putting together our results obtained thus far in Eqs. (VI-25.4) and (VI-25.6) and (VI-25.7), we find that we can meet all requirements of the problem except the prescribed temperature distribution to be maintained on the upper base by a function of the form

$$T = 40 + U = 40 + A_{nj}J_n(k_{nj}r) \cdot \sin n\theta \cdot \sinh(k_{nj}z). \qquad \text{(VI-25.8)}$$

But no single function T of this kind can equal the prescribed temperature $T_2 = r^2\theta$ on the upper base. This means that here, as in Prob. IV-9 for example, the final step in the solution of our problem is to construct an infinite series of solutions of the type found in Eq. (VI-25.8), determining the coefficients so that the series converges to $T_2 = r^2\theta$ on the upper base where $z = 20$. In the present problem we shall have to construct a doubly infinite series, since T_2 is a function of two independent arguments. However, in Eq. (VI-25.8) we have two indices over which we can sum, namely n and j.

Our task is to determine a doubly infinite series such that for $0 < r < 10$, $\quad 0 < \theta < \pi$

$$r^2\theta - 40 = \sum_{n=1}^{\infty} \left[A_n \sin n\theta \sum_{j=1}^{\infty} B_{nj} J_n(k_{nj}r) \sinh (20k_{nj}) \right].$$
(VI-25.9)

We can accomplish the desired expansion (VI-25.9) by making it to be the sum of two such expansions as follows. In the first one of them we make the summation with respect to j equal r^2 for each and every n. We make the trigonometric sum $\sum A_n \sin n\theta$ by itself equal to θ. In the second one we make the inner j-sum equal to unity for each and every n. This time we make the sine series by itself equal to unity. Let us denote the coefficients in the first sum by B'_{nj} and A'_n. In the second sum we denote the coefficients by B''_{nj} and A''_n. Then the solution to our problem is given by

$$T = 40 + \sum_{n=1}^{\infty} \left[A'_n \sin n\theta \sum_{j=1}^{\infty} B'_{nj} \sinh (k_{nj}z) J_n(k_{nj}r) \right]$$

$$- 40 \sum_{n=1}^{\infty} \left[A''_n \sin n\theta \sum_{j=1}^{\infty} B''_{nj} \sinh (k_{nj}z) J_n(k_{nj}r) \right]. \quad \text{(VI-25.10)}$$

The right side of Eq. (VI-25.10) equals $T_1 = 40$ on the lower base of R and on the curved surface $r = 10$ and on the vertical plane face and along the rim of the upper base. On the rest of the upper base the right side of Eq. (VI-25.10) equals $T_2 = r^2\theta$. At every interior point of R the function T defined by Eq. (VI-25.10) satisfies

Laplace's equation $\nabla^2 T = 0$, since each series is termwise differentiable inasmuch as each series is uniformly convergent for $0 \leqq r \leqq 10$, $0 \leqq z \leqq 20$, $0 < a \leqq \theta \leqq b < \pi$.

The formulas for the coefficients in the respective series in Eq. (VI-25.10) are as follows. Since we want to have

$$\sum_{n=1}^{\infty} A_n' \sin n\theta = \theta, \qquad \text{for } 0 \leqq \theta < \pi,$$

then by the well-known formula for coefficients in a Fourier sine series of an odd function we have

$$A_n' = \frac{2}{\pi} \int_0^\pi \theta \sin n\theta \, d\theta = \frac{2(-1)^{n+1}}{n}. \qquad \text{(VI-25.11)}$$

Similarly, since we require

$$\sum_{n=1}^{\infty} A_n'' \sin n\theta = 1, \qquad 0 < \theta < \pi,$$

then, by the Fourier series expansion for the odd function

$$f(\theta) = \begin{cases} -1, & -\pi < \theta < 0 \\ 0, & \theta = 0 \\ 1, & 0 < \theta < \pi, \end{cases}$$

we have

$$A_n'' = \frac{2}{\pi} \int_0^\pi \sin n\theta \, d\theta = \begin{cases} \dfrac{4}{n\pi}, & n \text{ odd} \\ 0, & n \text{ even.} \end{cases} \qquad \text{(VI-25.12)}$$

Since we require

$$\sum_{n=1}^{\infty} B_{nj}' \sinh (20 k_{nj}) J_n(k_{nj} r) = r^2,$$

then, by Eq. (VI-25.6a) together with Eq. (V-27.4), we have

$$B_{nj}' = \frac{2 \displaystyle\int_0^{10} r^3 J_n(k_{nj} r) dr}{10^2 [J_{n+1}(10 k_{nj})]^2 \sinh (20 k_{nj})}. \qquad \text{(VI-25.13)}$$

Similarly, since we require

$$\sum_{n=1}^{\infty} B''_{nj} \sinh (20k_{nj})J_n(k_{nj}r) = 1$$

we have

$$B''_{nj} = \frac{2\int_0^{10} rJ_n(k_{nj}r)dr}{10^2[J_{n+1}(10k_{nj})]^2 \sinh (20k_{nj})}. \qquad \text{(VI-25.14)}$$

REMARKS. 1. Each of the "duplex" series in the solution Eq. (VI-25.10) can be written as a simple series as follows. Let us indicate symbolically the terms in either series by their coefficient letters A and B without the primes. Then we can write out either "duplex" series as a simple series as follows:

$$A_1B_{1,1} + A_1B_{1,2} + A_1B_{1,3} + A_2B_{2,1} + A_1B_{1,4} + A_2B_{2,2}$$
$$+ A_1B_{1,5} + A_2B_{2,3} + A_3B_{3,1} + A_1B_{1,6} + A_2B_{2,4}$$
$$+ A_3B_{3,2} + \cdots.$$

The scheme just indicated is to take first the term for which $n + (n + j) = 3$, then add the one term for which $n + (n + j) = 4$, then add all the terms (2 in number) for which $n + (n + j) = 5$, and so on ad infinitum.

Let us see how the first "duplex" series in Eq. (VI-25.10) would look written out to three terms of such a simple series in case we wanted to evaluate T at the point $(5, \pi/4, 10)$. We use Eqs. (VI-25.11) and (VI-25.13) for the coefficients. And we recall from Eq. (VI-25.6a) that

$$k_{nj} = \text{one-tenth of the } j^{\text{th}} \text{ positive zero of } J_n.$$

We take the first and second positive zero of J_n from Table V-4. The terms we choose, indicated symbolically by their coefficients only, are $A_1' B_{1,1}'$, $A_1' B_{1,2}'$, $A_2' B_{2,1}'$. For the sum of these three terms, taking $r = 5$, $\theta = \pi/4$, and $z = 10$, we would have

$$\sqrt{2}\ \sinh(3.83)J_1(1.92)\frac{\int_0^{10} r^3 J_1([.383]r)dr}{50[J_2(3.83)]^2 \sinh(7.66)}$$

$$+ \sqrt{2}\ \sinh(7.02)J_1(3.51)\frac{\int_0^{10} r^3 J_1([.702]r)dr}{50[J_2(7.02)]^2 \sinh(14.04)}$$

$$- \sinh(5.14)J_2(2.57)\frac{\int_0^{10} r^3 J_2([.514]r)dr}{50[J_3(5.14)]^2 \sinh(10.28)}.$$

The values of the hyperbolic sine and the values of J_1 and J_2 for the arguments indicated can be taken from appropriate tables.

2. Variations of Prob. VI-25 can be solved by the method of superposition by virtue of the fact that the sum of a finite number of functions each harmonic (satisfying Laplace's equation) in a region R is also harmonic in R. Suppose, for example, that the prescribed boundary temperature distributions to be maintained on the face of the solid in Prob. VI-25 shall be one and the same constant, say 40, on only two of the faces, namely on the curved lateral face C and also on the plane lateral face L, while the prescribed temperature to be maintained on the bases shall be $f(r, \theta)$ on the upper base H and $g(r, \theta)$ on the lower base B. In such a case the required steady-state temperature T can be obtained as a tripartite sum as follows. We take a function U as determined in Prob. VI-25 and call it U_1. This function U_1 is such that $\nabla^2 U_1 = 0$ interior to R while U_1 takes on the boundary values

$$U_1 = \begin{cases} 0 & \text{on } C \text{ and } L \text{ and } B, \\ f(r, \theta) - 40 & \text{on } H. \end{cases} \tag{VI-25.15}$$

Then we determine in the manner of Prob. VI-25 a function U_2 such that $\nabla^2 U_2 = 0$ interior to R and such that

$$U_2 = \begin{cases} 0 & \text{on } C \text{ and } L \text{ and } H, \\ g(r, \theta) - 40 & \text{on } B. \end{cases} \tag{VI-25.16}$$

Then the solution to the new problem is

$$T = U_1 + U_2 + 40.$$

For, interior to R we have

$$\nabla^2 T = \nabla^2 U_1 + \nabla^2 U_2 + \nabla^2(40) = 0.$$

And by virtue of Eqs. (VI-25.15) and (VI-25.16) we have

$$T = \begin{cases} 0 & \text{on } C \text{ and } L, \\ f(r, \theta) & \text{on } H, \\ g(r, \theta) & \text{on } B. \end{cases}$$

In similar manner one can obtain solution of the variation of Prob. VI-25 in which the prescribed boundary temperature distribution shall be constant on only one face and on each of the other three faces shall be a nonconstant function of the two variables concerned. Also solvable by the method of superposition is the problem when the prescribed temperature to be maintained on the boundary shall be on each of the four faces a nonconstant function of the two variables concerned. In this latter case we first determine four functions U_1, U_2, U_3, U_4 each satisfying Laplace's equation interior to R and such that

$$U_1 = 0 \quad \text{on } B \text{ and } C \text{ and } L, \quad U_1 = f(r, \theta) \text{ on } H,$$
$$U_2 = 0 \quad \text{on } L \text{ and } C \text{ and } H, \quad U_2 = g(r, \theta) \text{ on } B,$$
$$U_3 = 0 \quad \text{on } H \text{ and } B \text{ and } C, \quad U_3 = h(r, z) \text{ on } L,$$
$$U_4 = 0 \quad \text{on } H \text{ and } B \text{ and } L, \quad U_4 = p(z, \theta) \text{ on } C.$$

Then the function

$$T = U_1 + U_2 + U_3 + U_4$$

is such that $\nabla^2 T = 0$ interior to R and such that

$$T = \begin{cases} f(r, \theta) & \text{on } H, \\ g(r, \theta) & \text{on } B, \\ h(r, z) & \text{on } L, \\ p(z, \theta) & \text{on } C. \end{cases}$$

Problem: Displacement of Vibrating Annular Membrane

VI-26. A stretched elastic membrane occupies the plane region bounded by two concentric circles having radii a and b, $b > a$ and lying in the (r, θ)-plane, where r and θ are polar coordinates with the pole at the common center of the two circles. The membrane is clamped along each circle (as in a drum head). It is vibrating as result of having been given an initial distortion (and then set free) such that each point of the membrane was initially displaced at right angles to the plane of the circles by a signed amount, dependent only on distance from the pole, namely

$$z_0 = f(r) \qquad \text{at time } t = 0, \qquad \text{(VI-26.1)}$$

where r is differentiable and of bounded variation for $a \leqq r \leqq b$. Under the approximating assumption that each point of the membrane moves in a straight line at right angles to the plane of the bounding circles (that is, parallel to the z-axis) the differential equation to be satisfied is

$$\frac{\partial^2 z}{\partial x^2} + \frac{\partial^2 z}{\partial y^2} = \frac{1}{c^2} \frac{\partial^2 z}{\partial t^2}, \qquad \text{(VI-26.2)}$$

where x and y are rectangular coordinates in the (r, θ)-plane, and where c is a constant, namely $\sqrt{T/m}$, T denoting the magnitude of tension (assumed constant) and m denoting the mass per unit area of the membrane. Obtain solution for Eq. (VI-26.2), assuming that the motion of each point of the membrane is a sinusoidal function of t.   ⋆

The statement of the problem calls for the use of cylindrical coordinates (r, θ, z). Taking the origin of xy-coordinates at the pole, we have

$$\frac{\partial^2 z}{\partial x^2} + \frac{\partial^2 z}{\partial y^2} = \frac{\partial^2 z}{\partial r^2} + \frac{1}{r^2} \frac{\partial^2 z}{\partial \theta^2} + \frac{1}{r} \frac{\partial z}{\partial r}. \qquad \text{(VI-26.3)}$$

But, the conditions of the problem are such that z is going to be independent of θ. This makes the middle term on the right in Eq. (VI-26.3) vanish. Accordingly, Eq. (VI-26.2) becomes

$$\frac{\partial^2 z}{\partial r^2} + \frac{1}{r}\frac{\partial z}{\partial r} = \frac{1}{c^2}\frac{\partial^2 z}{\partial t^2}. \qquad \text{(VI-26.4)}$$

Since z is then a function of r and t and is assumed sinusoidal in t for each r, we may hope to obtain solution for Eq. (VI-26.4) in the form of a composite of particular solutions of Eq. (VI-26.4), each particular solution being a product solution of the form

$$z = G(r)\Big[A\cos(\omega t + \lambda)\Big], \qquad \text{(VI-26.5)}$$

where A and ω and λ are constants to be determined by boundary and initial conditions. At time $t = 0$ we must have by condition (VI-26.1)

$$f(r) = G(r)\Big[A\cos\lambda\Big], \qquad a \leqq r \leqq b,$$

so that we may take $A = 1$ and $\lambda = 0$. Thus, we seek particular solutions for Eq. (VI-26.4) of the form

$$z = G(r)\cos\omega t, \qquad \omega \neq 0. \qquad \text{(VI-26.6)}$$

Substitution for z from Eq. (VI-26.6) into Eq. (VI-26.4) yields

$$(\cos\omega t)\Big[G'' + \frac{1}{r}G' + \frac{\omega^2}{c^2}G\Big] = 0, \qquad \text{(VI-26.7)}$$

where the primes on G denote first and second derivative with respect to r. Since $\cos\omega t$ will not be identically zero, satisfaction of Eq. (VI-26.7) requires that

$$G'' + \frac{1}{r}G' + \frac{\omega^2}{c^2}G = 0,$$

or

$$r^2 G'' + rG' + \lambda^2 r^2 G = 0, \qquad \lambda = \omega/c. \qquad \text{(VI-26.8)}$$

Solution of Eq. (VI-26.8) is immediately obtained by way of Eq. (VI-1.6) taking $r = x$, $D = \lambda$, $p = 0$. Thus, by Eqs. (VI-1.5) and (V-0.8) the general solution of Eq. (VI-26.8) is

$$G = C_1 J_0(\lambda r) + C_2 Y_0(\lambda r), \qquad \text{(VI-26.9)}$$

where C_1 and C_2 are arbitrary constants. There are, however, two boundary conditions to be met by virtue of the fact that the membrane is clamped along the two circles. These require by Eqs. (VI-26.6) and (VI-26.9) that

$$\left. \begin{aligned} C_1 J_0(\lambda a) + C_2 Y_0(\lambda a) &= 0 \\ C_1 J_0(\lambda b) + C_2 Y_0(\lambda b) &= 0 \end{aligned} \right\}. \qquad \text{(VI-26.10)}$$

Equations (VI-26.10) constitute two simultaneous equations in three unknowns, namely C_1, C_2, and λ. We solve the second of these equations for C_2:

$$C_2 = -\frac{J_0(\lambda b)}{Y_0(\lambda b)} C_1.$$

Then we substitute this for C_2 in the first equation:

$$C_1 \left[J_0(\lambda a) - \frac{J_0(\lambda b)}{Y_0(\lambda b)} Y_0(\lambda a) \right] = 0. \qquad \text{(VI-26.11)}$$

Eq. (VI-26.11) will be satisfied for arbitrary choice of C_1 if $\lambda = \lambda_n$ is a solution of the equation

$$J_0(\lambda a) - \frac{J_0(\lambda b)}{Y_0(\lambda b)} Y_0(\lambda a) = 0. \qquad \text{(VI-26.12)}$$

Assuming that Eq. (VI-26.12) has a set of infinitely many distinct solutions $\lambda_1, \lambda_2, \lambda_3, \cdots, \lambda_n, \cdots$, we see that all the conditions of the problem *except one* can be met by a function of the form

$$z = C_n \left[J_0(\lambda_n r) - \frac{J_0(\lambda_n b)}{Y_0(\lambda_n b)} Y_0(\lambda_n r) \right] \cos (\lambda_n ct).$$

$$\text{(VI-26.13)}$$

It is condition (1) which cannot be satisfied at time $t = 0$ by any single function of form (VI-26.13) or finite sum of such functions except in the special case that $f(r)$ is itself prescribed as such a sum. It is possible, however, by appropriate generalization of the expansion considered in Prob. V-26 to express a function of the sort described in connection with condition (VI-26.1) as an infinite series, namely

$$f(r) = \sum_{n=1}^{\infty} C_n \left[J_0(\lambda_n r) - \frac{J_0(\lambda_n b)}{Y_0(\lambda_n b)} Y_0(\lambda_n r) \right]. \quad \text{(VI-26.14)}$$

One finds by arguments similar to those used in Probs. V-24, V-25, V-26 (see also Byerly, *Fourier's Series and Spherical Harmonics*) that the family of functions

$$L_0(\lambda_n r) = \sqrt{r} \left[J_0(\lambda_n r) - \frac{J_0(\lambda_n b)}{Y_0(\lambda_n b)} Y_0(\lambda_n r) \right], \qquad n = 1, 2, 3, \cdots$$
$$\text{(VI-26.15)}$$

is orthogonal on the interval $a \leqq r \leqq b$ and that the integral over this interval of a squared member of the family is given by

$$\int_a^b \left[L_0(\lambda_n r) \right]^2 dr = \frac{1}{2} \left\{ b^2 \left[L_0'(\lambda_n b) \right]^2 - a^2 \left[L_0'(\lambda_n a) \right]^2 \right\}$$
$$\text{(VI-26.16)}$$

and the coefficients C_n in Eq. (VI-26.14) are given by

$$C_n = \frac{2}{b^2 [L_0'(\lambda_n b)]^2 - a^2 [L_0'(\lambda_n a)]^2} \int_a^b L_0(\lambda_n r) r f(r) dr.$$
$$\text{(VI-26.17)}$$

The solution of our problem is thus found to be

$$z = \sum_{n=1}^{\infty} C_n \left[J_0(\lambda_n r) - \frac{J_0(\lambda_n b)}{Y_0(\lambda_n b)} Y_0(\lambda_n r) \right] \cos(\lambda_n c t),$$
$$\text{(VI-26.18)}$$

where the coefficients C_n, $n = 1, 2, 3, \cdots$, are given by Eq. (VI-26.17).

Problem: Alternating Current Density

VI-27. Determine the current density in a straight homogeneous wire of circular cross-section lying in the z-axis and carrying alternating current, given the following:

(a) the current density vector $\vec{i}$ is continuous and is always parallel to the z-axis;

(b) the scalar component $i = i_z$ of $\vec{i}$ parallel to the z-axis is given by

$$i = u(r, t) = \mathrm{Re}\Big[g(r)e^{j\omega t} \Big], \qquad (\text{VI-27.1})$$

where t denotes time, $j = \sqrt{-1}$, ω is a constant, and $\mathrm{Re}[w]$ denotes the real part of the complex number $w = u + jv$;

(c) the current density i satisfies the equation

$$\nabla^2 i = \lambda\frac{\partial i}{\partial t}, \qquad (\text{VI-27.2})$$

where λ is a constant and $\nabla^2 i$ denotes the Laplacian of i. ⋆

We have placed the wire in the z-axis in order that we may use cylindrical space coordinates (r, θ, z). Since $i = u(r, t)$ is independent of θ and z, its Laplacian (compare Prob. VI-23), namely

$$\nabla^2 i = \frac{\partial^2 i}{\partial r^2} + \frac{1}{r}\frac{\partial i}{\partial r} + \frac{1}{r^2}\frac{\partial^2 i}{\partial \theta^2} + \frac{\partial^2 i}{\partial z^2},$$

reduces to two terms, so that Eq. (VI-27.2) becomes

$$\frac{\partial^2 i}{\partial r^2} + \frac{1}{r}\frac{\partial i}{\partial r} = \lambda\frac{\partial i}{\partial t}. \qquad (\text{VI-27.3})$$

Now if we can find a function $g(r)$, real or complex, such that the complex function

$$w = g(r)e^{j\omega t} \qquad (\text{VI-27.4})$$

satisfies the equation

$$\nabla^2 w = \lambda\frac{\partial w}{\partial t}, \qquad (\text{VI-27.5})$$

then by separation of reals and pure imaginaries, it will follow that the real part thereof, namely $i = u(r, t) = \mathrm{Re}[g(r)e^{j\omega t}]$, will satisfy

Eq. (VI-27.3), thus satisfying Eq. (VI-27.2); and our problem will be solved.

Substitution for w from Eq. (VI-27.4) in Eq. (VI-27.5) yields

$$e^{j\omega t}\left[\frac{\partial^2 g}{\partial r^2} + \frac{1}{r}\frac{\partial g}{\partial r}\right] = (\lambda g)\left[j\omega e^{j\omega t}\right].$$

The nonvanishing factor $e^{j\omega t}$ may be divided out:

$$g'' + \frac{1}{r}g' - j\omega\lambda g = 0,$$

where the primes now denote first and second derivatives with respect to r. Multiplying through by r^2 and letting $k^2 = -j\omega\lambda$, we have

$$r^2 g'' + rg' + k^2 r^2 g = 0. \tag{VI-27.6}$$

Identification of Eq. (VI-27.6) with Eq. (VI-1.6), with g and r playing the roles of y and x respectively, tells us by Eqs. (VI-1.5) and (V-0.8) that g shall be a function of the sort

$$g = C_1 J_0(kr) + C_2 Y_0(kr). \tag{VI-27.7}$$

The requirement of continuity of $\vec{i}$ in **(a)** means we shall take $C_2 = 0$ in Eq. (VI-27.7), since $Y_0(kr)$ becomes infinite at $r = 0$. We have put $k^2 = -j\omega\lambda$. This makes $k = \sqrt{-1}\,j^{1/2}\sqrt{\omega\lambda} = j^{3/2}\sqrt{\omega\lambda}$. Let us put $m = \sqrt{\omega\lambda}$. Then we have

$$g = C_1 J_0(j^{3/2}mr), \tag{VI-27.8}$$

which by Eq. (V-0.15) may be written

$$g = C_1[\text{ber}(mr) + j\,\text{bei}\,(mr)]. \tag{VI-27.9}$$

Thus, the solution to our problem, with C_1 yet to be determined, is

$$i = \text{Re}\left[C_1 J_0(j^{3/2}mr)e^{j\omega t}\right]$$

$$= \text{Re}\left[C_1 J_0(j^{3/2}mr)\left\{\cos\,\omega t + j\sin\,\omega t\right\}\right]. \tag{VI-27.10}$$

To determine C_1, let A denote the amplitude (maximum absolute value of i) at a point in the axis of the wire (A being the same for all such points) where $r = 0$. Now at $r = 0$ we have $J_0(j^{3/2}mr) = J_0(0) = 1$ by Eq. (V-0.4). Thus by Eq. (VI-27.10)

$$A = \max\left| \text{Re}\left[C_1 \left\{ \cos \omega t \ + j \sin \omega t \right\} \right] \right|.$$

If we take C_1 real, then

$$A = \max| C_1 \cos \omega t \ |$$
$$= |C_1|(\max|\cos \omega t \ |)$$
$$= |C_1|(1).$$

We may take $C_1 = A$ or $C_1 = - A$. We choose the former. Finally, then the solution of our problem is

$$i = A\text{Re}\left[J_0(j^{3/2}\sqrt{\omega\lambda}\, r)e^{j\omega t} \right], \qquad \text{(VI-27.11)}$$

that is,

$$i = A\left[\text{ber}\,(\sqrt{\omega\lambda}\, r) \cos \omega t \ - \text{bei}\,(\sqrt{\omega\lambda}\, r) \sin \omega t \ \right].$$
$$\text{(VI-27.12)}$$

To find the total current i_T in a wire of radius R we integrate the current density i in Eq. (VI-27.11) over the circular cross-section S of the wire:

$$i_T = \text{Re}\left[Ae^{j\omega t}\iint\limits_{S} J_0(j^{3/2}\sqrt{\omega\lambda}\, r)dS \right]$$

$$= \text{Re}\left[Ae^{j\omega t}\int_0^{2\pi} d\theta \int_0^R J_0(j^{3/2}\sqrt{\omega\lambda}\, r)r\, dr \right]$$

$$= \text{Re}\left[Ae^{j\omega t}2\pi \int_0^R rJ_0(j^{3/2}\sqrt{\omega\lambda}\, r)dr \right].$$

An integral of this type was evaluated in Prob. V-28. Thus, we have

$$i_T = \text{Re}\left[\frac{2\pi ARJ_1(j^{3/2}\sqrt{\omega\lambda}\, R)}{j^{3/2}\sqrt{\omega\lambda}}e^{j\omega t} \right]. \qquad \text{(VI-27.13)}$$

Problems: Eddy Current Density and Power Loss

VI-28. Determine the eddy current density *i* induced in the copper core of a solenoid under the following conditions. The core is a long solid copper circular cylinder of radius *R*. The solenoid wound around the core has *N* turns per cm. The solenoid is excited by a current having magnitude $I \cos 2\pi ft$ abamp., where *I* is the amplitude constant, *f* is the number of cycles per second, and *t* is time in seconds. The solenoid and core are sufficiently long (in ratio to *R*) that end effects may be neglected. The following are given:

(a) the magnetic field strength $H = H(r, t)$, where *r* denotes distance from the axis of the core, satisfies the equation

$$r \frac{\partial^2 H}{\partial r^2} + \frac{\partial H}{\partial r} = \lambda^2 r \frac{\partial H}{\partial t}, \qquad (\text{VI-28.1})$$

where $\lambda^2 = 4\pi/\rho$ and ρ is the resistivity of copper in abohms per cu. cm.,

(b) at the surface of the core the function *H* takes on the boundary value

$$H(R, t) = 4\pi NI \cos 2\pi ft, \qquad (\text{VI-28.2})$$

(c) the eddy current density *i* is related to the magnetic field strength by the equation

$$i = -\frac{1}{4\pi} \frac{\partial H}{\partial r}. \qquad \star \qquad (\text{VI-28.3})$$

If Eq. (VI-28.1) be divided through by *r*, the left side is then none other than the first two terms of the Laplacian

$$\nabla^2 H = \frac{\partial^2 H}{\partial r^2} + \frac{1}{r} \frac{\partial H}{\partial r} + \frac{\partial^2 H}{\partial z^2} + \frac{1}{r^2} \frac{\partial^2 H}{\partial \theta^2} \qquad (\text{VI-28.4})$$

when *H* is considered as a function of all three cylindrical coordinates (r, θ, z). Since, in the present problem, *H* is a function of only one of the three cylindrical coordinates, the first two terms on the right in

Eq. (VI-28.4) may actually be regarded as the Laplacian of H. Now it is well-known that partial differential equations in which the Laplacian of the desired function equals either zero or a constant times its partial derivative with respect to time may often be solved via a product function, or a composite of product functions, in which each factor in the product is a function of just one of the variables concerned. Accordingly, we shall try to find solution for Eq. (VI-28.1) as a function of the form

$$H = F(r)G(t) = FG \qquad \text{(VI-28.5)}$$

or as a composite of such functions.

Substitution for H from Eq. (VI-28.5) into Eq. (VI-28.1) yields

$$rF''G + F'G = \lambda^2 rFG', \qquad \text{(VI-28.6)}$$

where the primes on F and G denote first and second derivative with respect to r or to t, as the case may be.

Equation (VI-28.6) may be written

$$\frac{F''}{F} + \frac{1}{r}\frac{F'}{F} = \lambda^2 \frac{G'}{G}. \qquad \text{(VI-28.7)}$$

One side of Eq. (VI-28.7) is a function of r alone, the other a function of t alone. Since r and t are independent variables, Eq. (VI-28.7) can hold only when both sides thereof are identically equal to the same constant C. Our task now is to find out what this separation constant will be. If we take $C = 0$, then we have

$$\frac{\lambda^2 G'}{G} = 0, \qquad \text{(VI-28.7a)}$$

which requires that $G(t)$ be a constant. This can readily be seen by integrating Eq. (VI-28.7a). This would make H a constant as far as time is concerned. But H certainly varies with time, since H is produced by an exciting current which varies with time. Thus, we cannot take $C = 0$, since it leads to a contradiction.

If we take C positive, say b^2 with b real and different from zero, then the requirement

$$\frac{\lambda^2 G'}{G} = b^2$$

calls for G to be an exponential function of t of the form

$$G = Ae^{(b^2/\lambda^2)t},$$

which would increase beyond all bounds as $t \to \infty$ and would, therefore, cause H to do likewise. This, too, is out of the question. So, we cannot take C positive.

Choice of a negative number for C would require G to be a decreasing exponential function of time. This would make $H \to 0$ as $t \to \infty$, contrary to the nature of H as produced by periodic exciting current of fixed amplitude. Thus, C cannot be taken negative.

It looks as if we are stumped in our search for a product function H of type (5), which will satisfy Eq. (VI-28.1). In other problems in this chapter we were always able to effect solution by way of pertinent selection of the separation constant C: positive, negative or zero. But, there is yet another possibility: take C to be a *complex* constant $a + bj$, where a and b are real and $j = \sqrt{-1}$. A moment's reflection, however, indicates that it will suffice to take C as a pure imaginary number, since the real part of a complex constant would involve only the untenable solutions of Eq. (VI-28.1) obtained by taking C real. We, therefore, try $C = jk$, where k is real and $j = \sqrt{-1}$. This means that we shall have

$$\frac{\lambda^2 G'}{G} = jk,$$

whence

$$G = Ae^{jkt/\lambda^2} = A\left[\cos\frac{kt}{\lambda^2} + j\sin\frac{kt}{\lambda^2}\right], \qquad \text{(VI-28.8)}$$

where A is an arbitrary constant. Now the given boundary condition Eq. (VI-28.2) requires the period of H, and thereby the period of G, to be the period of the exciting current, namely $2\pi f$. Recalling that we put $\lambda^2 = 4\pi/\rho$, we find, therefore, that

$$k = \frac{8\pi^2 f}{\rho}. \qquad (\text{VI-28.9})$$

Thus, by Eqs. (VI-28.8) and (VI-28.9) we have

$$G = Ae^{j2\pi ft} = A\Big[\cos 2\pi ft + j\sin 2\pi ft\Big]. \qquad (\text{VI-28.10})$$

So much for G. And now, when $C = jk$ with k defined by Eq. (VI-28.9), the left side of Eq. (VI-28.7) requires that

$$\frac{F''}{F} + \frac{1}{r}\frac{F'}{F} = jk,$$

or

$$r^2 F'' + rF' - jkr^2 F = 0. \qquad (\text{VI-28.11})$$

We can solve Eq. (VI-28.11) by application of Eq. (VI-2.1), identifying r with x and F with y. We have

$$1 - 2A = 1, \quad 2E = 2, \quad D^2 E^2 = -jk, \quad A^2 - E^2 p^2 = 0,$$

whence

$$A = 0, \quad E = 1, \quad D = \sqrt{-jk}, \quad p = 0.$$

We express the constant D in a slightly different form as follows:

$$D = (-1)^{1/2} j^{1/2} k^{1/2} = jj^{1/2}k^{1/2} = j^{3/2}\sqrt{k}.$$

Then, by Eqs. (VI-2.2) and (V-0.8) and (V-0.15) and (V-0.17), the general solution of Eq. (VI-28.11) is

$$F = J_0(j^{3/2}\sqrt{k}\,r) = \text{ber}(\sqrt{k}\,r) + j\,\text{bei}(\sqrt{k}\,r) = M_0(\sqrt{k}\,r)e^{j\theta_0(\sqrt{k}\,r)}. \qquad (\text{VI-28.12})$$

Now, F as defined by Eq. (VI-28.12) is a complex function, which makes $H = FG$ a complex function. But, if a complex function H

satisfies Eq. (VI-28.1), then it is seen by separation of real and imaginary parts, that the real part alone of H will satisfy Eq. (VI-28.1) as will also the imaginary part (the coefficient of j). Accordingly, we may take for the magnetic field strength either of these two parts of the complex H. Let us, however, for convenience continue with the complex H defined by the product FG, where F is given by Eq. (VI-28.12) and G by Eq. (VI-28.10). We do this with the understanding that, when we come to determine the eddy current density i, we shall apply Eq. (VI-28.3) to the real part of H. With this understanding we may also determine the constant A in Eq. (VI-28.10) by writing both G and the boundary condition Eq. (VI-28.2) in the complex form. Thus, Eq. (VI-28.2) requires

$$H(R, t) = 4\pi NI e^{j2\pi ft},$$

that is,

$$A e^{j2\pi ft} J_0(j^{3/2}\sqrt{k}\,R) = 4\pi NI e^{j2\pi ft},$$

whence

$$A = \frac{4\pi NI}{J_0(j^{3/2}\sqrt{k}\,R)}.$$

We now have our complex solution of Eq. (VI-28.1) completely determined:

$$H = H_1 + jH_2 = \frac{4\pi NI}{J_0(j^{3/2}\sqrt{k}\,R)} J_0(j^{3/2}\sqrt{k}\,r)e^{j2\pi ft}$$

$$\text{(VI-28.13)}$$

with the understanding that the actual magnetic field strength is the real part of this H. (We remark that these functions H_1 and H_2 are not to be confused with the Hankel functions nor with the Struve functions.) One can also express this complex H via the $M_0(z)$ function as mentioned in the introduction to Chapter V. We have

$$H = \frac{4\pi NI}{M_0(\sqrt{k}\,R)e^{j\theta_0(\sqrt{k}\,R)}} M_0(\sqrt{k}\,r)e^{j\theta_0(\sqrt{k}\,r) + j2\pi ft}$$

where $M_0(\sqrt{k}\,r)$ and $\theta_0(\sqrt{k}\,r)$ are respectively the modulus and the amplitude of $J_0(j^{3/2}\sqrt{k}\,r)$, that is, the modulus and the amplitude of

$\text{ber}(\sqrt{k}\, r) + j\,\text{bei}(\sqrt{k}\, r)$. Upon separating real and imaginary parts, we find from Eq. (VI-28.13) that the actual real magnetic field strength H_1 which satisfies Eqs. (VI-28.1) and (VI-28.2) is

$$H_1 = \frac{4\pi NI}{M_0(\sqrt{k}R)} M_0(\sqrt{k}\, r)\cos\left[2\pi ft + \theta_0(\sqrt{k}\, r) - \theta_0(\sqrt{k}\, R) \right].$$

$$(\text{VI-28.14})$$

We are now finally in a position to determine the eddy current density i as given by Eq. (VI-28.3). We could do this directly by partial differentiation of Eq. (VI-28.14). However, it seems it will be more convenient to differentiate the complex H defined by Eq. (VI-28.13) and then take the real part of the result. Accordingly, by Eqs. (VI-28.3) and (VI-28.13), we get (for the complex i)

$$i = -\frac{NI}{J_0(j^{3/2}\sqrt{k}\, R)}\left[\frac{d}{dr} J_0(j^{3/2}\sqrt{k}\, r) \right] e^{j2\pi ft}.$$

$$(\text{VI-28.15})$$

The derivative indicated within the brackets is obtained immediately via formula (C) in Table V-1 by taking $x = r$, $p = 0$, $q_p = J_p$, $a = j^{3/2}\sqrt{k}$:

$$\frac{d}{dr} J_0(j^{3/2}\sqrt{k}\, r) = -j^{3/2}\sqrt{k}\, J_1(j^{3/2}\sqrt{k}\, r).$$

Thus the complex i is

$$i = i_1 + ji_2 = \frac{NI j^{3/2}\sqrt{k}}{J_0(j^{3/2}\sqrt{k}\, R)} J_1(j^{3/2}\sqrt{k}\, r)e^{j2\pi ft},$$

$$(\text{VI-28.16})$$

or

$$i = i_1 + ji_2 = \frac{NI e^{j(3\pi/4)}\sqrt{k}}{M_0(\sqrt{k}\, R)e^{j\theta_0(\sqrt{k}\, R)}} M_1(\sqrt{k}\, r)e^{j\theta_1(\sqrt{k}\, r) + j2\pi ft}.$$

$$(\text{VI-28.17})$$

Separation of real and imaginary parts yields the actual real eddy current density i_1, namely

$$i_1 = \frac{NI\sqrt{k}}{M_0(\sqrt{k}\, R)} M_1(\sqrt{k}\, r)\cos\left[2\pi ft + \theta_1(\sqrt{k}\, r) - \theta_0(\sqrt{k}\, R) + \frac{3\pi}{4} \right].$$

$$(\text{VI-28.18})$$

VI-29. Using formula (VI-28.18) in Prob. VI-28 for the eddy current density i in the copper core of a long (compared to radius) solenoid excited by alternating current of constant amplitude, compute the power loss P (ergs per second per centimeter of length) due to the eddy currents produced in the core. It is given that P equals the integral, taken over the circular area A of a cross-section of the core, of the product of the resistivity ρ (abohms per cubic centimeter) of copper and the mean square of i.   ★

It will be noticed that we have dropped the subscript 1 on the i in the formula (VI-28.18) of Prob. VI-28. There is no need to retain that subscript in the present problem, since there will be no call to use complex quantities.

Although the formula for the mean square of i can be found in appropriate texts, let us compute it for ourselves as an exercise. It is found by the well-known formula for the mean of the values taken by an integrable function $f(t)$ on an interval $t_1 \leq t \leq t_2$, namely

$$\text{mean } f = \frac{\int_{t_1}^{t_2} f(t)dt}{t_2 - t_1}. \qquad \text{(VI-29.1)}$$

Now, in the present problem, the value of i^2 at any chosen fixed point of the core is of the form

$$i^2 = B^2 \cos^2(2\pi f t + \lambda), \qquad \text{(VI-29.2)}$$

where

$$B = \frac{NI\sqrt{k}\, M_1(\sqrt{k}\, r)}{M_0(\sqrt{k}\, R)} \qquad \text{(VI-29.3)}$$

and does not change with t and where λ also is a constant as far as t is concerned. Now, since i^2 assumes on the interval defined by

$$t_1 \leq t \leq t_2, \quad 2\pi f t_1 + \lambda = 0, \quad 2\pi f t_2 + \lambda = \pi/2 \qquad \text{(VI-29.4)}$$

all the values it can possibly have, the mean of i^2 is given by Eq. (VI-29.1) where t_1 and t_2 are defined by (VI-29.4):

$$\mu = \text{mean } i^2 = \frac{\int_{t_1}^{t_2} B^2 \cos^2(2\pi ft + \lambda)dt}{t_2 - t_1}$$

$$= \frac{B^2}{t_2 - t_1} \int_{t_1}^{t_2} \left\{ \tfrac{1}{2} + \tfrac{1}{2}\cos(2[2\pi ft + \lambda]) \right\} dt$$

$$= \frac{B^2}{t_2 - t_1} \left[\tfrac{1}{2}(t_2 - t_1) + \tfrac{1}{2}\int_{t_1}^{t_2} \cos(2[2\pi ft + \lambda])dt \right].$$
(VI-29.5)

It is readily seen that, by virtue of (VI-29.4), the integral remaining to be evaluated in Eq. (VI-29.5) will have the value

$$(\text{constant})(\sin \pi - \sin 0) = 0.$$

Thus, Eq. (VI-29.5) yields

$$\mu = \text{mean } i^2 = \tfrac{1}{2}B^2.$$
(VI-29.6)

Then, by Eqs. (VI-29.6) and (VI-29.3) and the formula for P given in the statement of the problem, we have

$$P = \iint_A \rho \tfrac{1}{2} B^2 \, dA$$

$$= \frac{\rho}{2} \int_0^{2\pi} d\theta \int_0^R B^2 r \, dr$$

$$= \pi\rho \int_0^R B^2 r \, dr$$

$$= \frac{\pi\rho N^2 I^2 k}{[M_0(\sqrt{k}\,R)]^2} \int_0^R r \left[M_1(\sqrt{k}\,r) \right]^2 dr.$$
(VI-29.7)

The integral in Eq. (VI-29.7) suggests that we make a slight change of variable of integration, namely $\sqrt{k}\,r = x$. Then

$$\int_0^R r \left[M_1(\sqrt{k}\,r) \right]^2 dr = \frac{1}{k}\int_0^{\sqrt{k}R} x \left[M_1(x) \right]^2 dx.$$
(VI-29.8)

Here we can make use of Eq. (V-12.3) in Prob. V-12, whereby

$$M_1(x) = |\text{ber}'x + i\,\text{bei}'x|$$
$$= [(\text{ber}'x)^2 + (\text{bei}'x)^2]^{1/2}. \qquad \text{(VI-29.9)}$$

It follows at once from Eq. (VI-29.9) that

$$[M_1(x)]^2 = (\text{ber}'x)^2 + (\text{bei}'x)^2. \qquad \text{(VI-29.10)}$$

Thus, by Eqs. (VI-29.8) and (VI-29.10), we find that Eq. (VI-29.7) becomes

$$P = \frac{\pi\rho N^2 I^2}{[M_0(\sqrt{k}\,R)]^2} \int_0^{\sqrt{k}\,R} x\left[(\text{ber}'x)^2 + (\text{bei}'x)^2\right]dx. \qquad \text{(VI-29.11)}$$

Evaluation of the integral in Eq. (VI-29.11) now requires merely direct application of Eq. (V-11.3) with $a = \sqrt{k}\,R$, so that

$$P = \frac{\pi\rho N^2 I^2 \sqrt{k}\,R}{[M_0(\sqrt{k}\,R)]^2}\left\{\text{ber}\sqrt{k}R\,\text{ber}'\sqrt{k}R + \text{bei}\sqrt{k}R\,\text{bei}'\sqrt{k}R\right\}.$$

Problem: Heat-Flow Temperature Distribution

VI-30. Determine the steady-state temperature distribution T in a homogeneous material cylindrical shell whose height is h and whose lateral surfaces are coaxial circular cylinders having radii a and b with $b > a$, if temperatures are maintained on the four bounding faces as follows. If cylindrical coordinates (r, θ, z) are taken with the z-axis in the common axis of the two lateral faces, with the origin in the plane of one base and with the other base in the plane $z = h$, then the prescribed boundary temperatures to be maintained are

$$T = \begin{cases} 100 & \text{on lower base and both lateral faces,} \\ f(r) & \text{on upper base} \end{cases}$$

where $f(r)$ is continuous and differentiable and of bounded variation for $a \leq r \leq b$, where $f(a) = f(b) = 100$ and $f(r) > 100$ for $a < r < b$. ★

We shall find it convenient in this problem, as in other problems of similar character (see, for example Probs. VI-24 and VI-25), to use an adjusted steady-state temperature function

$$U = T - 100. \tag{VI-30.1}$$

At every interior point of the region occupied by the shell the function U must satisfy Laplace's equation $\nabla^2 U = 0$, which in cylindrical coordinates is

$$\frac{\partial^2 U}{\partial r^2} + \frac{\partial^2 U}{\partial z^2} + \frac{1}{r^2} \frac{\partial^2 U}{\partial \theta^2} + \frac{1}{r} \frac{\partial U}{\partial r} = 0. \tag{VI-30.2}$$

But the symmetry with respect to the z-axis of the prescribed boundary temperatures means that U will be independent of θ, so that Laplace's equation for this problem reduces to

$$\frac{\partial^2 U}{\partial r^2} + \frac{\partial^2 U}{\partial z^2} + \frac{1}{r} \frac{\partial U}{\partial r} = 0. \tag{VI-30.3}$$

We, therefore, seek a function U which satisfies Eq. (VI-30.3) at each interior point R and which takes on boundary values as follows:

$$U = \begin{cases} 0 & \text{on lower base and both lateral surfaces,} \\ f(r) - 100 & \text{on upper base.} \end{cases} \tag{VI-30.4}$$

It is well-known that, in problems requiring satisfaction of Laplace's equation (or a similar equation) in a region and also requiring the solution to take on prescribed boundary values, solution may be effected via a function which is a composite (an infinite series) of particular solutions of the differential equation, each particular solution being a product function of the form

$$G(u)H(v)M(w),$$

where u, v and w denote the space coordinates of the coordinate system that is to be used in the problem. Let us, then, see if we can solve the present problem via a composite of particular solutions of Eq. (VI-30.3), each particular solution being of the form

$$U = G(r)H(z) = GH.$$ (VI-30.5)

Substitution for U from Eq. (VI-30.5) into Eq. (VI-30.3) requires that

$$G''H + GH'' + \frac{1}{r} G'H = 0,$$ (VI-30.6)

where the primes on G and H denote first and second derivative with respect to argument r or z, as the case may be. Eq. (VI-30.6) may be written as

$$\frac{H''}{H} = - \frac{G''}{G} - \frac{1}{r} \frac{G'}{G}.$$ (VI-30.7)

The left side of Eq. (VI-30.7) is independent of r while the right side of Eq. (VI-30.7) is independent of z. Now, the only way in which a function of z alone can be equal identically (in a region) to a function of r alone is for each of the two functions to be identically equal to the same constant. Thus, the requirement that a product function $U = G(r)H(z)$ shall satisfy Eq. (VI-30.3) is equivalent to requiring simultaneous satisfaction of two equations

$$\frac{H''}{H} = C, \qquad - \frac{G''}{G} - \frac{1}{r} \frac{G'}{G} = C,$$ (VI-30.8)

where C is the same constant in both equations.

Let us now determine which kind of constant can be used for C, positive or negative or zero. If we take $C = 0$, then the second equation in (VI-30.8) becomes

$$rG'' + G' = 0,$$

whose solution is

$$G = A \log_e r + B,$$

where A and B are constants. And the first equation in (VI-30.8) becomes $H'' = 0$, whose solution is

$$H = Dz + E,$$

where D and E are arbitrary constants. We now have

$$U = (A \log_e r + B)(Dz + E). \tag{VI-30.9}$$

On the lower base, where $z = 0$, we must have

$$U = 0 = (A \log_e r + B)(E),$$

which requires that $E = 0$, since the function $A \log_e r + B$ cannot vanish identically for $a \leqq r \leqq b$, unless $A = 0$ and $B = 0$. Thus,

$$U = (A' \log_e r + B')z,$$

where $A' = AD$ and $B' = BD$.

We must also have $U = 0$ on both lateral surfaces. Since z does not vanish identically on the lateral faces, the requirement that U shall vanish thereon means that

$$\left.\begin{array}{l} A' \log_e b + B' = 0 \\ A' \log_e a + B' = 0 \end{array}\right\}.$$

Since $a \neq b$, the solution of this pair of simultaneous equations is $A' = 0$ and $B' = 0$, which makes $U(r, z)$ vanish identically. But U certainly does not vanish identically because the prescribed value of U on the upper base is greater than zero for $a < r < b$. We must, then, reject the assumption that led to this contradiction, namely the assumption that $C = 0$ in Eq. (VI-30.8).

If we take C negative in Eq. (VI-30.8), then the first equation in (VI-30.8) becomes

$$H'' = -k^2 H, \qquad C = -k^2 \neq 0,$$

whose solution is

$$H = A \cos kz + B \sin kz,$$

where A and B are arbitrary constants. This requires on the lower base, where $z = 0$, that

$$0 = G(r)[A], \qquad a \leqq r \leqq b.$$

Now $G(r)$ is certainly not identically zero for $a \leqq r \leqq b$. So, we must take $A = 0$. Then we have

$$U = G(r)B \sin kz. \qquad \text{(VI-30.10)}$$

Taking $C = -k^2$, $k \neq 0$, we find by the second equation in (VI-30.8) that G shall satisfy the equation

$$r^2G'' + rG' - k^2r^2G = 0,$$

whose general solution, by Eqs. (VI-1.6), (VI-1.5), (V-0.12), is

$$G = C_1I_0(kr) + C_2K_0(kr),$$

so that Eq. (VI-30.10) becomes

$$U = \left[C_1I_0(kr) + C_2K_0(kr) \right] \sin kz, \qquad \text{(VI-30.11)}$$

where the constant B in Eq. (VI-30.10) has been absorbed into the arbitrary constants C_1 and C_2.

Comparison of Eq. (VI-30.11) with the corresponding stage of development in Prob. VI-26 indicates that a solution of the present problem could be obtained in the form of a series of particular solutions of type (VI-30.11), namely

$$U = \sum_{n=1}^{\infty} C_n \left[I_0(\lambda_n r) - \frac{I_0(\lambda_n b)}{K_0(\lambda_n b)} K_0(\lambda_n r) \right] \frac{\sin(\lambda_n z)}{\sin(\lambda_n h)},$$

$$\text{(VI-30.12)}$$

where $\lambda = \lambda_n$ would be a solution of

$$I_0(\lambda a) - \frac{I_0(\lambda b)}{K_0(\lambda b)} K_0(\lambda a) = 0$$

and where the coefficients C_n would be given by a formula corresponding to the formula for the C_n in Prob. VI-26. The constant factor $1/\sin(\lambda_n h)$ is needed in Eq. (VI-30.12) because the coefficients C_n shall be determined so that on upper base, where $z = h$, we can have

$$f(r) - 100 = \sum_{n=1}^{\infty} C_n \left[I_0(\lambda_n r) - \frac{I_0(\lambda_n b)}{K_0(\lambda_n b)} K_0(\lambda_n r) \right].$$

(VI-30.13)

Thus, once the expansion (VI-30.13) has been obtained, the function U defined by Eq. (VI-30.12) meets all the requirements imposed on U by Eqs. (VI-30.3) and (VI-30.4), in particular on the upper base. The actual solution of the problem as stated is $T = U + 100$, where U is given by Eq. (VI-30.12).

Let us see if a solution can also be obtained when we take the separation constant C positive in Eq. (VI-30.8), say $C = k^2$, $k \neq 0$. Then $H'' = k^2 H$, whose general solution is

$$H = Ae^{kz} + Be^{-kz},$$

where A and B are arbitrary constants. Then, on the lower base where $U = 0$ and $z = 0$, we must have

$$0 = G(r)(A + B),$$

so that $B = -A$. We thus have

$$U = G(r)[A(e^{kz} - e^{-kz})].$$

Since A is arbitrary, we can let $M = A/2$ and write

$$U = MG(r) \sinh (kz). \qquad (VI\text{-}30.14)$$

When we take $C = k^2 \neq 0$ in (VI-30.8), we must have

$$r^2 G'' + rG' + k^2 r^2 G = 0,$$

whose general solution by Eqs. (VI-1.6), (VI-1.5), (V-0.8) is

$$G = C_1 J_0(kr) + C_2 Y_0(kr). \qquad (VI\text{-}30.15)$$

Consequently, by virtue of Eqs. (VI-30.14) and (VI-30.15), we can have particular solutions of the type

$$U = [C_1 J_0(kr) + C_2 Y_0(kr)] \sinh(kz). \qquad \text{(VI-30.16)}$$

And we can meet all the boundary conditions imposed on T as well as have T satisfy Laplace's equation at all interior points by a series solution similar to the solution obtained in Prob. VI-26, namely

$$T = 100 + U = 100 + \sum_{n=1}^{\infty} C_n \left[J_0(\lambda_n r) - \frac{J_0(\lambda_n b)}{Y_0(\lambda_n b)} Y_0(\lambda_n r) \right] \frac{\sinh(\lambda_n z)}{\sinh(\lambda_n h)},$$

where $\lambda = \lambda_n$ is a solution of

$$J_0(\lambda a) - \frac{J_0(\lambda b)}{Y_0(\lambda b)} Y_0(\lambda a) = 0$$

and where the coefficients C_n are determined (as in Prob. VI-26) so that

$$f(r) - 100 = \sum_{n=1}^{\infty} C_n \left[J_0(\lambda_n r) - \frac{J_0(\lambda_n b)}{Y_0(\lambda_n b)} Y_0(\lambda_n r) \right], \qquad a \leqq r \leqq b.$$

BIBLIOGRAPHY

Byerly, W. E. *Fourier's Series and Spherical Harmonics*, Ginn, 1893.

Eckert, E. R. G., and Drake, R. M., Jr. *Heat and Mass Transfer*, McGraw-Hill, 1959.

Franklin, P. *Treatise on Advanced Calculus*, Wiley, 1940.

——. *Methods of Advanced Calculus*, McGraw-Hill, 1944.

Gray, A., Mathews, G. B., MacRobert, T. M. *Bessel Functions*, Macmillan, 1931.

Hildebrand, F. B. *Advanced Calculus for Engineers*, Prentice-Hall, 1958.

Hobson, E. W. *Theory of Functions of a Real Variable*, Cambridge, 1926.

——. *The Theory of Spherical and Ellipsoidal Harmonics*, Cambridge, 1931.

Kaplan, W. *Advanced Calculus*, Addison-Wesley, 1952.

Lowan, A. N., Davids, N., and Levenson, A. *Table of the Zeros of the Legendre Polynomials of Order 1–16 and the Weight Coefficients for Gauss's Mechanical Quadrature*, Bulletin of American Mathematical Society, Volume 48, 1942.

MacRobert, T. M. *Spherical Harmonics*, Dover, 1948.

Margenau, H., and Murphy, G. M. *The Mathematics of Physics and Chemistry*, Van Nostrand, 1956.

McLachlan, N. W. *Bessel Functions for Engineers*, Oxford, 1955.

Murray, R. L. *Nuclear Reactor Physics*, Prentice-Hall, 1957.

National Bureau of Standards, *Tables of Functions and Zeros of Functions*, Applied Mathematics Series 37, 1954.

Pipes, L. A. *Applied Mathematics for Engineers and Physicists*, McGraw-Hill, 1958.

Rainville, E. D. *Special Functions*, Macmillan, 1960.

——. *Elementary Differential Equations*, Macmillan, 1958.

Reddick, H. W., and Miller, F. H. *Advanced Mathematics for Engineers*, Oxford, 1955.

Relton, F. E. *Applied Bessel Functions*, Blackie and Son, Ltd., 1946.

Seely, F. B., and Smith, J. O. *Resistance of Materials*, Wiley, 1956.

——, and ——. *Advanced Mechanics of Materials*, Wiley, 1952.

Sneddon, I. N. *Special Functions of Mathematical Physics and Chemistry*, Interscience, 1956.

Sokolnikoff, I. S., and Redheffer, R. M. *Mathematics of Physics and Modern Engineering*, McGraw-Hill, 1958.

Taylor, A. E. *Advanced Calculus*, Ginn, 1955

Watson, G. N. *Theory of Bessel Functions*, Cambridge, 1944.

Whittaker, E. T., and Watson, G. N. *Modern Analysis*, Cambridge, 1927.

Wiley, C. R. *Advanced Engineering Mathematics*, McGraw-Hill, 1960.

INDEX

SOME DOVER SCIENCE BOOKS

SOME DOVER SCIENCE BOOKS

WHAT IS SCIENCE?,
Norman Campbell
This excellent introduction explains scientific method, role of mathematics, types of scientific laws. Contents: 2 aspects of science, science & nature, laws of science, discovery of laws, explanation of laws, measurement & numerical laws, applications of science. 192pp. 5⅜ x 8.   60043-2 Paperbound $1.25

FADS AND FALLACIES IN THE NAME OF SCIENCE,
Martin Gardner
Examines various cults, quack systems, frauds, delusions which at various times have masqueraded as science. Accounts of hollow-earth fanatics like Symmes; Velikovsky and wandering planets; Hoerbiger; Bellamy and the theory of multiple moons; Charles Fort; dowsing, pseudoscientific methods for finding water, ores, oil. Sections on naturopathy, iridiagnosis, zone therapy, food fads, etc. Analytical accounts of Wilhelm Reich and orgone sex energy; L. Ron Hubbard and Dianetics; A. Korzybski and General Semantics; many others. Brought up to date to include Bridey Murphy, others. Not just a collection of anecdotes, but a fair, reasoned appraisal of eccentric theory. Formerly titled *In the Name of Science*. Preface. Index. x + 384pp. 5⅜ x 8.
                                                        20394-8 Paperbound $2.00

PHYSICS, THE PIONEER SCIENCE,
L. W. Taylor
First thorough text to place all important physical phenomena in cultural-historical framework; remains best work of its kind. Exposition of physical laws, theories developed chronologically, with great historical, illustrative experiments diagrammed, described, worked out mathematically. Excellent physics text for self-study as well as class work. Vol. 1: Heat, Sound: motion, acceleration, gravitation, conservation of energy, heat engines, rotation, heat, mechanical energy, etc. 211 illus. 407pp. 5⅜ x 8. Vol. 2: Light, Electricity: images, lenses, prisms, magnetism, Ohm's law, dynamos, telegraph, quantum theory, decline of mechanical view of nature, etc. Bibliography. 13 table appendix. Index. 551 illus. 2 color plates. 508pp. 5⅜ x 8.
                        60565-5, 60566-3 Two volume set, paperbound $5.50

THE EVOLUTION OF SCIENTIFIC THOUGHT FROM NEWTON TO EINSTEIN,
A. d'Abro
Einstein's special and general theories of relativity, with their historical implications, are analyzed in non-technical terms. Excellent accounts of the contributions of Newton, Riemann, Weyl, Planck, Eddington, Maxwell, Lorentz and others are treated in terms of space and time, equations of electromagnetics, finiteness of the universe, methodology of science. 21 diagrams. 482pp. 5⅜ x 8.
                                                        20002-7 Paperbound $2.50

CHANCE, LUCK AND STATISTICS: THE SCIENCE OF CHANCE,
Horace C. Levinson
Theory of probability and science of statistics in simple, non-technical language.
Part I deals with theory of probability, covering odd superstitions in regard to
"luck," the meaning of betting odds, the law of mathematical expectation,
gambling, and applications in poker, roulette, lotteries, dice, bridge, and other
games of chance. Part II discusses the misuse of statistics, the concept of statistical probabilities, normal and skew frequency distributions, and statistics applied to various fields—birth rates, stock speculation, insurance rates, advertis-
ing, etc. "Presented in an easy humorous style which I consider the best kind of
expository writing," Prof. A. C. Cohen, Industry Quality Control. Enlarged
revised edition. Formerly titled *The Science of Chance.* Preface and two new
appendices by the author. xiv + 365pp. 5⅜ x 8.   21007-3 Paperbound $2.00

BASIC ELECTRONICS,
prepared by the U.S. Navy Training Publications Center
A thorough and comprehensive manual on the fundamentals of electronics.
Written clearly, it is equally useful for self-study or course work for those with
a knowledge of the principles of basic electricity. Partial contents: Operating
Principles of the Electron Tube; Introduction to Transistors; Power Supplies
for Electronic Equipment; Tuned Circuits; Electron-Tube Amplifiers; Audio
Power Amplifiers; Oscillators; Transmitters; Transmission Lines; Antennas and
Propagation; Introduction to Computers; and related topics. Appendix. Index.
Hundreds of illustrations and diagrams. vi + 471pp. 6½ x 9¼.
61076-4 Paperbound $2.95

BASIC THEORY AND APPLICATION OF TRANSISTORS,
prepared by the U.S. Department of the Army
An introductory manual prepared for an army training program. One of the
finest available surveys of theory and application of transistor design and
operation. Minimal knowledge of physics and theory of electron tubes required.
Suitable for textbook use, course supplement, or home study. Chapters: Introduction; fundamental theory of transistors; transistor amplifier fundamentals;
parameters, equivalent circuits, and characteristic curves; bias stabilization;
transistor analysis and comparison using characteristic curves and charts; audio
amplifiers; tuned amplifiers; wide-band amplifiers; oscillators; pulse and switch-
ing circuits; modulation, mixing, and demodulation; and additional semi-
conductor devices. Unabridged, corrected edition. 240 schematic drawings,
photographs, wiring diagrams, etc. 2 Appendices. Glossary. Index. 263pp.
6½ x 9¼.   60380-6 Paperbound $1.75

GUIDE TO THE LITERATURE OF MATHEMATICS AND PHYSICS,
N. G. Parke III
Over 5000 entries included under approximately 120 major subject headings of
selected most important books, monographs, periodicals, articles in English,
plus important works in German, French, Italian, Spanish, Russian (many
recently available works). Covers every branch of physics, math, related engineering. Includes author, title, edition, publisher, place, date, number of
volumes, number of pages. A 40-page introduction on the basic problems of
research and study provides useful information on the organization and use of
libraries, the psychology of learning, etc. This reference work will save you
hours of time. 2nd revised edition. Indices of authors, subjects, 464pp. 5⅜ x 8.
60447-0 Paperbound $2.75

THE RISE OF THE NEW PHYSICS (formerly THE DECLINE OF MECHANISM), *A. d'Abro*
This authoritative and comprehensive 2-volume exposition is unique in scientific publishing. Written for intelligent readers not familiar with higher mathematics, it is the only thorough explanation in non-technical language of modern mathematical-physical theory. Combining both history and exposition, it ranges from classical Newtonian concepts up through the electronic theories of Dirac and Heisenberg, the statistical mechanics of Fermi, and Einstein's relativity theories. "A must for anyone doing serious study in the physical sciences," *J. of Franklin Inst.* 97 illustrations. 991pp. 2 volumes.
20003-5, 20004-3 Two volume set, paperbound $5.50

THE STRANGE STORY OF THE QUANTUM, AN ACCOUNT FOR THE GENERAL READER OF THE GROWTH OF IDEAS UNDERLYING OUR PRESENT ATOMIC KNOWLEDGE, *B. Hoffmann*
Presents lucidly and expertly, with barest amount of mathematics, the problems and theories which led to modern quantum physics. Dr. Hoffmann begins with the closing years of the 19th century, when certain trifling discrepancies were noticed, and with illuminating analogies and examples takes you through the brilliant concepts of Planck, Einstein, Pauli, de Broglie, Bohr, Schroedinger, Heisenberg, Dirac, Sommerfeld, Feynman, etc. This edition includes a new, long postscript carrying the story through 1958. "Of the books attempting an account of the history and contents of our modern atomic physics which have come to my attention, this is the best," H. Margenau, Yale University, in *American Journal of Physics.* 32 tables and line illustrations. Index. 275pp. 5⅜ x 8.
20518-5 Paperbound $2.00

GREAT IDEAS AND THEORIES OF MODERN COSMOLOGY, *Jagjit Singh*
The theories of Jeans, Eddington, Milne, Kant, Bondi, Gold, Newton, Einstein, Gamow, Hoyle, Dirac, Kuiper, Hubble, Weizsäcker and many others on such cosmological questions as the origin of the universe, space and time, planet formation, "continuous creation," the birth, life, and death of the stars, the origin of the galaxies, etc. By the author of the popular *Great Ideas of Modern Mathematics.* A gifted popularizer of science, he makes the most difficult abstractions crystal-clear even to the most non-mathematical reader. Index. xii + 276pp. 5⅜ x 8½. 20925-3 Paperbound $2.50

GREAT IDEAS OF MODERN MATHEMATICS: THEIR NATURE AND USE, *Jagjit Singh*
Reader with only high school math will understand main mathematical ideas of modern physics, astronomy, genetics, psychology, evolution, etc., better than many who use them as tools, but comprehend little of their basic structure. Author uses his wide knowledge of non-mathematical fields in brilliant exposition of differential equations, matrices, group theory, logic, statistics, problems of mathematical foundations, imaginary numbers, vectors, etc. Original publications, appendices. indexes. 65 illustr. 322pp. 5⅜ x 8. 20587-8 Paperbound $2.25

THE MATHEMATICS OF GREAT AMATEURS, *Julian L. Coolidge*
Great discoveries made by poets, theologians, philosophers, artists and other non-mathematicians: Omar Khayyam, Leonardo da Vinci, Albrecht Dürer, John Napier, Pascal, Diderot, Bolzano, etc. Surprising accounts of what can result from a non-professional preoccupation with the oldest of sciences. 56 figures. viii + 211pp. 5⅜ x 8½. 61009-8 Paperbound $2.00

COLLEGE ALGEBRA, *H. B. Fine*

Standard college text that gives a systematic and deductive structure to algebra; comprehensive, connected, with emphasis on theory. Discusses the commutative, associative, and distributive laws of number in unusual detail, and goes on with undetermined coefficients, quadratic equations, progressions, logarithms, permutations, probability, power series, and much more. Still most valuable elementary-intermediate text on the science and structure of algebra. Index. 1560 problems, all with answers. x + 631pp. 5⅜ x 8. 60211-7 Paperbound $2.75

HIGHER MATHEMATICS FOR STUDENTS OF CHEMISTRY AND PHYSICS, *J. W. Mellor*

Not abstract, but practical, building its problems out of familiar laboratory material, this covers differential calculus, coordinate, analytical geometry, functions, integral calculus, infinite series, numerical equations, differential equations, Fourier's theorem, probability, theory of errors, calculus of variations, determinants. "If the reader is not familiar with this book, it will repay him to examine it," *Chem. & Engineering News*. 800 problems. 189 figures. Bibliography. xxi + 641pp. 5⅜ x 8. 60193-5 Paperbound $3.50

TRIGONOMETRY REFRESHER FOR TECHNICAL MEN, *A. A. Klaf*

A modern question and answer text on plane and spherical trigonometry. Part I covers plane trigonometry: angles, quadrants, trigonometrical functions, graphical representation, interpolation, equations, logarithms, solution of triangles, slide rules, etc. Part II discusses applications to navigation, surveying, elasticity, architecture, and engineering. Small angles, periodic functions, vectors, polar coordinates, De Moivre's theorem, fully covered. Part III is devoted to spherical trigonometry and the solution of spherical triangles, with applications to terrestrial and astronomical problems. Special time-savers for numerical calculation. 913 questions answered for you! 1738 problems; answers to odd numbers. 494 figures. 14 pages of functions, formulae. Index. x + 629pp. 5⅜ x 8. 20371-9 Paperbound $3.00

CALCULUS REFRESHER FOR TECHNICAL MEN, *A. A. Klaf*

Not an ordinary textbook but a unique refresher for engineers, technicians, and students. An examination of the most important aspects of differential and integral calculus by means of 756 key questions. Part I covers simple differential calculus: constants, variables, functions, increments, derivatives, logarithms, curvature, etc. Part II treats fundamental concepts of integration: inspection, substitution, transformation, reduction, areas and volumes, mean value, successive and partial integration, double and triple integration. Stresses practical aspects! A 50 page section gives applications to civil and nautical engineering, electricity, stress and strain, elasticity, industrial engineering, and similar fields. 756 questions answered. 556 problems; solutions to odd numbers. 36 pages of constants, formulae. Index. v + 431pp. 5⅜ x 8. 20370-0 Paperbound $2.25

INTRODUCTION TO THE THEORY OF GROUPS OF FINITE ORDER, *R. Carmichael*

Examines fundamental theorems and their application. Beginning with sets, systems, permutations, etc., it progresses in easy stages through important types of groups: Abelian, prime power, permutation, etc. Except 1 chapter where matrices are desirable, no higher math needed. 783 exercises, problems. Index. xvi + 447pp. 5⅜ x 8. 60300-8 Paperbound $3.00

FIVE VOLUME "THEORY OF FUNCTIONS" SET BY KONRAD KNOPP

This five-volume set, prepared by Konrad Knopp, provides a complete and readily followed account of theory of functions. Proofs are given concisely, yet without sacrifice of completeness or rigor. These volumes are used as texts by such universities as M.I.T., University of Chicago, N. Y. City College, and many others. "Excellent introduction . . . remarkably readable, concise, clear, rigorous," *Journal of the American Statistical Association.*

ELEMENTS OF THE THEORY OF FUNCTIONS,
Konrad Knopp
This book provides the student with background for further volumes in this set, or texts on a similar level. Partial contents: foundations, system of complex numbers and the Gaussian plane of numbers, Riemann sphere of numbers, mapping by linear functions, normal forms, the logarithm, the cyclometric functions and binomial series. "Not only for the young student, but also for the student who knows all about what is in it," *Mathematical Journal.* Bibliography. Index. 140pp. 5⅜ x 8. 60154-4 Paperbound $1.50

THEORY OF FUNCTIONS, PART I,
Konrad Knopp
With volume II, this book provides coverage of basic concepts and theorems. Partial contents: numbers and points, functions of a complex variable, integral of a continuous function, Cauchy's integral theorem, Cauchy's integral formulae, series with variable terms, expansion of analytic functions in power series, analytic continuation and complete definition of analytic functions, entire transcendental functions, Laurent expansion, types of singularities. Bibliography. Index. vii + 146pp. 5⅜ x 8. 60156-0 Paperbound $1.50

THEORY OF FUNCTIONS, PART II,
Konrad Knopp
Application and further development of general theory, special topics. Single valued functions. Entire, Weierstrass, Meromorphic functions. Riemann surfaces. Algebraic functions. Analytical configuration, Riemann surface. Bibliography. Index. x + 150pp. 5⅜ x 8. 60157-9 Paperbound $1.50

PROBLEM BOOK IN THE THEORY OF FUNCTIONS, VOLUME 1.
Konrad Knopp
Problems in elementary theory, for use with Knopp's *Theory of Functions,* or any other text, arranged according to increasing difficulty. Fundamental concepts, sequences of numbers and infinite series, complex variable, integral theorems, development in series, conformal mapping. 182 problems. Answers. viii + 126pp. 5⅜ x 8. 60158-7 Paperbound $1.50

PROBLEM BOOK IN THE THEORY OF FUNCTIONS, VOLUME 2,
Konrad Knopp
Advanced theory of functions, to be used either with Knopp's *Theory of Functions,* or any other comparable text. Singularities, entire & meromorphic functions, periodic, analytic, continuation, multiple-valued functions, Riemann surfaces, conformal mapping. Includes a section of additional elementary problems. "The difficult task of selecting from the immense material of the modern theory of functions the problems just within the reach of the beginner is here masterfully accomplished," *Am. Math. Soc.* Answers. 138pp. 5⅜ x 8. 60159-5 Paperbound $1.50

NUMERICAL SOLUTIONS OF DIFFERENTIAL EQUATIONS,
H. Levy & E. A. Baggott
Comprehensive collection of methods for solving ordinary differential equations
of first and higher order. All must pass 2 requirements: easy to grasp and
practical, more rapid than school methods. Partial contents: graphical integration of differential equations, graphical methods for detailed solution. Numerical solution. Simultaneous equations and equations of 2nd and higher orders.
"Should be in the hands of all in research in applied mathematics, teaching,"
Nature. 21 figures. viii + 238pp. 5⅜ x 8.   60168-4 Paperbound $1.85

ELEMENTARY STATISTICS, WITH APPLICATIONS IN MEDICINE AND THE
BIOLOGICAL SCIENCES, F. E. Croxton
A sound introduction to statistics for anyone in the physical sciences, assuming no prior acquaintance and requiring only a modest knowledge of math.
All basic formulas carefully explained and illustrated; all necessary reference
tables included. From basic terms and concepts, the study proceeds to frequency
distribution, linear, non-linear, and multiple correlation, skewness, kurtosis,
etc. A large section deals with reliability and significance of statistical methods.
Containing concrete examples from medicine and biology, this book will prove
unusually helpful to workers in those fields who increasingly must evaluate,
check, and interpret statistics. Formerly titled "Elementary Statistics with Applications in Medicine." 101 charts. 57 tables. 14 appendices. Index. vi +
376pp. 5⅜ x 8.   60506-X Paperbound $2.25

INTRODUCTION TO SYMBOLIC LOGIC,
S. Langer
No special knowledge of math required — probably the clearest book ever
written on symbolic logic, suitable for the layman, general scientist, and philos-
opher. You start with simple symbols and advance to a knowledge of the
Boole-Schroeder and Russell-Whitehead systems. Forms, logical structure, classes,
the calculus of propositions, logic of the syllogism, etc. are all covered. "One
of the clearest and simplest introductions," Mathematics Gazette. Second enlarged, revised edition. 368pp. 5⅜ x 8.   60164-1 Paperbound $2.25

A SHORT ACCOUNT OF THE HISTORY OF MATHEMATICS,
W. W. R. Ball
Most readable non-technical history of mathematics treats lives, discoveries of
every important figure from Egyptian, Phoenician, mathematicians to late 19th
century. Discusses schools of Ionia, Pythagoras, Athens, Cyzicus, Alexandria,
Byzantium, systems of numeration; primitive arithmetic; Middle Ages, Renais-
sance, including Arabs, Bacon, Regiomontanus, Tartaglia, Cardan, Stevinus,
Galileo, Kepler; modern mathematics of Descartes, Pascal, Wallis, Huygens,
Newton, Leibnitz, d'Alembert, Euler, Lambert, Laplace, Legendre, Gauss,
Hermite, Weierstrass, scores more. Index. 25 figures. 546pp. 5⅜ x 8.
                                        20630-0 Paperbound $2.75

INTRODUCTION TO NONLINEAR DIFFERENTIAL AND INTEGRAL EQUATIONS,
Harold T. Davis
Aspects of the problem of nonlinear equations, transformations that lead to
equations solvable by classical means, results in special cases, and useful
generalizations. Thorough, but easily followed by mathematically sophisticated
reader who knows little about non-linear equations. 137 problems for student
to solve. xv + 566pp. 5⅜ x 8½.   60971-5 Paperbound $2.75

An INTRODUCTION TO THE GEOMETRY OF N DIMENSIONS,
D. H. Y. Sommerville
An introduction presupposing no prior knowledge of the field, the only book
in English devoted exclusively to higher dimensional geometry. Discusses
fundamental ideas of incidence, parallelism, perpendicularity, angles between
linear space; enumerative geometry; analytical geometry from projective and
metric points of view; polytopes; elementary ideas in analysis situs; content of
hyper-spacial figures. Bibliography. Index. 60 diagrams. 196pp. 5⅜ x 8.
60494-2 Paperbound $1.50

ELEMENTARY CONCEPTS OF TOPOLOGY, *P. Alexandroff*
First English translation of the famous brief introduction to topology for the
beginner or for the mathematician not undertaking extensive study. This unusually useful intuitive approach deals primarily with the concepts of complex,
cycle, and homology, and is wholly consistent with current investigations.
Ranges from basic concepts of set-theoretic topology to the concept of Betti
groups. "Glowing example of harmony between intuition and thought," David
Hilbert. Translated by A. E. Farley. Introduction by D. Hilbert. Index. 25
figures. 73pp. 5⅜ x 8.   60747-X Paperbound $1.25

ELEMENTS OF NON-EUCLIDEAN GEOMETRY,
D. M. Y. Sommerville
Unique in proceeding step-by-step, in the manner of traditional geometry.
Enables the student with only a good knowledge of high school algebra and
geometry to grasp elementary hyperbolic, elliptic, analytic non-Euclidean geom-
etries; space curvature and its philosophical implications; theory of radical
axes; homothetic centres and systems of circles; parataxy and parallelism;
absolute measure; Gauss' proof of the defect area theorem; geodesic representation; much more, all with exceptional clarity. 126 problems at chapter endings
provide progressive practice and familiarity. 133 figures. Index. xvi + 274pp.
5⅜ x 8.   60460-8 Paperbound $2.00

INTRODUCTION TO THE THEORY OF NUMBERS, *L. E. Dickson*
Thorough, comprehensive approach with adequate coverage of classical literature, an introductory volume beginners can follow. Chapters on divisibility,
congruences, quadratic residues & reciprocity. Diophantine equations, etc. Full
treatment of binary quadratic forms without usual restriction to integral coefficients. Covers infinitude of primes, least residues. Fermat's theorem. Euler's
phi function, Legendre's symbol, Gauss's lemma, automorphs, reduced forms,
recent theorems of Thue & Siegel, many more. Much material not readily
available elsewhere. 239 problems. Index. I figure. viii + 183pp. 5⅜ x 8.
60342-3 Paperbound $1.75

MATHEMATICAL TABLES AND FORMULAS,
compiled by Robert D. Carmichael and Edwin R. Smith
Valuable collection for students, etc. Contains all tables necessary in college
algebra and trigonometry, such as five-place common logarithms, logarithmic
sines and tangents of small angles, logarithmic trigonometric functions, natural
trigonometric functions, four-place antilogarithms, tables for changing from
sexagesimal to circular and from circular to sexagesimal measure of angles, etc.
Also many tables and formulas not ordinarily accessible, including powers,
roots, and reciprocals, exponential and hyperbolic functions, ten-place logarithms of prime numbers, and formulas and theorems from analytical and
elementary geometry and from calculus. Explanatory introduction. viii +
269pp. 5⅜ x 8½.   60111-0 Paperbound $1.50

A SOURCE BOOK IN MATHEMATICS,
D. E. Smith
Great discoveries in math, from Renaissance to end of 19th century, in English translation. Read announcements by Dedekind, Gauss, Delamain, Pascal, Fermat, Newton, Abel, Lobachevsky, Bolyai, Riemann, De Moivre, Legendre, Laplace, others of discoveries about imaginary numbers, number congruence, slide rule, equations, symbolism, cubic algebraic equations, non-Euclidean forms of geometry, calculus, function theory, quaternions, etc. Succinct selections from 125 different treatises, articles, most unavailable elsewhere in English. Each article preceded by biographical introduction. Vol. I: Fields of Number, Algebra. Index. 32 illus. 338pp. 5⅜ x 8. Vol. II: Fields of Geometry, Probability, Calculus, Functions, Quaternions. 83 illus. 432pp. 5⅜ x 8.
60552-3, 60553-1 Two volume set, paperbound $5.00

FOUNDATIONS OF PHYSICS,
R. B. Lindsay & H. Margenau
Excellent bridge between semi-popular works & technical treatises. A discussion of methods of physical description, construction of theory; valuable for physicist with elementary calculus who is interested in ideas that give meaning to data, tools of modern physics. Contents include symbolism; mathematical equations; space & time foundations of mechanics; probability; physics & continua; electron theory; special & general relativity; quantum mechanics; causality. "Thorough and yet not overdetailed. Unreservedly recommended," *Nature* (London). Unabridged, corrected edition. List of recommended readings. 35 illustrations. xi + 537pp. 5⅜ x 8. 60377-6 Paperbound $3.50

FUNDAMENTAL FORMULAS OF PHYSICS,
ed. by D. H. Menzel
High useful, full, inexpensive reference and study text, ranging from simple to highly sophisticated operations. Mathematics integrated into text—each chapter stands as short textbook of field represented. Vol. 1: Statistics, Physical Constants, Special Theory of Relativity, Hydrodynamics, Aerodynamics, Boundary Value Problems in Math, Physics, Viscosity, Electromagnetic Theory, etc. Vol. 2: Sound, Acoustics, Geometrical Optics, Electron Optics, High-Energy Phenomena, Magnetism, Biophysics, much more. Index. Total of 800pp. 5⅜ x 8.
60595-7, 60596-5 Two volume set, paperbound $4.75

THEORETICAL PHYSICS,
A. S. Kompaneyets
One of the very few thorough studies of the subject in this price range. Provides advanced students with a comprehensive theoretical background. Especially strong on recent experimentation and developments in quantum theory. Contents: Mechanics (Generalized Coordinates, Lagrange's Equation, Collision of Particles, etc.), Electrodynamics (Vector Analysis, Maxwell's equations, Transmission of Signals, Theory of Relativity, etc.), Quantum Mechanics (the Inadequacy of Classical Mechanics, the Wave Equation, Motion in a Central Field, Quantum Theory of Radiation, Quantum Theories of Dispersion and Scattering, etc.), and Statistical Physics (Equilibrium Distribution of Molecules in an Ideal Gas, Boltzmann Statistics, Bose and Fermi Distribution. Thermodynamic Quantities, etc.). Revised to 1961. Translated by George Yankovsky, authorized by Kompaneyets. 137 exercises. 56 figures. 529pp. 5⅜ x 8½.
60972-3 Paperbound $3.50

MATHEMATICAL PHYSICS, *D. H. Menzel*
Thorough one-volume treatment of the mathematical techniques vital for classical mechanics, electromagnetic theory, quantum theory, and relativity. Written by the Harvard Professor of Astrophysics for junior, senior, and graduate courses, it gives clear explanations of all those aspects of function theory, vectors, matrices, dyadics, tensors, partial differential equations, etc., necessary for the understanding of the various physical theories. Electron theory, relativity, and other topics seldom presented appear here in considerable detail. Scores of definition, conversion factors, dimensional constants, etc. "More detailed than normal for an advanced text . . . excellent set of sections on Dyadics, Matrices, and Tensors," *Journal of the Franklin Institute*. Index. 193 problems, with answers. x + 412pp. 5⅜ x 8.   60056-4 Paperbound $2.50

THE THEORY OF SOUND, *Lord Rayleigh*
Most vibrating systems likely to be encountered in practice can be tackled successfully by the methods set forth by the great Nobel laureate, Lord Rayleigh. Complete coverage of experimental, mathematical aspects of sound theory. Partial contents: Harmonic motions, vibrating systems in general, lateral vibrations of bars, curved plates or shells, applications of Laplace's functions to acoustical problems, fluid friction, plane vortex-sheet, vibrations of solid bodies, etc. This is the first inexpensive edition of this great reference and study work. Bibliography, Historical introduction by R. B. Lindsay. Total of 1040pp. 97 figures. 5⅜ x 8.   60292-3, 60293-1 Two volume set, paperbound $6.00

HYDRODYNAMICS, *Horace Lamb*
Internationally famous complete coverage of standard reference work on dynamics of liquids & gases. Fundamental theorems, equations, methods, solutions, background, for classical hydrodynamics. Chapters include Equations of Motion, Integration of Equations in Special Gases, Irrotational Motion, Motion of Liquid in 2 Dimensions, Motion of Solids through Liquid-Dynamical Theory, Vortex Motion, Tidal Waves, Surface Waves, Waves of Expansion, Viscosity, Rotating Masses of Liquids. Excellently planned, arranged; clear, lucid presentation. 6th enlarged, revised edition. Index. Over 900 footnotes, mostly bibliographical. 119 figures. xv + 738pp. 6⅛ x 9¼.   60256-7 Paperbound $4.00

DYNAMICAL THEORY OF GASES, *James Jeans*
Divided into mathematical and physical chapters for the convenience of those not expert in mathematics, this volume discusses the mathematical theory of gas in a steady state, thermodynamics, Boltzmann and Maxwell, kinetic theory, quantum theory, exponentials, etc. 4th enlarged edition, with new material on quantum theory, quantum dynamics, etc. Indexes. 28 figures. 444pp. 6⅛ x 9¼.
60136-6 Paperbound $2.75

THERMODYNAMICS, *Enrico Fermi*
Unabridged reproduction of 1937 edition. Elementary in treatment; remarkable for clarity, organization. Requires no knowledge of advanced math beyond calculus, only familiarity with fundamentals of thermometry, calorimetry. Partial Contents: Thermodynamic systems; First & Second laws of thermodynamics; Entropy; Thermodynamic potentials: phase rule, reversible electric cell; Gaseous reactions: van't Hoff reaction box, principle of LeChatelier; Thermodynamics of dilute solutions: osmotic & vapor pressures, boiling & freezing points; Entropy constant. Index. 25 problems. 24 illustrations. x + 160pp. 5⅜ x 8.   60361-X Paperbound $2.00

CELESTIAL OBJECTS FOR COMMON TELESCOPES,
Rev. T. W. Webb
Classic handbook for the use and pleasure of the amateur astronomer. Of inestimable aid in locating and identifying thousands of celestial objects. Vol I, The Solar System: discussions of the principle and operation of the telescope, procedures of observations and telescope-photography, spectroscopy, etc., precise location information of sun, moon, planets, meteors. Vol. II, The Stars: alphabetical listing of constellations, information on double stars, clusters, stars with unusual spectra, variables, and nebulae, etc. Nearly 4,000 objects noted. Edited and extensively revised by Margaret W. Mayall, director of the American Assn. of Variable Star Observers. New Index by Mrs. Mayall giving the location of all objects mentioned in the text for Epoch 2000. New Precession Table added. New appendices on the planetary satellites, constellation names and abbreviations, and solar system data. Total of 46 illustrations. Total of xxxix + 606pp. 5⅜ x 8.   20917-2, 20918-0 Two volume set, paperbound $5.00

PLANETARY THEORY,
E. W. Brown and C. A. Shook
Provides a clear presentation of basic methods for calculating planetary orbits for today's astronomer. Begins with a careful exposition of specialized mathematical topics essential for handling perturbation theory and then goes on to indicate how most of the previous methods reduce ultimately to two general calculation methods: obtaining expressions either for the coordinates of planetary positions or for the elements which determine the perturbed paths. An example of each is given and worked in detail. Corrected edition. Preface. Appendix. Index. xii + 302pp. 5⅜ x 8½.   61133-7 Paperbound $2.25

STAR NAMES AND THEIR MEANINGS,
Richard Hinckley Allen
An unusual book documenting the various attributions of names to the individual stars over the centuries. Here is a treasure-house of information on a topic not normally delved into even by professional astronomers; provides a fascinating background to the stars in folk-lore, literary references, ancient writings, star catalogs and maps over the centuries. Constellation-by-constellation analysis covers hundreds of stars and other asterisms, including the Pleiades, Hyades, Andromedan Nebula, etc. Introduction. Indices. List of authors and authorities. xx + 563pp. 5⅜ x 8½.   21079-0 Paperbound $3.00

A SHORT HISTORY OF ASTRONOMY, *A. Berry*
Popular standard work for over 50 years, this thorough and accurate volume covers the science from primitive times to the end of the 19th century. After the Greeks and the Middle Ages, individual chapters analyze Copernicus, Brahe, Galileo, Kepler, and Newton, and the mixed reception of their discoveries. Post-Newtonian achievements are then discussed in unusual detail: Halley, Bradley, Lagrange, Laplace, Herschel, Bessel, etc. 2 Indexes. 104 illustrations, 9 portraits. xxxi + 440pp. 5⅜ x 8.   20210-0 Paperbound $2.75

SOME THEORY OF SAMPLING, *W. E. Deming*
The purpose of this book is to make sampling techniques understandable to and useable by social scientists, industrial managers, and natural scientists who are finding statistics increasingly part of their work. Over 200 exercises, plus dozens of actual applications. 61 tables. 90 figs. xix + 602pp. 5⅜ x 8½.
          61755-6 Paperbound $3.50

PRINCIPLES OF STRATIGRAPHY,
A. W. Grabau
Classic of 20th century geology, unmatched in scope and comprehensiveness. Nearly 600 pages cover the structure and origins of every kind of sedimentary, hydrogenic, oceanic, pyroclastic, atmoclastic, hydroclastic, marine hydroclastic, and bioclastic rock; metamorphism; erosion; etc. Includes also the constitution of the atmosphere; morphology of oceans, rivers, glaciers; volcanic activities; faults and earthquakes; and fundamental principles of paleontology (nearly 200 pages). New introduction by Prof. M. Kay, Columbia U. 1277 bibliographical entries. 264 diagrams. Tables, maps, etc. Two volume set. Total of xxxii + 1185pp. 5⅜ x 8.   60686-4, 60687-2 Two volume set, paperbound $6.25

SNOW CRYSTALS, *W. A. Bentley and W. J. Humphreys*
Over 200 pages of Bentley's famous microphotographs of snow flakes—the product of painstaking, methodical work at his Jericho, Vermont studio. The pictures, which also include plates of frost, glaze and dew on vegetation, spider webs, windowpanes; sleet; graupel or soft hail, were chosen both for their scientific interest and their aesthetic qualities. The wonder of nature's diversity is exhibited in the intricate, beautiful patterns of the snow flakes. Introductory text by W. J. Humphreys. Selected bibliography. 2,453 illustrations. 224pp. 8 x 10¼.   20287-9 Paperbound $3.25

THE BIRTH AND DEVELOPMENT OF THE GEOLOGICAL SCIENCES,
F. D. Adams
Most thorough history of the earth sciences ever written. Geological thought from earliest times to the end of the 19th century, covering over 300 early thinkers & systems: fossils & their explanation, vulcanists vs. neptunists, figured stones & paleontology, generation of stones, dozens of similar topics. 91 illustrations, including medieval, renaissance woodcuts, etc. Index. 632 footnotes, mostly bibliographical. 511pp. 5⅜ x 8.   20005-1 Paperbound $2.75

ORGANIC CHEMISTRY, *F. C. Whitmore*
The entire subject of organic chemistry for the practicing chemist and the advanced student. Storehouse of facts, theories, processes found elsewhere only in specialized journals. Covers aliphatic compounds (500 pages on the properties and synthetic preparation of hydrocarbons, halides, proteins, ketones, etc.), alicyclic compounds, aromatic compounds, heterocyclic compounds, organophosphorus and organometallic compounds. Methods of synthetic preparation analyzed critically throughout. Includes much of biochemical interest. "The scope of this volume is astonishing," *Industrial and Engineering Chemistry.* 12,000-reference index. 2387-item bibliography. Total of x + 1005pp. 5⅜ x 8.   60700-3, 60701-1 Two volume set, paperbound $4.50

THE PHASE RULE AND ITS APPLICATION,
Alexander Findlay
Covering chemical phenomena of 1, 2, 3, 4, and multiple component systems, this "standard work on the subject" (*Nature*, London), has been completely revised and brought up to date by A. N. Campbell and N. O. Smith. Brand new material has been added on such matters as binary, tertiary liquid equilibria, solid solutions in ternary systems, quinary systems of salts and water. Completely revised to triangular coordinates in ternary systems, clarified graphic representation, solid models, etc. 9th revised edition. Author, subject indexes. 236 figures. 505 footnotes, mostly bibliographic. xii + 494pp. 5⅜ x 8.
60091-2 Paperbound $2.75

A COURSE IN MATHEMATICAL ANALYSIS,
Edouard Goursat

Trans. by E. R. Hedrick, O. Dunkel, H. G. Bergmann. Classic study of fundamental material thoroughly treated. Extremely lucid exposition of wide range of subject matter for student with one year of calculus. Vol. 1: Derivatives and differentials, definite integrals, expansions in series, applications to geometry. 52 figures, 556pp. 60554-X Paperbound $3.00. Vol. 2, Part I: Functions of a complex variable, conformal representations, doubly periodic functions, natural boundaries, etc. 38 figures, 269pp. 60555-8 Paperbound $2.25. Vol. 2, Part II: Differential equations, Cauchy-Lipschitz method, nonlinear differential equations, simultaneous equations, etc. 308pp. 60556-6 Paperbound $2.50. Vol. 3, Part I: Variation of solutions, partial differential equations of the second order. 15 figures, 339pp. 61176-0 Paperbound $3.00. Vol. 3, Part II: Integral equations, calculus of variations. 13 figures, 389pp. 61177-9 Paperbound $3.00 60554-X, 60555-8, 60556-6 61176-0, 61177-9 Six volume set, paperbound $13.75

PLANETS, STARS AND GALAXIES,
A. E. Fanning

Descriptive astronomy for beginners: the solar system; neighboring galaxies; seasons; quasars; fly-by results from Mars, Venus, Moon; radio astronomy; etc. all simply explained. Revised up to 1966 by author and Prof. D. H. Menzel, former Director, Harvard College Observatory. 29 photos, 16 figures. 189pp. 5⅜ x 8½.
21680-2 Paperbound $1.50

GREAT IDEAS IN INFORMATION THEORY, LANGUAGE AND CYBERNETICS,
Jagjit Singh

Winner of Unesco's Kalinga Prize covers language, metalanguages, analog and digital computers, neural systems, work of McCulloch, Pitts, von Neumann, Turing, other important topics. No advanced mathematics needed, yet a full discussion without compromise or distortion. 118 figures. ix + 338pp. 5⅜ x 8½.
21694-2 Paperbound $2.25

GEOMETRIC EXERCISES IN PAPER FOLDING,
T. Sundara Row

Regular polygons, circles and other curves can be folded or pricked on paper, then used to demonstrate geometric propositions, work out proofs, set up well-known problems. 89 illustrations, photographs of actually folded sheets. xii + 148pp. 5⅜ x 8½.
21594-6 Paperbound $1.00

VISUAL ILLUSIONS, THEIR CAUSES, CHARACTERISTICS AND APPLICATIONS,
M. Luckiesh

The visual process, the structure of the eye, geometric, perspective illusions, influence of angles, illusions of depth and distance, color illusions, lighting effects, illusions in nature, special uses in painting, decoration, architecture, magic, camouflage. New introduction by W. H. Ittleson covers modern developments in this area. 100 illustrations. xxi + 252pp. 5⅜ x 8.
21530-X Paperbound $1.50

ATOMS AND MOLECULES SIMPLY EXPLAINED,
B. C. Saunders and R. E. D. Clark

Introduction to chemical phenomena and their applications: cohesion, particles, crystals, tailoring big molecules, chemist as architect, with applications in radioactivity, color photography, synthetics, biochemistry, polymers, and many other important areas. Non technical. 95 figures. x + 299pp. 5⅜ x 8½.
21282-3 Paperbound $1.50

The Principles of Electrochemistry,
D. A. MacInnes

Basic equations for almost every subfield of electrochemistry from first principles, referring at all times to the soundest and most recent theories and results; unusually useful as text or as reference. Covers coulometers and Faraday's Law, electrolytic conductance, the Debye-Hueckel method for the theoretical calculation of activity coefficients, concentration cells, standard electrode potentials, thermodynamic ionization constants, pH, potentiometric titrations, irreversible phenomena. Planck's equation, and much more. 2 indices. Appendix. 585-item bibliography. 137 figures. 94 tables. ii + 478pp. 5⅜ x 8⅜.
60052-1 Paperbound $3.00

Mathematics of Modern Engineering,
E. G. Keller and R. E. Doherty

Written for the Advanced Course in Engineering of the General Electric Corporation, deals with the engineering use of determinants, tensors, the Heaviside operational calculus, dyadics, the calculus of variations, etc. Presents underlying principles fully, but emphasis is on the perennial engineering attack of set-up and solve. Indexes. Over 185 figures and tables. Hundreds of exercises, problems, and worked-out examples. References. Total of xxxiii + 623pp. 5⅜ x 8. 60734-8, 60735-6 Two volume set, paperbound $3.70

Aerodynamic Theory: A General Review of Progress,
William F. Durand, editor-in-chief

A monumental joint effort by the world's leading authorities prepared under a grant of the Guggenheim Fund for the Promotion of Aeronautics. Never equalled for breadth, depth, reliability. Contains discussions of special mathematical topics not usually taught in the engineering or technical courses. Also: an extended two-part treatise on Fluid Mechanics, discussions of aerodynamics of perfect fluids, analyses of experiments with wind tunnels, applied airfoil theory, the nonlifting system of the airplane, the air propeller, hydrodynamics of boats and floats, the aerodynamics of cooling, etc. Contributing experts include Munk, Giacomelli, Prandtl, Toussaint, Von Karman, Klemperer, among others. Unabridged republication. 6 volumes. Total of 1,012 figures, 12 plates, 2,186pp. Bibliographies. Notes. Indices. 5⅜ x 8½. 61709-2, 61710-6, 61711-4, 61712-2, 61713-0, 61715-9 Six volume set, paperbound $13.50

Fundamentals of Hydro- and Aeromechanics,
L. Prandtl and O. G. Tietjens

The well-known standard work based upon Prandtl's lectures at Goettingen. Wherever possible hydrodynamics theory is referred to practical considerations in hydraulics, with the view of unifying theory and experience. Presentation is extremely clear and though primarily physical, mathematical proofs are rigorous and use vector analysis to a considerable extent. An Engineering Society Monograph, 1934. 186 figures. Index. xvi + 270pp. 5⅜ x 8.
60374-1 Paperbound $2.25

Applied Hydro- and Aeromechanics,
L. Prandtl and O. G. Tietjens

Presents for the most part methods which will be valuable to engineers. Covers flow in pipes, boundary layers, airfoil theory, entry conditions, turbulent flow in pipes, and the boundary layer, determining drag from measurements of pressure and velocity, etc. Unabridged, unaltered. An Engineering Society Monograph. 1934. Index. 226 figures, 28 photographic plates illustrating flow patterns. xvi + 311pp. 5⅜ x 8. 60375-X Paperbound $2.50

APPLIED OPTICS AND OPTICAL DESIGN,
A. E. Conrady

With publication of vol. 2, standard work for designers in optics is now complete for first time. Only work of its kind in English; only detailed work for practical designer and self-taught. Requires, for bulk of work, no math above trig. Step-by-step exposition, from fundamental concepts of geometrical, physical optics, to systematic study, design, of almost all types of optical systems. Vol. 1: all ordinary ray-tracing methods; primary aberrations; necessary higher aberration for design of telescopes, low-power microscopes, photographic equipment. Vol. 2: (Completed from author's notes by R. Kingslake, Dir. Optical Design, Eastman Kodak.) Special attention to high-power microscope, anastigmatic photographic objectives. "An indispensable work," *J., Optical Soc. of Amer.* Index. Bibliography. 193 diagrams. 852pp. 6⅛ x 9¼.

60611-2, 60612-0 Two volume set, paperbound $8.00

MECHANICS OF THE GYROSCOPE, THE DYNAMICS OF ROTATION,
R. F. Deimel, Professor of Mechanical Engineering at Stevens Institute of Technology

Elementary general treatment of dynamics of rotation, with special application of gyroscopic phenomena. No knowledge of vectors needed. Velocity of a moving curve, acceleration to a point, general equations of motion, gyroscopic horizon, free gyro, motion of discs, the damped gyro, 103 similar topics. Exercises. 75 figures. 208pp. 5⅜ x 8.

60066-1 Paperbound $1.75

STRENGTH OF MATERIALS,
J. P. Den Hartog

Full, clear treatment of elementary material (tension, torsion, bending, compound stresses, deflection of beams, etc.), plus much advanced material on engineering methods of great practical value: full treatment of the Mohr circle, lucid elementary discussions of the theory of the center of shear and the "Myosotis" method of calculating beam deflections, reinforced concrete, plastic deformations, photoelasticity, etc. In all sections, both general principles and concrete applications are given. Index. 186 figures (160 others in problem section). 350 problems, all with answers. List of formulas. viii + 323pp. 5⅜ x 8.

60755-0 Paperbound $2.50

HYDRAULIC TRANSIENTS,
G. R. Rich

The best text in hydraulics ever printed in English . . . by former Chief Design Engineer for T.V.A. Provides a transition from the basic differential equations of hydraulic transient theory to the arithmetic integration computation required by practicing engineers. Sections cover Water Hammer, Turbine Speed Regulation, Stability of Governing, Water-Hammer Pressures in Pump Discharge Lines, The Differential and Restricted Orifice Surge Tanks, The Normalized Surge Tank Charts of Calame and Gaden, Navigation Locks, Surges in Power Canals—Tidal Harmonics, etc. Revised and enlarged. Author's prefaces. Index. xiv + 409pp. 5⅜ x 8½.

60116-1 Paperbound $2.50

Prices subject to change without notice.

Available at your book dealer or write for free catalogue to Dept. Adsci, Dover Publications, Inc., 180 Varick St., N.Y., N.Y. 10014. Dover publishes more than 150 books each year on science, elementary and advanced mathematics, biology, music, art, literary history, social sciences and other areas.